Water Recycled Earlier

From Classical to Modern Chemistry

The Instrumental Revolution

Dedication

To Frank Greenaway, Keeper of Chemistry, Science Museum, London, 1967–1980, who by his judicious acquisition of modern instrumentation, his exacting scholarship and his generous encouragement of young curators and historians laid the foundations for the history of the instrumental revolution.

From Classical to Modern Chemistry

The Instrumental Revolution

Edited by

Peter J. T. Morris
Science Museum, London SW7 2DD, UK

RS•C
ROYAL SOCIETY OF CHEMISTRY

science museum

ISBN 0-85404-479-5

A catalogue record for this book is available from the British Library

Published by The Royal Society of Chemistry in association with the Science Museum, London and the Chemical Heritage Foundation
Thomas Graham House, Science Park, Milton Road, Cambridge CB4 0WF, UK

Registered Charity Number 207890

For further information see our web site at www.rsc.org

Typeset in Great Britain by Vision Typesetting, Manchester
Printed in Great Britain by TJ International Ltd, Padstow, Cornwall

Foreword

Diary Entry (11 August 1983)

We sit in Copenhagen:
Chemists from a dozen countries.
The talk is heavy; the words are long.

Male contraception,
Cures for cancer,
Morphine substitutes,
Drugs from the sea,
Medicines for the year 2000.

We've mouthed these words for many years,
Formulae hiding the chemists.
Who are these colleagues, students, strangers?
What do they do besides chemistry?

If this were the Holiday Inn,
Not the Royal Danish Academy,
Would I guess who they are?

A convention of grocers? Too Serious.
Car salesmen? Too little polyester.
Bankers? Lawyers? No vests.
Clergymen? Wrong collars.
Poets? Nobody smokes.

How did they come to chemistry?
What do they do besides chemistry?
What do I do besides what I do
Besides Chemistry?

(From C. Djerassi, *The Clock Runs Backward*, Story Line Press, 1991)

When I embarked on my research career in 1942 the only physical method used by organic chemistry was ultraviolet spectroscopy, involving laborious point-by-point measurements. The University of Wisconsin Chemistry Department, where I did my graduate work, did not even have a Beckman DU ultraviolet spectrophotometer.

I have always been interested in structure elucidation. Before the Second World War, structure elucidation was a painstaking process of doing experimental chemistry and piecing the often puzzling answers into a coherent framework. It was similar to solving a jigsaw puzzle, but one in which not all the pieces were on the table. Carrying out the right reactions would yield another piece, which had to be fitted into the partly completed puzzle. More often than not, the puzzle would have to be taken apart and reassembled almost from scratch. The method of trial and error could be frustrating, but it was also intellectually exciting.

As a structure elucidation chemist I was emotionally receptive to new physical methods. In contrast to many of my natural products colleagues, I also became interested in research on the physical methods themselves, notably optical rotatory dispersion and mass spectroscopy – eventually even extending it to computer-aided artificial intelligence approaches. At the same time, other techniques, notably nuclear magnetic resonance also began to make their mark. Steroid chemistry played a key rôle in this process. Had steroids not been so important in the 1940s and 1950s – or had their structures been less complex – the development of instrumental methods might well have occurred more slowly.

The excitement of finding out more about the structure of compounds from their chemical reactions had almost completely receded when the above introductory poem was written – when degradative chemical structure elucidation had turned into applied spectroscopy or X-ray crystallography. While the actual process of structure elucidation had become less challenging on the intellectual level, the scientific gain has been enormous. In the process, the emphasis has moved from straightforward elucidation of structure to biosynthesis and function. Using modern methods, natural product chemists can frequently establish the complete structure of a complex molecule with only micrograms of material and do so in record time. We have now reached the stage where we can find out more about the structure of a compound with an amount of material that is often too small for a melting point determination compared to the gram quantities required decades ago.

The papers in this volume are of interest to me, because I took part in the 'instrumental revolution' myself. I don't agree with everything they say, but they demonstrate the enormous impact that new physical instrumentation had on chemistry and biochemistry. My own bias is clearly visible if I highlight mass spectrometry among all other methods covered.

However, you don't need to have been working in chemistry around the middle of the twentieth century to find these papers interesting; they tell a fascinating story about the transformation of a mature science into an entirely new way of investigating nature. Nowadays, the natural product chemist is not an organic chemist anymore, but an analytical chemist or a spectroscopist. Laboratory glassware and reagents have been replaced by 'black boxes' – and expensive ones at that! It is important that we understand how this happened and what it did to a branch of chemistry. And given how little attention research chemists pay to the actual history of their science, the present set of contributed papers sheds some light on the great transformation that swept across the chemical sciences during the twentieth century.

Carl Djerassi
Professor of Chemistry
Stanford University
Stanford, CA 94305-5080

Preface

This volume stems from a conference on the history of chemical instrumentation, 'From the Test-tube to the Autoanalyzer: The Development of Chemical Instrumentation in the Twentieth Century', which was held at Imperial College, London, in August 2000. The conference was organised by the Commission on the History of Modern Chemistry, a constituent organisation of the Division of History of Science of the International Union of the History and Philosophy of Science (IUHPS); it was co-sponsored by the Scientific Instruments Commission of the IUHPS and the Historical Group of the Royal Society of Chemistry. The papers from this conference have been completely revised for publication and edited to attain a degree of uniformity that is not normally found in conference proceedings. I selected the photographs and wrote the captions. The individual authors bear no responsibility for any errors or misjudgements with respect to these images.

The Commission on the History of Modern Chemistry was set up in 1997 to promote the history of twentieth century chemistry in the face of the pre-occupation of many historians of chemistry with the eighteenth and nineteenth century and the failure of many contemporary chemists to take the history of chemistry seriously. This split between practising chemists and the history of their field is of relatively recent origin. Traditionally, most chemists had learnt the history of chemistry (however imperfectly and however laden with 'Whiggish' preconceptions[1]) as part of their training. Furthermore, many leading historians of chemistry were also working scientists, notably James Riddick Partington. Leaving aside the usual (and doubtlessly valid) excuses about increasing specialisation and pressures of other duties, one reason for this divergence must surely be the failure of historians of chemistry to address the development of modern chemistry. The basic questions for any professional group are 'how did we get here?' and 'why do we do what we do?' These questions have not been addressed adequately by contemporary historians of chemistry, who are increasingly, professionally trained *historians* rather than historically inclined *chemists*. Twenty-five years ago, when I was an undergraduate chemist at Oxford, I eagerly bought a book entitled *From Classical to Modern Chemistry*,[2] taking it for granted that it would tell me about the development of modern chemistry, how

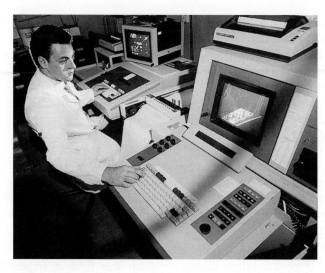

Few techniques have had more impact on the development of modern chemistry than nuclear magnetic resonance (NMR) spectroscopy. NMR is not only a powerful probe of the molecular environment, but is also very versatile, useful for a large number of chemical investigations and applicable to a wide variety of elements. Courtesy of Geoff Thompkinson/Science Photo Library

organic reaction mechanisms and physical instrumentation completely transformed chemistry. To my deep disappointment, it was entirely about the nineteenth century and very early twentieth century.[3] This gap has remained unfilled for many years. Many professional historians of chemistry are uneasy with modern chemistry; it is complex and there are too many sources of information. Instead, they have preferred to concentrate on issues that will win them credibility with other professional historians.

Fortunately, historians of chemistry have come to realise that they cannot completely ignore the development of chemistry over the span of some eight decades. Moreover, there has been a realisation that the introduction of physical instrumentation into chemistry was a crucial element in its evolution. Having tackled the boundaries of chemistry with other sciences in its first major conference, held in Munich in 1999,[4] the Commission on the History of Modern Chemistry decided to devote its second annual meeting to this theme.

To be sure, this conference did not take place in an intellectual vacuum. Although Bernard Cohen (to give just one illustrious example) completely ignored the 'instrumental revolution' in his *Revolution in Science*,[5] published in 1985, the subject has been growing slowly but surely over the last two decades. The first serious attempt to analyse the introduction of the 'new' physical methods into chemistry appears to have been an Oxford University undergraduate (Chemistry Part Two) thesis submitted in 1973.[6] Unfortunately, like most Part Twos, it then sank without leaving any intellectual trace. In *The Techniques of Analytical Chemistry*,[7] published in the following year, Harry Irving and Frank Greenaway stressed the continuity of chemical analysis, although they were aware of the radical changes taking place. Then, in 1977, the American Chemical

Society's Division of Analytical Chemistry published a collection of histories of instrumental techniques.[8] Written by experts in the various fields, it is still a valuable information source, but suffers from a number of drawbacks, notably the lack of any citations and a narrow (if understandable) focus on chemical analysis. John Stock and Mary Virginia Orna in 1986 published the proceedings of an American Chemical Society symposium on the history of chemical instrumentation.[9] The four papers (out of eighteen) that covered twentieth century techniques were all personal recollections without any attempt at historical analysis. In the same year, the Tarbells published the first serious attempt to analyse the introduction of physical instrumentation into organic chemistry and it would seem that they coined the term 'instrumental revolution'.[10] They were able to speak with authority from their personal experiences as chemists, but perhaps unwisely, they attempted to cover the entire revolution in a brief survey of some fifteen pages.

The publication in 1987 of Yakov Rabkin's excellent paper in *Isis*,[11] on the adoption of infrared spectroscopy by chemists, marked a watershed in this field. He broke decisively with earlier internalist accounts[12] by showing the importance of social networks, how physicists and chemists at American Cyanamid's Stamford research laboratories collaborated in the early 1940s with the neighbouring Perkin–Elmer company. The two groups carried out collaborative research on the applications of spectrophotometers in chemistry and then promoted their use by both academic and industrial chemists. He also demonstrated that the rival firm of Beckman Instruments had developed similar instruments, but had been hindered by wartime secrecy. After 1945, both Perkin–Elmer and Beckman Instruments enjoyed considerable commercial success. This kind of rich interpretation had not hitherto existed for the history of modern chemical instrumentation. It is indeed fortunate that we have been able to reprint Rabkin's pathbreaking paper in this volume.

As an émigré from the Soviet Union, Rabkin came from outside the Western tradition in the history of chemistry and his example was not emulated until Davis Baird published his paper on the 'big scientific instrumentation revolution' in *Annals of Science* six years later.[13] Baird surveyed the instrumentation changes in analytical chemistry, but more importantly he pointed out that there was a *philosophical* change as well. Between 1920 and the 1960s, he argued, analytical chemists had switched from identifying unknowns by their chemical properties to distinguishing chemicals by their physical properties. Coming into the field from philosophy, and his interest in the subject was provoked by family papers rather than a non-existent historiography, Baird was also an outsider.

In the same year, I was invited by John Krige to write a chapter on the impact of physical instrumentation on organic chemistry in *Science in the Twentieth Century*.[14] Given the enormity of the topic, I restricted my chapter to structural organic chemistry, and then invited Anthony Travis to assist me. Drawing on the valuable first-hand accounts that were then appearing thanks to Jeffrey Seeman's stimulating series, *Profiles, Pathways and Dreams*,[15] we showed how the instrumental revolution radically altered what chemists did. We also stressed the importance of textbooks as means of converting chemists to the new way of doing things. Not least because our paper provides an overview of the main techniques, the first complete version of our chapter is reproduced in this volume. By reprinting these three pioneering papers alongside the contributions from the conference, I hope this book will be an excellent starting point, and indeed a significant resource, for anyone interested in 'how we got here' in modern chemistry.[16]

The history of modern chemical instrumentation is a play with many actors, ranging from venture capitalists and instrument manufacturers to single-minded inventors and intrepid chemists. It is a play in which the actors change the scenery, and in his introduction to the papers on instrument development in social, economic and political context, David Knight reminds us of the enormous changes that have been made to the intellectual scenery of chemistry. Drawing on his earlier research into development of physical instrumentation, Terry Shinn describes the play's stage to us, side-stepping the time-worn debates about science *versus* technology by introducing the concept of an autonomous field of 'research-technology'. The key features of this zone, according to Shinn, are 'interstitiality', 'genericity' and metrology. Using Shinn's scheme as a framework, Charlotte Bigg illustrates the rôle played by the instrument maker, by examining how Frank Twyman of the London firm of Adam Hilger tried to persuade chemists to use ultraviolet spectroscopy by simplifying the apparatus. Since Adam Hilger marketed its 'constant deviation wavelength spectrometer' in 1903, Bigg also reminds us that the 'instrumental revolution' can be traced back to the beginning of the twentieth century. As the son of Walter Baird, a pioneering instrument manufacturer, Davis Baird is in a unique position to give us an insight into the development of modern instrumentation. Using his father's classification of 'thing, idea, people and money', he reveals how the different actors collaborated to introduce new types of instrumentation. Stuart Bennett shows how two groups of actors often regarded as peripheral, process control companies and the chemical industry, had much in common with the other actors. Like Hilger and Baird Associates, the process control firms saw themselves as enablers operating in a generic field: 'tell us your problem and we will find the solution' to paraphrase an early advertisement.

One only has to visit a modern chemistry laboratory to see how the new instrumentation has wrought immense physical changes. The philosophical impact is less obvious, and often passes completely unrecognised by practising chemists. The philosophy of chemistry is experiencing a revival at present, and we were indeed fortunate that several philosophers took part in this conference. In his introduction to the papers on the impact of instrumentation on chemistry, Pierre Laszlo, who is both a chemist and a philosopher, sets the agenda for the future by listing thirteen *desiderata*. He then treats us to a wide-ranging discussion of the impact of the instrumentation revolution, with particular reference to its philosophical and sociological implications. In his paper on the impact of instrumentation on chemical species identity, Joachim Schummer boldly ventures where chemists fear to tread by examining the ontological[17] implications of the new instrumentation, specifically spectroscopy, for species identity. He explores the changing practices of synthetic organic chemists, who moved during the latter half of the twentieth century from simple constitutional formulas to complete configurational structures. Schummer argues that there has been a fundamental ontological shift from pure chemical substances to frequently un-isolated quasi-molecular species. In his paper on 'the reification of chemical structures', Leo Slater takes a different starting point, taking the example of the leading synthetic organic chemist Robert Burns Woodward to show us how the new instrumentation converted chemical structures from being abstractions which summarised chemical knowledge into true representations. He also demonstrates that Woodward was a powerful advocate for the new physical techniques, though in many respects he was a traditional chemist. Carsten Reinhardt's paper reveals how organic chemists grappled with Schummer's quasi-molecular species, in this case produced by the mass spectrometer. A small but determined group of chemists converted apparently random molecular fragmentations into a powerful set of predictable rearrangements. This was a triumph of the relatively novel field of organic reaction mechanisms, logical thinking, sheer doggedness and of course, a new generation of mass spectrometers.

In his introduction to the papers on environmental and biomedical technology, Nicolas Rasmussen seeks to discover whether instruments can revolutionise knowledge. He concludes from a close reading of the three papers in this section that on rare occasions instruments can 'escape' from their original setting and have a revolutionary impact. He also notes in passing that the life sciences played a major rôle in the development of separation techniques and instruments that were eventually more important elsewhere (such as the electron capture detector) nonetheless had their origins within the life sciences. Luigi Cerruti opens his paper with a short but valuable history of paper chromatography, paper electrophoresis and polyacrylamide gel electrophoresis (PAGE). He then shows that the development of these techniques enabled biochemists and physicians to study abnormal haemoglobins, which in turn assisted the development of medical genetics. As chemical factories produced more and more organic compounds in the twentieth century, they had to monitor the impact of these chemicals (and their by-products) on the environment, in particular, rivers and groundwater. Some companies turned to instrument manufacturers for assistance (in keeping with the Baird–Bennett characterisation of such companies as problem-solvers), and as Anthony Travis illustrates in his paper, this interaction accelerated the introduction of spectrophotometry and other techniques into analytical chemistry. Travis argues that developments in chemical technology, and consequently the need for improved environmental monitoring, remain a major driving force for innovation in the field of chemical instrumentation. However, these innovations are not always a result of directed research. One of the most sensitive instruments for detecting pollutants, the electron capture detector (ECD), arose within the same biochemical arena described by Cerruti. In my paper about the development of the ECD and its consequences, I place its inventor, James Lovelock (better known now for his development of the Gaia theory[18]), firmly in Shinn's zone of research-technology, as a scientist situated between organic chemistry, physiology, biology and physics. The ECD solved one of the most pressing problems facing pesticide analysts in the late 1950s, the detection of DDT at the level of parts per billion, but thereby (and unwittingly) created another conundrum; how to set safe levels for food contaminants when the hitherto comforting concept of 'zero tolerance' had been rendered meaningless.

Given the still immature state of the history of modern chemical instrumentation, this volume can only be a starting point for future research. Without wishing to add to the burdens of future historians and philosophers, already charged with thirteen tasks by Pierre Laszlo, I would like nonetheless to conclude this preface by highlighting three areas for potential development. I agree with Joachim Schummer[19] that there is a need to define rigorously the concepts and terms used, even such basic ones as 'instrument' and, *pace* Laszlo, 'tool'. Although several papers in this volume have considered the economic and social context of instrument manufacture, much more work is needed to map the social and economic environment in which chemical instrumentation evolves. Above all, as historians and museum curators, we need to show that the instruments are part of the historical record, that a rounded history of chemical instrumentation cannot be based on written records and oral history alone, but must compass a close analysis of the instruments themselves. Without detailed studies of the innards of our black boxes, the history of chemistry will remain tantalisingly incomplete.

Acknowledgements

I would like thank my fellow members of the organising committee – Professor Charles Rees FRS, Anthony S. Travis, Luigi Cerruti, Carsten Reinhardt and Stuart Bennett – for their considerable assistance in arranging the conference's programme, and Professors Charles Rees and William Griffith for their help with the local arrangements. The organisers gratefully acknowledge the generous funding provided by the Hans Jenemann-Stiftung, the Chemical Heritage Foundation, and the Angela and Tony Fish Bequest of the Royal Society of Chemistry. My thanks also go to the speakers and commentators who made the conference a success and this volume possible. The efforts of Jane Davies and Alice Nicholls during the conference were much appreciated by all the participants. At the editing stage, I was aided by Anthony Travis, Pierre Laszlo and Carsten Reinhardt. I am particularly grateful to Joachim Schummer for his help at various stages. The path of this volume's publication was smoothened by the professionalism of Ela Ginalska (Science Museum) and Janet Freshwater (RSC Publications). I am grateful for the constant support of Robert Bud (Science Museum), Professor Christoph Meinel (CHMC) and David Betteridge (RSC Historical Group). Finally, my warmest thanks must go to Magda Wheatley for her sterling efforts during the conference and her assistance with editing of the papers.

Peter J. T. Morris
Picnic Day, 2001

Notes

1 The term 'Whiggish' was originally given to the traditional political histories, written by Liberals, that saw the constitutional history of England as a smooth transition, *via* the Glorious Revolution and the Reform Act of 1832, to a perfect democracy. In the history of science it is applied, usually pejoratively, to accounts of scientific developments that assume that past events led inexorably to the current state of the field, and consequently only those activities that lead to the present paradigm (to use Kuhn's term) are historically important.

2 A. J. Berry, *From Classical to Modern Chemistry: Some Historical Sketches*, Dover Publications, New York, 1968.

3 I had not realised, of course, that it was a reprint of a book first published by Cambridge University Press in 1954 and based on an even earlier work, *Modern Chemistry – Some Sketches of its Historical Development*, published by Cambridge University Press in 1946. To Arthur Berry, born in 1886 and educated at the turn of the twentieth century, the chemistry of G. N. Lewis and Walter Nernst was indeed 'modern chemistry'.

4 The papers were subsequently published, with an additional chapter on organic chemistry, in Carsten Reinhardt (ed.), *Chemical Sciences in the 20th Century: Bridging Boundaries*, Wiley-VCH, Weinheim, 2001.

5 I. B. Cohen, *Revolution in Science*, Belknap Press of Harvard University Press, Cambridge, MA, 1985. I am indebted to Davis Baird's paper on analytical chemistry (see below) for this example.

6 J. M. D. Symes, 'The Development of Two Physical Methods in Chemistry: Infrared Spectroscopy and Nuclear Magnetic Resonance', Chemistry Part II thesis, Oxford University, 1973, cited in R. G. W. Anderson, 'Instruments and Apparatus', Chapter Ten of C. A. Russell (ed.), *Recent Developments in the History of Chemistry*, Royal Society of Chemistry, London, 1985, p. 230. Stored partly in the History Faculty Library and partly in the Museum of the History of Science, the unpublished Oxford Part II theses are perhaps the largest untapped resource in the history of chemistry.

7 H. M. N. H. Irving, *The Techniques of Analytical Chemistry*, with an introduction by Frank Greenaway, HMSO, London, 1974. This small book was published to accompany a Science Museum exhibition. Frank Greenaway also published 'Instruments' in *A History of Technology*, Vol. VII, *The Twentieth Century*, Part II, T.I. Williams (ed.), Clarendon Press, Oxford, 1978, pp. 1204–1219. This was a broad survey of twentieth century instrumentation with only brief comments on pH measurement, X-ray crystallography and NMR.

8 Herbert A. Laitinen and Galen W. Ewing (eds.), *A History of Analytical Chemistry*, Division of Analytical Chemistry, American Chemical Society, Washington DC, 1977.

9 John T. Stock and Mary Virginia Orna, OSU (eds.), *The History and Preservation of Chemical Instrumentation*, D. Reidel, Dordrecht, 1986.

10 Dean Stanley Tarbell and Ann Tracy Tarbell, *Essays on the History of Organic Chemistry in the United States, 1875–1955*, Folio Publishers, Nashville, TN, 1986, p. 336, so cited in Leo Slater's paper in this volume. Personally, I believe the term must have been used earlier, and I would suspect before the more grandiose 'second chemical revolution' credited by Davis Baird to Jon Eklund of the National Museum of American History.

11 Yakov M. Rabkin, 'Technological Innovation in Science: The Adoption of Infrared Spectroscopy by Chemists', *Isis*, 1988, **78**, 31–54.

12 The best of these internalist histories, with many interesting details, is R. Norman

Jones, 'Analytical Applications of Vibrational Spectroscopy – A Historical Review' in James R. Durig (ed.), *Chemical, Biological and Industrial Applications of Infrared Spectroscopy*, John Wiley and Sons, Chichester, 1985.

13 Davis Baird, 'Analytical Chemistry and the "Big" Scientific Instrumentation Revolution', *Annals of Science*, 1993, **50**, 267–290.

14 Peter J. T. Morris and Anthony S. Travis, 'The Rôle of Physical Instrumentation in Structural Organic Chemistry', Chapter 37 of John Krige and Dominique Pestre (eds.), *Science in the Twentieth Century*, Harwood Academic, Amsterdam, 1997.

15 A series of twenty autobiographies of chemists, chiefly organic chemists, edited by Jeffrey Seeman and published by the American Chemical Society, Washington DC, between 1990 and 1998. Shorter multiple first-hand accounts can also be found in L. S. Ettre and A. Zlatkis, 75 *Years of Chromatography: A Historical Dialogue*, Volume 17 of the Journal of Chromatography Library, Elsevier, Amsterdam and Oxford, 1979; and David M. Grant and Robin K. Harris (eds.), *Encyclopedia of Nuclear Magnetic Resonance*, Volume 1, *Historical Perspectives*, John Wiley and Sons, Chichester, 1996.

16 Two other multi-author volumes on the history of instrumentation have been published recently, but, useful though they are, neither covers the history of the instrumental revolution in any depth: see Frederic L. Holmes and Trevor H. Levere (eds.), *Instruments and Experimentation in the History of Chemistry*, MIT Press, Cambridge, MA, 2000; and Bernward Joerges and Terry Shinn (eds.), *Instrumentation Between Science, State and Industry*, Volume XXII of *Sociology of the Sciences*, Kluwer, Dordrecht, 2001. At a late stage, I also became aware of the interesting website 'The Tools of the Second Chemical Revolution: Instrumental Design and History' by Christopher J. Brubaker and David L. Powell at http://www.wooster.edu/chemistry/is/brubaker/default.html (accessed 29 November 2001). It does, however, confine itself to four key instruments (Beckman Model G pH meter, Beckman DU and Perkin–Elmer 21 spectrometers, and the Varian A-60 NMR machine).

17 In case you are wondering, ontology was traditionally a branch of metaphysics that examines the nature of being. In modern terms, it covers the specification of a conceptualisation, such as the concept of identity.

18 James Lovelock, *Gaia: A New Look at Life on Earth*, revised edition, Oxford University Press, Oxford, 1995.

19 An issue he raises in his excellent report on this conference in *Hyle*, 2001, **7**(1), 78–81.

Contents

Impact of Instrumentation on Chemistry

Authors

Davis Baird is Professor and Chair of the Department of Philosophy at the University of South Carolina. His research focuses on the history and philosophy of scientific instruments, particularly those developed for analytical chemistry during the twentieth century. Here he pursues a familial interest being the son of a co-founder of an early developer of spectrographic instrumentation, Baird Associates. He is the author of *Thing Knowledge*: *A Philosophy of Scientific Instruments* (forthcoming from University of California Press) and currently is working on a history of the firm his father founded. He edits *Techné*: *Technology in Culture and Concept*, Journal of the Society for Philosophy and Technology.

Stuart Bennett trained as a mechanical engineer and has taught at the University of Sheffield since 1968. He is now a Reader in the Department of Automatic Control and Systems Engineering. He has published extensively on the history of industrial measuring instruments. His current interests relate to the impact of instruments on the organisational and operational structure of the process industries. His non-historical research is on the socio-technical modelling of manufacturing processes.

Charlotte Bigg is completing a PhD at the University of Cambridge on the history of spectroscopy c. 1880–1920 in chemistry, physics and astrophysics. She is currently a research fellow of the Max Planck Institute for the History of Science in Berlin.

Luigi Cerruti graduated in chemistry at the Università di Torino in 1964. He was a physical chemist by training and profession until 1979, since when he has worked on the history of chemistry and physics in the nineteenth and twentieth centuries. His principal fields of interest are the contributions of British physicists on the border between chemistry and physics, the chemical atomic–molecular theories, and the history of the Italian chemical community. He is associate professor of the history of chemistry at the Università di Torino, Dipartimento di Chimica Generale ed Organica Applicata.

David Knight has taught the history of science at the University of Durham since 1964. He has published *Ideas in Chemistry*: *A History of the Science* (2nd ed., 1995), *Humphry Davy*: *Science and Power*, (2nd ed., 1998), and edited *The Making of the Chemist*: *The Social History of Chemistry in Europe, 1789–1914*, (1998, with

Helge Kragh). From 1994 to 1996 he was President of the British Society for the History of Science.

Pierre Laszlo is emeritus from both the École polytechnique and the University of Liège. He published *Organic Spectroscopy* with Peter Stang in 1971. A member of the editorial board of *The Journal of Magnetic Resonance*, he taught spectroscopic methods for elucidation of the structure of organic molecules: also as guest professor in various academic settings and as instructor in numerous workshops. His hands-on experience was with nuclear magnetic resonance. In the early 1960s, he was involved in signal-to-noise enhancement using multi-channel analysers. In the 1970s, he pioneered sodium-23 and cobalt-59 NMR, and later edited the two-volume *NMR of Newly Accessible Nuclei* (1983).

Peter J. T. Morris is Senior Curator at the Science Museum, London, in charge of the chemistry collections and a visiting lecturer at the chemistry department of Imperial College of Science, Technology and Medicine. He has previously edited *Milestones in 150 Years of the Chemical Industry* (1991 with Hugh Roberts and W. Alec Campbell), *The Development of Plastics* (1994, with Susan Mossman), *Determinants of the Evolution of the European Chemical Industry, 1900–1939* (1998, with Anthony S. Travis, Harm G. Schröter and Ernst Homburg), and *Robert Burns Woodward: Architect and Artist in the World of Molecules* (2001, with Theodor Benfey). At present, he is working on the relationship between the chemical industry and the environment, and edits *Ambix*, the journal of the Society for the History of Alchemy and Chemistry.

Yakov M. Rabkin has taught the history of science at the University of Montreal since 1973. He has worked in two broad fields: (a) the study of relations between science and technology, and (b) cultural and political history of science. His paper is a good example of his work in the former field. His works in the latter field include books *Science between the Superpowers* (1988) and *Interaction between Scientific and Jewish Cultures in Modern Times* (1995) as well as over a hundred articles, including quite a few on the history of Russian and Soviet science.

Nicolas Rasmussen has written extensively on the rôle of technique in scientific change in the life sciences, including an award-winning book on the origins and impact of the electron microscope in cell and molecular biology, *Picture Control: The Electron Microscope and the Transformation of Biology in America, 1940–1960* (Stanford University Press, 1997). His current research deals with the interplay between physiological knowledge and pharmaceutical development during the early and middle twentieth century.

Carsten Reinhardt is assistant professor for history of science at the University of Regensburg, Germany. His research interests include the history of industrial research in the nineteenth and twentieth centuries and the history of physical instrumentation in the chemical and biochemical sciences, 1945–1985.

Joachim Schummer studied chemistry, philosophy, sociology and history of art at the Universities of Bonn and Karlsruhe where he received his diploma in chemistry (1990), MA in philosophy and sociology (1991) and PhD in philosophy (1994). He is currently a lecturer of the philosophy and history of science at the University of Karlsruhe and is undertaking a research project in philosophy of chemistry, with the financial assistance of the Deutsche Forschungsgemein-schaft. He has published *Realismus und Chemie* (1996), *Philosophie der Chemie* (1996, co-editor), *Glück und Ethik* (1998, editor) and 40 papers on various philosophical topics. Since 1995, he has also been the editor of *Hyle: International Journal for Philosophy of Chemistry*.

Terry Shinn is Directeur de Recherche at the CNRS and based at the Paris Maison des Sciences de l'Homme. His recent work focuses on the emergence and operations of a new form of instrumentation research, 'research-technology'. He has also published extensively on science education, the organisation and dynamics of laboratory research, and on changing industry/science relations.

Leo B. Slater is currently a Senior Research Historian and the John C. Haas Fellow in the History of the Chemical Industries at the Chemical Heritage Foundation in Philadelphia. In addition to holding a masters degree in chemistry from Stanford University and a doctorate in history from Princeton University, he worked in drug discovery at the Schering-Plough Research Institute for four years. His current research deals with the chemical history of malaria control.

Anthony S. Travis is deputy director of the Sidney M. Edelstein Center for the History and Philosophy of Science, Technology, and Medicine at The Hebrew University of Jerusalem, where he is Senior Researcher in the history of technology. He is author of *The Rainbow Makers: The Origins of the Synthetic Dyestuffs Industry in Western Europe* (1993), and co-author (with Carsten Reinhardt) of *Heinrich Caro and the Creation of Modern Chemical Industry* (2000). He is currently researching the history of the former American Cyanamid dye-manufacturing plant at Bound Brook, New Jersey.

Setting the Scene

Technological Innovation in Science: The Adoption of Infrared Spectroscopy by Chemists[*]

YAKOV M. RABKIN

Department of History, University of Montreal, Montreal H3C 3J7, Canada,
Email: yakov.rabkin@umontreal.ca

After World War II chemical research changed significantly under the influence of new physical research methods. Among the most important new methods was infrared spectroscopy, which acquired remarkable popularity during the 1950s and 1960s. The number of infrared instruments, a handful before the war, rose to 700 in 1947, to 3000 in 1958, and to 20000 in 1969.[1] The technique's use in scientific research, as recorded in a 1965 report issued by the National Academy of Sciences in Washington, DC, skyrocketed correspondingly (see Figure 1).[2] There are several important reasons for the popularity of infrared spectroscopy.

The technique significantly expanded the limits of organic and inorganic chemical research and enabled chemists to tackle problems whose scope and nature had hitherto eluded fruitful investigation.[3] The qualitative information conveyed by an infrared spectrum (*i.e.* its structural diagnosis) contained more indicators of the nature of the molecule being examined than more laborious and less precise techniques (boiling-point measurements, refractometry, densitometry) used earlier. The compound could be analysed by infrared spectroscopy in any phase or condition; the versatility of the new method was such that liquids, solutions, gases, solid materials, powders, coating films and finishes could all be turned into samples for an infrared spectrophotometer. It became possible to distinguish between two very similar molecules and to assign certain bands in the spectrum to particular bonds or groups of atoms in the molecule. Although the molecule vibrates as a whole, certain groups may vibrate some-

*Reprinted with permission from Y. M. Rabkin, *Isis*, 1987, **78**, 31–54.

what independently of the rest because of symmetry or because of widely separated frequencies. Recognition of the recurrence of the same groups in organic compounds led to the emergence of group frequency analysis, which soon became a standard identification method in organic chemistry. This, in turn, opened new and exciting opportunities for empirical research on the mechanisms of organic reactions.

Infrared spectroscopy also enriched the chemist's understanding of molecular symmetry and structure. Standard sources of spectral information, made available thanks to the standardization of equipment, largely eliminated the technical difficulty of identifying compounds in complex mixtures. Use of the technique also enabled the chemist to calculate thermodynamic functions of chemical equilibria and to study weak chemical interactions, such as hydrogen bonding. Moreover, the application of the technique was relatively simple, requiring no specialized training beyond the usual expertise of a research chemist. This factor, plus the rapid decrease in the price of the equipment, turned infrared spectroscopy into a routine tool used by undergraduate students of chemistry alongside the pH meter and analytical balance. Indeed, by 1960 'every undergraduate chemistry major at the University of California had the opportunity to record and interpret infrared spectra in two laboratory courses'.[4]

The consensus among chemists and historians of chemistry is that infrared spectroscopy was a crucial factor in the development of chemical research in the 1950s and 1960s. Adoption of the technique took place rapidly shortly before that time. According to Jean Lecomte, a Paris spectroscopist, 'avant 1939, la prise d'un spectre d'absorption ou de réflexion infrarouge constituait une opération délicate, reservée seulement à un petit nombre d'initiés'. In 1966, however, Norman Wright, a veteran of the field, wrote: 'From a little-known and relatively unused method two and a half decades ago, infrared spectroscopy has now reached a status of general acceptance and utility unequalled among the analytical methods of organic chemistry'.[5] Indeed, even as early as 1948 the technique had obviously become a routine analytical procedure, as one of the first textbooks on 'practical spectroscopy', published that year, makes clear:

> The accomplishments achieved by scientists' use of the spectroscope form a list so as to leave no doubt that this instrument is one of the most powerful now available for investigating the natural universe. But spectroscopy is valuable *not only* to the research scientist: it finds everyday and increasing use in technological laboratories. Today directors of such varied enterprises as factories, assay offices, arsenals, mines, crime detection bureaux, public health departments, hospitals, museums and technical research institutes consider access to spectroscopic equipment essential to the proper functioning of their laboratories.[6]

The passage just quoted suggests a path of diffusion, from the research scientist to the director of a factory, that abounds in the literature on the relationship between science and technology. Closer study of the actual course of adoption of infrared spectroscopy will test the adequacy of this construct and

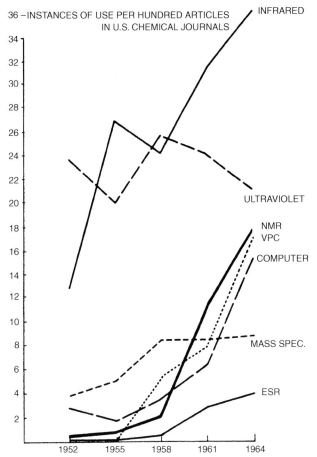

The use rate, as defined by citations of use in journal articles, of seven types of instruments commonly used in chemical research in the United States. Note the striking increase in the use of infrared spectrometers

From Chemistry: Opportunity and Needs, *National Academy of Sciences, Washington, DC, 1965 (Figure 16)*

illuminate the role of scientific instruments in the advancement of science. That role is not yet fully explored. The debate on the relation of science to technology is animated but factually deficient.[7] Most existing historical studies of scientific instruments are largely oriented toward antiquarian interests. Philosophers of science are quite sensitive to the importance of instruments for the development of science, recognizing, for example, that 'the experience of modern natural science is *apparative* experience, and experience necessarily structured by instrumentation. The instruments used for the collection of empirical results ... are themselves constitutive of these results'.[8] But philosophical statements on the subject rarely refer to the empirical historical data presumably used as a basis for their formulation.

This article seeks to remedy that deficiency by treating infrared spectroscopy as a case study in the introduction of new instruments into science, examining how it originated and diffused into chemical research laboratories. In identifying the events that accounted for the drastic change from an exotic technique to a commonplace research method, I will rely on interviews with protagonists in the wartime development of infrared spectroscopy. In general, scientific articles, which can be informative and even revealing on issues related to conceptual advances, are relatively poor sources of information about advances in instrumentation. Moreover, the spread of the infrared method occurred during the war, which imposed severe limitations on scientific publishing, especially of research that was intimately connected with the conduct of the war. After a discussion of infrared spectroscopy in chemistry in a comparative perspective, I will outline the implications of this example for our understanding of the interface between science and technology.

Beginnings

Infrared usually denotes the invisible part of the spectrum ranging from 0.75 to 1000 microns. The discovery of infrared radiation can be traced to an experiment undertaken by William Herschel in 1800. Herschel observed two maxima in the sun's spectrum, one visible and the other invisible, and assumed that two kinds of radiation were responsible for them: light radiation for the visible maximum and heat radiation for the invisible one.[9] Herschel's work followed logically from the discovery of the spectrum by Isaac Newton over one hundred years before and was therefore in the mainstream of the expanding scientific literature on the phenomenon of light. Some historians date the discovery of infrared radiation somewhat earlier. The concept of radiant heat was formulated by Carl Scheele in the famous treatise of 1777 in which he introduced his discovery of oxygen. Thirteen years later Marc-Auguste Pictet reported an experiment with two parallel mirrors, in which invisible radiation was reflected from one mirror to the other, causing an increase of temperature in its focus. The effect became more pronounced when the tip of the thermometer was blackened. Pictet's work was published in 1790 in Geneva but apparently remained unknown to those who later developed the field.[10]

After Herschel's experiment, infrared radiation was, and remained for over a century, an object of research rather than a research tool. How did this object of physical research become a chemist's tool? First, physicists developed the instrumentation capable of investigations into the nature of the invisible rays of heat. Advances in the field were to be contingent mainly on the development of radiation detection methods.

In Florence in 1829 Leopoldo Nobili was the first to report an early infrared detector: the thermopile. Thermopiles are series of pairs of dissimilar metals in which electric current is generated when temperatures differ between the junctions. Later Nobili used blackening (which he either reinvented or borrowed from Pictet without acknowledgment) as a means of increasing the sensitivity of the new detector. Nobili's friend, compatriot and collaborator Macedonio Melloni extended the range of infrared detectors when he discovered that rock salt was much more transparent to infrared radiation than glass and therefore made better prisms for the detectors. This discovery came after years of experiments showing that 'colors' exist in the infrared range and that different prisms behave toward the transmission of these rays just as transparent-colored materials behave toward visible light.[11] Once it was ascertained that infrared radiation was in fact an extension of the visible part of the spectrum and, later, a bridge to electromagnetic radiation, properties of the infrared rays were studied with respect to their absorption by various transparent bodies. The object of these investigations remained the radiation, not the bodies it was made to traverse.

The progress of infrared techniques slowed in the 1840s; work resumed some forty years later when the future secretary of the Smithsonian Institution, Samuel Langley, who was an investigator of solar radiation, turned to studies of the infrared region. He introduced a more precise heat detector called a bolometer, a sensitive resistance thermometer connected with a galvanometer which enabled him to measure the relatively weak grating spectra and determine wavelengths in the infrared range.[12] (Gratings were made by ruling a large number of closely spaced scratches; they produced narrower spectral lines. The advance in precision, however, was offset by a loss in the intensity of the resulting spectra.) This first accurate measurement of absolute wavelengths opened a way to analytical uses of infrared radiation: it became possible to identify absorption bands independently of the method of dispersion or of detection. These applications were, however, outside Langley's interests in solar radiation.

Once regularities in the absorptive behavior of infrared radiation were correlated with the nature and structure of the absorbents, then absorbents and their properties became objects of research, as well as the infrared rays and their properties. The research remained in the domain of physics at first because only physicists possessed the requisite knowledge of the infrared phenomenon and its complex instrumentation. Also working in the 1880s, two British officers, William Abney and E. R. Festing, examined the infrared spectra of nearly fifty organic compounds and speculated about the spectral identification of carbon–hydrogen vibrations. They found that aliphatic radicals produce similar bands in infrared spectra even when linked with different elements; the conclusion of one of their papers contained an explicit recommendation to use infrared spectra in research on radicals and functional groupings in organic molecules. As R. Norman Jones, a Canadian pioneer in spectroscopy, later observed, 'no organic chemists of the day were sufficiently aroused to follow it up; again the boat was missed'. Yet their work was not completely unknown: in 1892 a Dutch paper reporting on infrared spectra of alcohols cited their data on aliphatic radicals as important empirical evidence. It was cited again by William Coblentz in 1905, when he took up spectral identification of organic compounds.[13]

By the turn of the century infrared studies had developed primarily in connection with investigations of the phenomena of light. At this point investigations of molecular structure by means of infrared spectra began to come slowly into their own, spurred in part by work at the crossroads of experimental and theoretical physics that led to the emergence of important conceptual problems concerning the interaction between matter and radiation. Most important was work on blackbody theory by such scientists as Wilhelm Wien and Max Planck. Ultimately Planck's work on the theory of black body radiation led him to formulate his radical hypothesis of the quantum, and the development of quantum mechanics in the 1920s in turn led to a new phase of experimental and theoretical work in infrared spectroscopy, that is, determination of vibration frequencies.[14]

Experimental work of importance to the organic chemist was also being carried out at the turn of the century, by William Coblentz at the National Bureau of Standards (NBS) in Washington, DC. Although Coblentz had been trained in physics and remained a physicist, he undertook the spectral identification of many organic compounds. Published in 1905, his *Investigations of Infrared Spectra* contained over one hundred spectra of such compounds and provided the first experimental evidence of the link between the molecular structure and the spectral characteristics of chemical substances. Coblentz defined the infrared region as one to sixteen microns and applied infrared spectroscopy to simple molecules. His equipment consisted of a rock-salt prism mounted on a mirror spectrometer. The source of radiation was a Nernst lamp mounted before the collimator slit. Coblentz chose a more sensitive radiometer, designed several years earlier, in preference to the more common bolometer. All of his equipment was of course custom-made and, as his description shows, required laborious calibrations and adjustments. He drew a number of significant conclusions from his experimental work: that configurations of the links between atoms in a molecule are reflected in the spectrum; that an increase in molecular weight does not lead to a shift of the absorption maximum; and that certain absorption frequencies remain constant for given molecular groupings even in the presence of other groupings in the same molecule. These fundamental principles laid the basis for a wide variety of analytical uses.[15]

Though primarily concerned with chemical applications of the infrared technique, Coblentz kept abreast of the work of those concerned with basic problems of radiation. In the introduction to his 1905 work he distinguished between these two motivations for engaging in infrared research and provided detailed discussions of the recent data obtained by J. J. Thomson, Johannes Stark, Georges Friedel and other prominent scientists. His reference to Planck is telling: 'To crown all this we have the electromagnetic theory of selective absorption in isotropic nonconductors of Planck, in which Stockl's observations that maxima of absorption bands depend upon the solvent are applied to the question of intramolecular resonance. This is of great interest, since it is the first theoretical recognition of the possible unification of the selective absorption of a solution and the selective absorption of the solvent'. Coblentz found the imperfections of physical theory no hindrance to his 'taxonomic' work. As he noted elsewhere in his book: 'Experimental observations always have some value. This is not always true of theories which are built, more or less, upon hypotheses and must stand or fall with them'.[16]

Despite the relevance of Coblentz's experiments to organic chemists, they did not lead to an explosion of chemical analyses on the same lines. As Norman Jones put it several decades later, 'there were no chemists ready to take over from where Coblentz left off'.[17] Jones blamed a communication gap between organic chemists and molecular spectroscopists for the delay in the diffusion of the technique into chemistry. But at least some potential users were apprised of the new method, notably Charles F. Mabery of the Case School of Applied Science in Cleveland, a recognized specialist in petroleum analysis in the United States. Mabery helped Coblentz procure samples of pure distillates of petroleum unavailable elsewhere, as Coblentz acknowledged: 'When one considers that it has taken seven years to prepare them, and that his [Mabery's] collection is the most complete in existence, the value of the accession of these samples to the list of compounds investigated becomes apparent'.[18] Mabery obviously saw the infrared spectra of the compounds he had put at Coblentz's disposal: their cooperation accounts for the tradition of uninterrupted, albeit limited, use of the technique in petroleum chemistry.

The slow diffusion of Coblentz's method appears to have been due chiefly to its technical difficulties. Co blentz himself was quite explicit on the point: 'The process of mapping infra-red spectra … is, at best, a slow and tedious one'. More important, at the time there was no great call for the results. Several decades later Coblentz wrote that 'the discovery of the spectral absorptive properties of materials which would have a practical application could not then be foreseen.... Furthermore, the real need did not arise until three or four decades later'.[19]

Infrared investigations remained the province of physicists for the first third of the twentieth century. As the nature of chemical bonds became clearer in the light of quantum physics, various centers of molecular spectroscopy developed that tackled the issue of the vibrational energy levels of molecules. They concentrated on a limited group of symmetrical molecules of low molecular weight, not exceeding five or six atoms. The advent of Raman spectroscopy refined the process, enabling scientists to use the two techniques jointly to predict and interpret the energetics of simple molecules in terms of their rotational and vibrational behavior.[20] While advancing the knowledge of the nature of chemical bonds, however, these developments were of little help in interpreting the vibrational spectra of the vast majority of molecules that were of central concern to most organic chemists. These molecules were too heavy to give vibrational bands with a resolvable rotational fine structure, and their complexity and asymmetry precluded meaningful interpretation of the vibrational frequencies.

Among these schools of vibrational spectroscopy were two groups in the United States that followed up on Coblentz's work: scientists working at the NBS, where Coblentz was head of the Physics Division, and graduates and faculty of the Johns Hopkins University in Baltimore, less than fifty miles away. The two groups of followers, though overlapping at times, had distinct research interests. The NBS scientists used infrared radiation as a complementary method of identification of hydrocarbons in their systematic work on the chemical composition of petroleum; the Johns Hopkins people aimed at the loftier objective of understanding infrared spectra in terms of molecular structure and, more generally, of theoretical chemistry. It was the NBS work that eventually led to routine adoption of infrared spectroscopy.

R. Bowling Barnes, a Hopkins graduate and a noted physical experimentalist working at Princeton, extended the application of the infrared technique beyond the simple molecules studied by Coblentz to other chemical problems. Barnes never worked with Coblentz but considered his contribution 'the foundation of every beginner's knowledge of infra-red spectra'. Studying the nature of chemical bonds, Frederick Bell, a Hopkins professor, made important observations about the infrared spectral characteristics of organic isomers.[21] However, neither Barnes nor Bell could extend the work to molecules larger than C_6; theoretical calculations were then impossible for larger organic molecules. Frequency assignment and thermodynamic computations were important research issues for physicists and theoretical chemists all over the world but were of little interest to the organic chemist, who ignored the technique.[22] Applications of infrared spectroscopy to organic problems developed slowly.

The next stage in this development occurred in the early 1930s, as fuel companies first began to learn of the technique and to consider its possibilities. Fuels were a particularly propitious subject for infrared spectroscopy because the spectra of isomers such as octanes showed pronounced differences, while their nearly identical boiling temperatures precluded physical separation and identification. Work on fuels was based on the empirically observed high degree of invariance of the vibrations controlled by motions of atoms forming a particular link or group.

The work on fuels built on studies at the NBS. In 1928 Frederick Brackett, a Hopkins graduate and a former employee of the NBS, reported distinctions in the absorption spectra of the CH, CH_2 and CH_3 radicals in the 1.2-micron range. Five years later a report from the NBS, written by Urner Liddel and Charles Kasper, shed light on the peculiarities of the infrared spectra of the same radicals in some aromatic hydrocarbons.[23] These findings prompted a small cooperative research group of the American Petroleum Institute (API) to apply the technique systematically in a study of the chemical composition of petroleum. The group was then located at the NBS in Washington, where Coblentz was still head of the Physics Division; this explains the easy diffusion of the technique he had developed into the group's work. The API group was doing research for the entire oil industry, and its expertise and findings were thus easily accessible for petroleum and chemical companies throughout the 1930s and 1940s.[24]

The head of the API group, Edward Washburn, published the first report of the analyses of petroleum fractions in 1933, the same year the Liddel and Kasper report appeared.[25] Washburn, a noted physical chemist and chief chemist at the NBS, who died less than a year after the publication of this paper, hailed the new method as a source of 'fingerprints' of various compounds in the gasoline fraction. The technique added nothing new to what was already known from other sources, but it confirmed the identities of hydrocarbons that had already been separated and characterized by more traditional methods.

In 1937 Frank W. Rose of the API–NBS team issued a report that made considerable impact. It was the first one to deal exclusively with the infrared properties of selected petroleum hycrocarbons. A year later Rose interpreted infrared spectra of petroleum compounds in a novel way by considering aliphatic

CH$_3$, CH$_2$ and CH groupings and the aromatic CH as independent variables.[26] He interpreted the infrared spectra of fifty-five pure hydrocarbons either synthesized by the API–NBS team or separated from petroleum at the NBS and in the laboratories of certain American oil companies. Two important ideas emerged in Rose's 1938 paper: that different structural groupings of hydrocarbons have absorption maxima at different frequencies; and that a given grouping has a constant absorption intensity at each of its characteristic frequencies. The latter conclusion allowed for a calculation of the molecule's absorption on the basis of the specific absorptions of the hydrocarbon groupings. In contradistinction to previous work on the infrared characteristics of hydrocarbons, the 1938 report enabled the identification of compounds or at least their attribution to a given class of series. The technique provided information about the presence of hydrocarbons in a mixture comprising certain structural groupings.

Because it was not impressive in terms of theoretical chemistry, the 1938 article by Rose has hardly been mentioned in written and oral histories of infrared technology. Yet it was this research that apparently opened the way to industrial uses of the technique. Chemical industries and petroleum companies, then undergoing a profound change (which I have elsewhere termed 'chemicalization'), found the method a fast and reliable means of monitoring complex organic reactions.[27] The 1938 article by Rose was the last published work from the API–NBS group that was devoted to infrared technology as a new research method. The group continued to use the technique in its work in the late 1930s, but the center of infrared research was shifting to industrial laboratories.

The infrared technique offered considerable advantages to researchers working with a limited range of similar compounds, as was the case of the API–NBS group and is usually the case with industrial companies involved in the mass production of often homologous substances. While traditional methods of analysis, such as measurements of melting point, boiling point or refraction index, yield information about one characteristic of the compound, an infrared spectrum offers dozens of physical indicators, each of which provides specific information about the composition and the structure of the compound. The empirical tradition of the infrared method, stemming largely from Coblentz's investigations at the NBS, involved 'comparing the spectra of the largest obtainable number of different molecules having a common atomic group, including molecules of as great dissimilarity as possible, provided that they contain the common atomic groupings'.[28] The presence of absorption bands whose frequencies remain reasonably constant throughout the array of molecules forms the basis of identification of this specific atomic grouping in the unknown material. Even in the 1930s, obtaining infrared spectra could be made routine if enough time and money were invested in the calibration and standardization of equipment. These were conditions that only the largest chemical and petroleum companies could satisfy.

What did those companies that made the investment get from it? Among the industrial applications of infrared spectroscopy were 'fingerprinting' of compounds (*i.e.* unequivocal identification of a substance without damaging it), recognition of specific chemical bonds, linkages, or groups (*i.e.* O–H, N–H,

C≡N, C=O, CH_3), and determination of gross structural features (*e.g.* spatial configuration of atoms in a molecule, *cis–trans-* or *ortho–para–meta*-isomerism, and tautomerism). This qualitative information was essential in manufacturing organic compounds. No less important were various quantitative data about industrial processes, such as measurement of reaction rates as functions of the pressure, the temperature, or the catalyst used. Samples could be taken from the reaction chamber at regular intervals, the reactants consumed, and the new products formed quickly determined. Moreover, the removal of such small samples (less than 1 gram) would have no significant effect on the course of the reaction. Correlations between infrared absorption and concentration would indicate the relative presence of various compounds in the mixture – which was, in the late 1930s, impossible to determine by any other means, physical or chemical.

These attractions can be found conveniently listed in a paper presented at the 1940 annual meeting of the American Chemical Society in Detroit by a chemist from Dow Chemical, Norman Wright. Wright assessed the technique's comparative advantages as an industrial production control method in six areas of application: identification of organic compounds, determination of organic traces and their quantitative evaluation, studies of mechanisms and kinetic characteristics of chemical reactions, studies of tautomers and isomers, studies of polymerization, and determination of dissociation constants.[29] The tone of the paper suggested that industry was finding these applications of significant practical value. While Wright drew on Rose's 1938 paper in some aspects of his presentation, the main ideas could be traced to Coblentz. But Rose's paper was significant in its very lack of theoretical innovation: rather than looking for new physical aspects of infrared radiation and contributing to research activities in physics. Rose and his colleagues at the NBS had shown how practical problems of trivial scientific significance could be tackled by the technique.

While it was mainly in industry that the empirical tradition of infrared technology developed, the more fundamental infrared-assisted studies of chemical structure continued at universities. By most accounts, the world center of such research by the late 1930s was the University of Michigan. There Harrison Randall extended infrared research to the difficult long-wave region beyond 30 microns. He also saw the challenge of improving the methods of measurement, which, he claimed, had undergone no 'fundamental change' for over forty years. Thus he was the first to use a grating spectrometer in effective combination with a rock salt spectrometer; this greatly improved the precision and resolution of infrared spectra. While the principal advantage of salt prisms as dispersive elements is the relative simplicity of the instruments that use them, gratings provide consistently higher and relatively constant dispersion, particularly in the short-wavelength portion of the infrared range. High-quality gratings (which were first successfully ruled by Henry Rowland at Johns Hopkins University in 1890) are also less sensitive to variations of temperature. Randall's work was therefore a creative breakthrough in instrumentation, initially important mainly for spectroscopists working on calculations of molecular structures but at a later stage essential for the improvement of on-line control instruments used in chemical processes. Randall himself was apparently unconcerned with uses of the infrared technique for chemical analysis: he was a physicist interested in basic questions of molecular structure.[30]

On the eve of World War II, then, infrared spectroscopy as an analytical tool was firmly in the domain of the major chemical and petroleum companies. In the universities, the technique was used by a few groups of physicists studying theoretical chemistry, but organic chemists were not interested in engaging in delicate and labor-intensive calibrations and adjustments that the technique still required. Organic compounds could be identified through a combination of traditional techniques of organic analysis that, although also time-consuming, were familiar, relatively simple, and possible to delegate to assistants. Not that petroleum and chemical companies resorted to the infrared technique because they could afford the time and money while the university chemists could not; to the contrary, industry found that infrared spectroscopy saved money, given the amount and the speed of analytical work required. It was an efficient method of analysing organic compounds, and so industry invested in its development. The technique was needed as a routine tool of production control, and at that time no other method or combination of methods offered comparable advantages. Research chemists, with the obvious exception of those interested in molecular vibration in the infrared range, could satisfy their analytical needs without the infrared spectrometer so long as it remained a nonstandardized, capricious, and difficult tool.

Wartime Developments

The metamorphosis of the spectrometer into a routine analytical tool occurred where it was needed – in industry – and not in the university, where it was not. Further, it occurred under the pressures of the war. Coblentz, that veteran of infrared spectroscopy, later expressed his ambivalence about the rapid progress his method made in the 1940s: 'The regrettable thing is that this sudden advance could not come in peaceful surroundings, and be used for wholly peaceful purposes'.[31]

Efforts in Britain

The initial impetus was felt, not surprisingly, in Britain. Before the war, as in the United States, infrared work was done by and for physicists; it was carried out primarily in universities, with the two main groups being established at Cambridge and at Oxford. The group at Cambridge comprised C. P. Snow, of metascientific fame, and Gordon Sutherland, known for his scientific pursuits in the field of infrared spectroscopy. Sutherland had studied two years with Randall at Michigan; at Cambridge he continued his work on the infrared characteristics of gases, of hydrogen bonding, and of solid HCl and CH_4.[32] His activities focused on issues of significant interest to physicists and theoretical chemists.

The Oxford group centered on Harold Thompson. Like Sutherland, Thompson was trained in physics and physical chemistry. As a young scientist he went

Early infrared spectrometer developed by Harold Thompson at Oxford University in the late 1930s. Science Museum collections, inventory number 1977-244. It was based on a Hilger D83 spectrometer, c. 1934. Science Museum/Science and Society Picture Library

to Germany where, in the late 1920s, he studied under such luminaries as Walther Nernst and Fritz Haber. Of no less import for Thompson's scientific development was a year he spent as a paying guest in the house of Max Planck. Back at Oxford he tackled the issue of molecular vibrations, which led him to become interested in ultraviolet spectroscopy and, later, in the infrared technology. 'A primitive peace of equipment', he told me, 'was purchased for IR research in 1934 thanks to a grant from the Royal Society. IR absorption spectroscopy was to be used to measure molecular vibration frequencies and to do calculations on simple force constants. It was of purely academic interest. I had always had a sort of inclination towards quantum theory here at Oxford since the late twenties'. Thompson had some links with the chemical giant Imperial Chemical Industries (ICI), which was peripherally interested in the potentialities of the technique, but, as he concisely summed up the status of infrared work in Britain: 'Until the war came along no one thought the work some others and I were doing at Oxford and Cambridge was worth a damn'.[33]

This indifference did not persist once Britain declared war on Nazi Germany. Sutherland and Thompson were summoned to the Ministry of Aircraft Production in London, where they were to help gather intelligence about Germany's petroleum supplies. More specifically, they were to answer the following questions: Were the Germans synthesizing isooctane and octane? Did the Germans use any other synthetic processes for the production of aviation fuel? Fuel from downed German bombers provided samples, and the two groups, one at Oxford and the other at Cambridge, set out to prepare pure hydrocarbons in order to calibrate their primitive infrared apparatus. 'Conceptually the work was quite easy', Thompson commented forty years later, 'but technically it certainly was not'.[34]

The API–NBS articles published just before the war provided the basic ideas for the British infrared intelligence gathering. Monetary restrictions did not seem to exist: 'How much money do you want? You can have what you like!' was the message from the Ministry of Aircraft Production.[35] Hundreds of pure hydrocarbons in the 20–200 °C range were synthesized at the two universities and at ICI, with which Thompson kept up contacts established before the war. Within a year answers were found: Germany was using various mixtures of East European and Baku oil with synthetic hydrocarbons (from September 1939 to June 1941 Russia was Germany's main supplier of petroleum). Some of the expertise resulting from the study was also channeled to the control of fuel production in Great Britain.

About this time the United States was drawn into the war, with most important repercussions for the development of infrared spectroscopy. 'Whatever happened in the IR in Britain was linked up with what was going on in the USA', reminisced Sir Harold Thompson four decades later.[36] Massive intervention of the federal government in the economic life of the United States opened the way for nationwide cooperation in two strategic production areas where the infrared technique was urgently needed. This concerted effort built on the expertise in the technique already developed by the chemical and oil companies.

The Synthetic Rubber Program

The US synthetic rubber program was one of the strategic production areas that relied on infrared spectroscopy. The program was based on the polymerization of butadiene, a chemical available only from petroleum refinery gases. But the gases contained dozens of different hydrocarbons whose identification by chemical means would have been unthinkable given the scope now required. Analyses could be done only by means of infrared spectroscopy. The Office of Rubber Reserve, which was in charge of synthetic rubber production, applied the legal and financial stimuli required for the speedy manufacturing of the equipment. It instructed those companies versed in the technique, American Cyanamid, Dow Chemical and Shell Development, to pass on their expertise to all corporate participants in the rubber production program.

American Cyanamid and Shell Development had already begun working with small neighboring instrument firms at the start of the rubber program. American Cyanamid had been one of the first companies to use infrared techniques. Since standard equipment did not exist, the company began in 1936 to build infrared apparatus.[37] When more instruments were needed to produce synthetic rubber, American Cyanamid contracted out production of those instruments to a small local optical company, Perkin–Elmer. American Cyanamid chose Perkin–Elmer because it specialized in high-quality optics, primarily for military uses, and because both were located in the Stamford, Connecticut, area. American Cyanamid first ordered Perkin–Elmer to build several off-axis collimating mirrors; then it requested a few infrared instruments for its plants elsewhere in the United States; finally Perkin–Elmer began producing infrared equipment for other industrial clients.[38] These wartime instruments eventually evolved into Model

12, the commercial version of the infrared spectrophotometer that Perkin–Elmer began manufacturing after the war.

Like American Cyanamid, the Shell Development Company of Emeryville, California, was expert at infrared analysis and even at building infrared instruments for in-house use. Robert Brattain, a physicist with experience using the technique to elucidate the structure of penicillin, was at the core of the Shell Development's infrared program.[39] The Office of Rubber Reserve instructed Brattain to organize mass production of standard infrared equipment for the analysis of C_4 hydrocarbons. The actual manufacture was begun at a nearby instrument company with experience making pH meters and ultraviolet spectrophotometers, National Technical Laboratories. The company, which later became Beckman Instruments, produced about seventy-five copies of Brattain's prototype by 1945. Beckman's expertise in measuring small electric currents generated in photoelectric cells became the decisive factor in its success with infrared equipment.

Beckman's IR-1 was produced under direct contract with the Office of Rubber Reserve. Its manufacture was facilitated by the top-level AAA priority rating that secured supplies of scarce raw materials, and the IR-1 could be sold only to AAA-rated customers, who were under total wartime blackout as to publication in technical and scientific journals. Perkin–Elmer, involved in a different part of the government program, operated under looser controls, and American Cyanamid's infrared specialists enjoyed a greater freedom to discuss the industrial uses of the technology even during the war.[40] This had significant repercussions for the diffusion of infrared technology and, to a certain degree, explains the advantage Perkin–Elmer enjoyed right after the war when its instruments, in contrast to the secrecy-bound Beckman IR-1, were more widely known and appreciated.

The infrared equipment built by Perkin–Elmer and that built by Beckman had little in common beside its uses. The Stamford equipment was called 'Breadboard': it was a bulky single-beam instrument with a slow thermocouple and a galvanometer to read out the thermocouple. Its strength was the high-quality dispersive element: Perkin–Elmer's expertise had traditionally been in the production of unique optical devices. Model 12, the first mass-produced instrument made by Perkin–Elmer, differed from the Breadboard only in that it was slightly more compact. Beckman's IR-1, built for the Office of Rubber Reserve in 1942, also used a rock salt prism, but the quality of its electrical and electronic design was higher than that of the Stamford instrument. The collaboration of Beckman and Shell was very close: the design came from Brattain, and the first IR-1 was shipped to Shell. Beckman's previous experience with ultraviolent spectroscopy enabled the company to design the IR-1 with such convenient features as turret stops for the prism arm, which permitted quick selection of over a dozen specified wavelengths. The infrared instruments that were produced in quantity, then, reflected the practical demands of the industrial users and the respective strengths of Perkin–Elmer and Beckman in optics and electronics, rather than different scientific ideas or theoretical concepts.

The Beckman IR-1 spectrophotometer
  Courtesy of Beckman Instruments, Inc

Petroleum Refining

The second strategic industrial area that contributed to the standardization of infrared spectroscopy and its diffusion was petroleum refining. Relatively simple uses of the techniques, mainly for analysis of C_4 hydrocarbons, sufficed for the rubber program. However, the oil companies, which embarked on full-scale production of gasoline by methods of alkylation and polymerization, required more versatile methods for on-line control. Infrared and Raman spectroscopy would satisfy that requirement.

The transfer of the infrared technology to the oil industry was directly connected with the work of the API–NBS group in Washington. The group had pioneered its application to petroleum chemistry; it had also helped diffuse the actual methods into practice by preparing the pure hydrocarbon samples indispensable for calibrating infrared equipment in dozens of oil refineries in the United States and, later, in Great Britain. The American war effort relaxed anti-trust laws for the duration of the hostilities. Thus the API–NBS group could serve as a research center for the entire oil industry not only in relatively fundamental scientific areas but also in applied work such as the synthesis and purification of standard hydrocarbons. The War Petroleum Board, whose main concern was to win the '100-octane race', especially emphasized dissemination of the sample hydrocarbons as a condition for standardizing the refining control method.[41]

In 1942 the API–NBS group began the large-scale production of infrared spectra of pure hydrocarbons. Supervised by a committee of infrared specialists from major oil companies, the group began distributing the calibration samples, accompanied by standard spectra of the sample hydrocarbons. Thus oil companies could promptly put their infrared equipment to efficient use. The API

Table 1 *Distribution of hydrocarbon calibration samples by the API, 1929–1952*

	Number of laboratories	Number of infrared spectra taken
Oil industry	56	48
Universities	31	9
Federal government	43	43

program was crucial in diffusing the technology even beyond the confines of the oil industry proper, because the samples and the spectral characteristics of pure hydrocarbons were also distributed to university and government laboratories in the United States and elsewhere. Conversely, both synthesis and spectral characterization of the pure compounds were carried out, under the auspices of the API–NBS group, in a variety of cooperating laboratories. By the mid-1940s these efforts were distributed as shown in Table 1. As late as 1948 the API collection of infrared spectra was still considered the best in the world.[42]

The Dissemination of Infrared Technology

For the first part of the twentieth century, the spread of information about infrared spectroscopy was not a highly organized affair. The diffusion of expertise was effected largely through diffusion of research personnel. As Paul Wilks put it: 'When the initial decision was made to manufacture serial IR instruments it was on the basis of knowledge of people who were in other, mainly chemical industries.... There were no graduates in the infrared techniques then. Before the 1950s no one came to Perkin–Elmer from the university' – that is, there were no freshly trained graduates in the technique. Thus American Cyanamid relied on the experience gained by physicists like Bowling Barnes and Urner Liddel from work carried out for the National Bureau of Standards. Industry also relied on high-level experience gained in research universities. Both Van Zandt Williams, who came to American Cyanamid in 1941 and went to Perkin–Elmer in 1948, and Robert Brattain at Shell, who made the designs that started Beckman in the infrared field, had done their graduate training at the spectroscopy laboratory at Princeton; Barnes had taught and carried out research there for three years before coming to Cyanamid. Their acquaintance with theoretical aspects of spectroscopy was an important asset when solving practical production problems, one that they put at the disposal of their industrial employers.

One tentative effort was made in the 1930s to educate the chemical profession at large, and especially students, about the uses of infrared spectroscopy. About the time he moved to Cyanamid, Barnes and Lyman Bonner, a research fellow in physics also at Princeton, published a detailed account of the technique suitable for students in 1937 in the *Journal of Chemical Education*.[43]

Once the war began, this situation changed. The cooperative nature of the wartime infrared work that characterized the US rubber program and the 100-octane race resulted in fast and effective diffusion of infrared spectroscopy.

When Harold Thompson visited the United States in 1943, he saw the first routine applications of the technique and one of the first standard infrared spectrophotometers, the Beckman IR-1, being used to monitor the polymerization of isobutene.[44] By the end of World War II infrared spectroscopy was used in dozens of industrial companies and federal laboratories in the United States and Great Britain. Just as important for wide use of the method, its diffusion to the universities was also well under way: infrared spectroscopy had become an accurate and reproducible way of analysing complex organic and inorganic mixtures.

The postwar diffusion did not occur 'naturally' or 'spontaneously': it came about because of a combination of business initiatives and scientific foresight. One of the crucial major promoters of infrared chemistry after the war was Van Zandt Williams, then at American Cyanamid. In 1943 Williams and several other infrared specialists at the company published an article describing their experience with the technique that found a wide audience. A relatively unknown physicist at the time, Williams already had considerable enthusiasm for the new method in 1943. A remark he and his co-authors made in the article, despite the constraints of military secrecy, was prophetic: 'Interest in the potentialities of this physical tool has become so widespread that infrared spectrometers may now be found in the laboratories of an impressive list of American companies, and the time will soon arrive when such instruments will be standard equipment in most laboratories and plants'.[45]

When Williams moved to Perkin–Elmer after the war, he began a campaign to make the company's infrared instruments cheaper, easier to use, and therefore more popular. Other observations made in that milestone paper reveal the direction he was to take in promoting the technology. Standardization was the first necessity, coupled with brand-name identification:

> Indicative of the difficulties of infrared technique is the fact that no standardized research spectrometers are available today. In other branches of spectroscopy, it is customary to mention a medium Hilger, a larger Bausch & Lomb quartz spectrograph, or a General Electric recording spectrophotometer and let a reference to the literature suffice as a description of the instrument. In contrast to this, the various infrared laboratories use instruments constructed according to individual ideas, designs and demands of each investigator.

Several other passages in the article identify priorities in no uncertain terms, especially ease of use and education in that use: 'It is necessary that commercial spectrometers be available, that there be sufficient background information in the literature so that an operator may be put on one of these instruments to produce results immediately without having to spend a long time acquiring basic data. The first person to recognize this demand was the infrared spectroscopist already at work in the field'.[46] While Williams may not have been the 'first spectroscopist' to recognize the new need, he certainly was the first entrepreneurial one to do so.

Among the novel initiatives that Williams undertook at Perkin–Elmer was the organization of summer schools as a means of educating the world of chemistry about infrared procedures. Universities were an important target of this campaign, which was led by Williams but enjoyed the collaboration of other companies, including Beckman. The importance of the universities lay not so much in the potential markets for infrared equipment that their research facilities constituted. Rather, as Paul Wilks put it, 'We were anxious to get our instruments into universities because they were training the students who would ultimately become our customers'. Universities were offered discounts, they used instruments free of charge, and university faculty were offered grants to go to summer schools to learn about infrared technology. 'We promoted the summer schools through our advertising literature. We really made these courses possible, that is, to all the instrument companies', observed Wilks. 'Each manufacturer would bring his instrument and provide instructors, so all the latest models would be included as part of the course'.[47] The marketing strategies of the instrument manufacturers thus helped spread the method.

As was pointed out at the beginning of this article, the spread of infrared spectroscopy as a routine tool widened the limits and improved the precision of analysis in both organic and inorganic chemistry. But standardization of the instruments has also had negative consequences. While they are much simpler to use, their manufacture has grown in complexity so that a chemist can no longer build tailor-made instruments for a particular research purpose. The manufacturer, assessing the potential to attract industrial as well as academic clients, decides what instruments will sell best and plans production accordingly. The researcher must then adapt projects to the equipment available and may even look for research problems to justify the expense of its purchase. 'That is why', commented Angelo Mangini, a spectroscopist from Italy, 'so much bad work is done today. People buy instruments and use them for the sake of it. They invent work that may have little significance to justify their acquisition'.[48]

Ironically, the commercialization of scientific instruments has also exacerbated the difficulty of comparing various spectroscopic techniques as tools for organic analysis and molecular structural determination. The high cost of instruments has made it virtually impossible for any academic laboratory to concentrate within its walls a complete range of spectroscopic equipment and to compare their performance on the same structural or analytical problem. Even the larger and wealthier industrial companies, which can afford to buy all the latest equipment, rarely engage in comparative studies since, as a veteran spectroscopist put it, 'human nature being what it is, there can be strong interdepartmental competition between analytical groups dedicated to different techniques'.[49]

Comparative Examples

One of the stated purposes of this article was to test a model, frequently cited in the literature, of the relation between science and technology: the model that traces the path of innovation from the research laboratory to an industrial setting. This brief history of the spread of infrared spectroscopy has touched on the role of developments in both areas. In assessing the relative importance of the theoretical breakthroughs in quantum physics, on the one hand, and of the military and economic factors, on the other, in the diffusion of the infrared technology into the practice of chemical research, the example of foreign, and especially German, developments may be quite instructive.

The importance of environmental factors is particularly well illustrated by the fate of work carried out by Jean Lecomte in Paris, work that built on Coblentz's. The Germans could conceivably have capitalized on it but, not surprisingly given its reception in France, failed to do so. Lecomte had carried out the infrared analysis of petroleum hydrocarbons even before the analyses undertaken at the American Petroleum Institute in the early 1930s. In an interview in 1977 Lecomte described using infrared spectroscopy 'to look at automobile and aviation fuels in 1928 – ten [*sic*] years before the Americans. We had 7000 spectra of the components of fuels and we published them, although I never published any interpretations'. (Lecomte's dating reflects the relative impact of the paper published by Washburn in 1933 and that published by Rose of the NBS–API team in 1937.)[50]

Lecomte's work acquired no following, although his article 'Spectres dans l'infrarouge' appeared in 1936 in a prestigious volume on organic chemistry edited by Victor Grignard. Lecomte even offered his work to the French air force, but it showed no interest, although he linked infrared spectroscopy directly with its technical needs. Indeed, his last prewar article appeared in the journal of the French *Armée de l'air* in 1939. By the time Paris fell to the Germans, Lecomte had accumulated infrared spectra of hundreds of organic compounds. A man of independent means, he quietly continued his work in Paris under the occupation, and his help was never solicited by the German military. He resumed publication of his results in 1945.

Lecomte's lack of impact begs an explanation. That offered by Sir Harold Thompson rests on his disciplinary affiliation: 'Lecomte's work never really made the impact it should have done. He didn't know enough chemistry. He was a physicist'.[51] But so were most Americans active in the infrared field: yet their efforts were eventually fruitful, while Lecomte's was not. It appears instead that Lecomte was strenuously 'pushing' his technical innovation toward practical uses at a time when there was no 'pull'. France's petroleum industry, unlike that in the United States, did not face the challenge of vastly different sources of crude oil. Nor did the French military get enough time to become concerned with Germany's industrial strategy, as did their British counterparts, who called in Sutherland and Thompson. Thus Lecomte's case may be a good example of how potentially relevant knowledge engendered by the internal logic of scientific pursuit and the researcher's practical foresight does not lead to innovation so long as there exists no external user of this knowledge. Lack of avenues for practical exploitation does not in itself hamper the emergence of applicable technical knowledge in science; it simply makes such knowledge sterile.

The situation in Germany was somewhat more complex. Germany had virtually all the ingredients for developing an infrared instruments industry well in

place: a highly sophisticated chemical industry with a long tradition of manufacturing synthetic fuels, a science-based instrument-making industry, and a vener-
able corps of world-renowned organic chemists and specialists in quantum
theory. Much of the prewar work in physics on the use of frequency values and
moments of inertia in the calculations of specific heats and other thermodynamic
constants of molecules was done in Germany. Moreover, German strength in
organic chemistry in general, and in its theoretical aspects in particular, could
hardly be matched. Yet when the war was over, German scientists found themselves well behind their British and American counterparts. 'They [the Ger-
mans] were staggered when they were told what could be done [in structural
diagnosis]', Thompson recollected of his postwar visit to Germany.[52] Most
surprising to him was that Karl Ziegler, a future Nobel Prize winner in organic
chemistry who attended one of his lectures, had no idea how simple it was to
distinguish between different radicals in complex organic molecules by means of
the infrared technique.

What factors can explain Germany's failure to develop chemical applications
of the technique? One likely factor is the general havoc caused by the Nazi
government in the academic world. The integration of academic scientists into
the war effort was different in both mode and scope from that in the United
States and Britain. German universities were brought into the war effort in a
direct and subordinate manner. Moreover, many creative scientists were alien-
ated, driven away, or killed. In fact, anti-intellectual trends in the Nazi movement
remained strong until the end.[53]

Second, the very reliance on scientific theory that had propelled German
industry to spectacular achievements during the preceding fifty years may have
become a negative factor in this case. Accumulating spectra that no theoretical
specialist in the infrared field could even begin to interpret had perhaps no place
in the social structure of German science and technology. In the United States,
however – and only in the United States – a significant demand for infrared
equipment already existed in the chemistry-based industries. This industrial
demand – not theoretical understanding that the technique could be applied –
created a spin-off effect among the instrument makers, who then found it
profitable to enlarge the market by spreading knowledge of infrared technology,
often at their own expense, into academia. Also important was the tradition in
the United States and Britain of scientists pursuing projects that were possibly of
little practical significance (like Thompson's prewar projects) or, conversely, of
little theoretical interest but considerable practical import (like Rose's 1938
work). These scientists' freedom to interact among themselves and with industrial companies was preserved in wartime under the aegis of a rather loose
organization of research. The resulting interplay between three sectors – universities, industry and government – rather than the impact of quantum mechanics,
should be credited for the diffusion of the infrared technique into chemistry. This
contradicts the often-held scientistic belief that 'one had to build a scientific
foundation – discover the principal laws of heat radiation, clarify its nature and
properties – in order to begin to use the infrared in practice'.[54]

The significance in this combination of factors of the preponderance of non-research users of the infrared technique becomes more evident when the growth of the infrared instruments industry after the war is examined. As with most standardized products, the United States took the lead in the commercialization of scientific instruments. The large number of users of such instruments outside basic research in the country, which resulted in a significant level of consumption of the equipment, may explain US predominance from the 1940s to the 1960s. The magnitude of the phenomenon becomes clear from a comparison of total sales and employment in the infrared instrument industry, which increased by seventeen and ten times, respectively, between 1940 and 1960. The uses of these instruments in basic research accounted for only 20 percent of the total.[55]

The British example points to another factor that led to the United States' being the only country to mass-produce infrared spectrometers. Like France and Germany, Britain had lacked the impetus to expand the infrared method beyond the usual theoretical realm before the war began. Even during the war, when Thompson and Sutherland successfully developed specific chemical applications of the technique, British industry had little capacity to innovate. Instead it availed itself of the industrial innovations and resulting products brought to Britain by lend-lease from the United States. European countries, when it came to the potential market for mass-produced infrared equipment, could not match the wealthy, war-stimulated American market.

Conclusion

The infrared analysis of organic chemical compounds took forty years to diffuse into scientific practice. This diffusion was made possible by a combination of military, legal and economic factors created by national emergency. The infrared technique became a reliable tool of scientific research only after it had been developed, refined, standardized and disseminated by industrial companies with important help from the federal government.

It was a physicist, Coblentz, who first used the infrared technique for a series of chemical analyses. Those who, through their isolated efforts, continued the practice in the 1930s and early 1940s were also physicists capable of using the infrared method and interested in understanding molecular structure. When physicists came to work for chemical and oil companies, the infrared technique became a routine, albeit difficult and capricious, tool. Its diffusion to the world of chemistry occurred only when the physicists joined instrument-making companies and the instruments became standardized and, eventually, easy for nonspecialists to operate. The urge to capitalize on the expertise in infrared instrumentation they had acquired during the war prompted these companies to embark on an educational campaign to create a need and thus expand the market to medium and small chemical plants and, ultimately, to university chemists. This consecrated the new role of the infrared technique, that of a potent and widespread tool of chemical analysis.

Chemists became responsive to the tool only when it became simple, reliable and relatively inexpensive. A rank-and-file chemist, whether in industry or in a university, would not build an instrument, grow a crystal to serve as its dispersive element, develop an electronic device to detect radiation, calibrate the instrument (after synthesizing pure chemicals for that purpose), and maintain constant humidity and temperature in the room that would house the spectrometer in order to use the infrared technique. The experience with infrared instrumentation suggests, then, that technical difficulties are a serious obstacle in the diffusion of a scientific idea. These have to be removed before the idea can be accepted as the basis for a new analytical method.

This 'detour' of a scientific idea is a significant instance of the impact of technology on science. Technology can routinize research techniques, which then diffuse into scientific practice. The infrared case can be compared to the diffusion of computers into everyday life. Once an exotic machine whose operation required the constant attention of a specialist, the computer has become a largely foolproof gadget accessible to the layperson as a result of application of advances in microelectronics. The story of infrared technology is similar in that expertise in spectroscopy was no longer a prerequisite for taking and interpreting spectra. So, too, operation of infrared equipment, mass-produced since the late 1940s, no longer requires a spectroscopist. The lay chemist no longer needs the priest-spectroscopist in order to commune with the hitherto-mysterious infrared spectra.

As the comparison with the computer shows, the history of infrared spectroscopy offers but one example of the impact on science of specialization in the production of scientific instrumentation. The impact of technology on science has intensified with the advent of a strong scientific instruments industry. Whereas during the war the chemical and oil companies only facilitated the diffusion of the infrared technique into scientific practice, the influence of instrumentation on science has since become more pronounced. And, as noted in the section on diffusion above, routinization, at least of infrared technology, has been a mixed blessing.

Historically documented cases like this one, which highlights the effect of technological and economic factors on the orientation of basic science, advance our understanding of the interface of science and technology, particularly when they are presented in relation to larger questions in the history of those fields.[56] Unfortunately, the data at the disposal of historians so far cannot support grand generalizations about the character of the science and technology interface. Peter Janich's claim, for example, that 'natural science is to be understood as a secondary consequence of technology rather than technology as an application of natural science' may be a stimulating contribution to our thinking on the subject, but it is hardly a statement of historical fact.[57] The nature of this interface may be too complex to lend itself to one-sided claims of primacy, particularly in view of the definitional imprecision of the two terms. This imprecision appears to arise from a lack of conceptual rigor rather than merely reflecting the empirical situation. While I appreciate his general description of the instruments industry, I find it difficult to agree with the conclusion drawn by Frank Greenaway from his material in the authoritative *History of Technology*: 'The instrument industry fitted into the general pattern of technology, both by providing for immediate requirements and, no less important, by making possible new technological advances. Many of the new instruments originated in scientific laboratories as an aid to fundamental research, but in many cases this industrial potential was quickly realized. The gradual disappearance of a distinction between science and technology is nowhere better illustrated than in this field'.[58]

It appears doubtful that our knowledge of the interaction between science and technology will increase if we turn away from cognitive, social and institutional distinctions between them as we admire the growing frequency of their interaction. In the case of infrared spectroscopy the demarcation between science and technology is self-evident – which, some would say, makes it easy for me to preach the virtues of conceptual rigor. But the experience of other branches of scientific instrumentation in chemistry seems to suggest a pattern of development close to the one found for infrared technology. 'Because the vitality of research depends ultimately on the practical importance of its applications, appropriate commercial instruments must be developed to sustain progress', one reads in a recent review of electrochemical instrumentation.[59] Apparently, this has become a truism among both the makers and the users of scientific instruments, albeit not yet among the historians of science. Historians of technology, traditionally more attentive to economic and other 'extraneous' forces, have noticed the role played by industry in the conduct of fundamental research. For example, Greenaway himself says that nuclear magnetic resonance, which was developed during World War II, was diffused into chemical research through the perseverance of

> electronic firms, encouraged by the potential market for rapid organic analysis offered by large chemical and oil companies.... It is clear that the development of this elaborate and expensive instrument could not have taken place unless it was manufactured on an industrial scale to serve investigations also carried out on an industrial scale. At the same time, it became available for academic research workers, concerned primarily with the advancement of fundamental knowledge.

Another British historian of technology, Donald S. L. Cardwell, has noticed that most works on the history of science neglect the role of scientific instruments and suggests that this may be due to certain philosophical preferences of historians of science, for example, their view of science as a 'pure' intellectual activity, positivism, and the 'immemorial academic snobbery of philosophers'. In Cardwell's opinion, 'it is possible to envisage a history of technology which is closely related to the history of science and to the history of ideas generally. Technology so regarded is not to be thought of as the dependent variable, drawing its ideas from and parasitic upon science; rather it is an equal partner contributing at least as much to the common stock as it draws out'.[60] I intend my study of the case of infrared spectroscopy to serve as an example of such an approach.

Acknowledgements

Research for this article was supported in part by the Social Sciences and Humanities Research Council of Canada. I wish to thank also Marika Ainley, and to recall the discussions I had on this subject with the late Derek De Solla Price.

Notes and References

1 Interview with Paul Wilks, Norwalk, Connecticut, Nov. 1981.
2 *Chemistry: Opportunity and Needs*, National Academy of Sciences, Washington, DC, 1965, Chapter 4.
3 See R. Norman Jones and Camille Sandorfy, *Technique of Organic Chemistry*, Wiley Interscience, New York, 1959, Vol. IX; and J. J. Turner, 'Inorganic Chemistry and Infrared Spectroscopy', *Chemistry and Industry*, 15 Jan. 1966, 109–114.
4 George C. Pimentel, 'Infrared Spectroscopy: A Chemist's Tool', *Journal of Chemical Education*, 1960, **37**, 651–657, on p. 651; see also Paul A. Wilks, 'Infrared Equipment for Teaching', *J. Chem. Educ.*, 1969, **46**, A9–A26.
5 Jean Lecomte, 'Conférence d'ensemble sur la spectrométrie infrarouge', *Spectrochimica Acta*, 1957, **11**, 479–480, on p. 479. Norman Wright is quoted in David N. Kendall, *Applied Infrared Spectroscopy*, Reinhold, New York, 1966, p. vii.
6 G. R. Harrison *et al.*, *Practical Spectroscopy*, Prentice-Hall, New York, 1948, p. 1 (emphasis added).
7 For a recent update of that debate see Yakov M. Rabkin, 'Science and Technology: Can One Hope to Find a Measurable Relationship?', *Fundamental Scientia*, 1981, **2**, 413–423.
8 Peter Janich, 'Physics – Natural Science or Technology', in *The Dynamics of Science and Technology*, Wolfgang Krohn *et al.* (eds.), Reidel, Dordrecht, 1978, p. 10.
9 William Herschel, 'Experiments of the Solar and Terrestrial Rays That Occasion Heat', *Philosophical Transactions of the Royal Society of London*, 1800, **90**, 293–326, 437–538.
10 Carl W. Scheele, *Chemische Abhandlung von der Luft und dem Feuer*, Uppsala/Leipzig, 1777; and Marc-Auguste Pictet, *Essai sur le feu*, Geneva, 1790. Among those mentioning this work before Herschel is V. A. Gurikov, in 'K predystorii infrakrasnoi tekhniki' (The early history of infrared technology), *Voprosy istorii estestvoznaniia i tekhniki* (Issues in the History of Natural Science and Technology), 1973, **3**(44), 53–56.
11 Leopoldo Nobili, 'Description d'un thermo-multiplicateur', *Bibliothèque Universelle des Sciences, Belles-Lettres et Arts*, 1830, **44**, 225–234; and Macedonio Melloni, 'Mèmoire sur la transmission libre de la chaleur rayonnante par différents corps solides et liquides', *Annales de Chimie*, 1833, **53**, 5–73.
12 Samuel Langley, 'The Bolometer and Radiant Energy', *Proceedings of the American Academy of Arts and Sciences*, 1881, **16**, 352–358.
13 W. Abney and E. R. Festing, 'On the Influence of the Atomic Grouping in the Molecules of Organic Bodies on Their Absorption in the Infrared Region of the Spectrum', *Phil. Trans. Roy. Soc. London*, 1882, **172**, 887–918; R. Norman Jones, 'Some Retrospective Thoughts on Vibrational Spectroscopy', *Canadian Journal of Spectroscopy*, 1981, **26**(1), 1–9, on p. 4; H. Julius, 'Bolometrisch onderzoek van Absorptiespectra', *Verhandelingen der Koninklijke Akademie van Wetenschappen*, 1892, **1**(1), 1–49; and William W. Coblentz, *Investigations of Infrared Spectra*, Carnegie Institu-

tion of Washington Publications, 35, Washington, DC, 1905, p. 5.

14 See Willy Wien, 'Uber die Energieverteilung im Emissionsspectrum eines schwarzen Korpers', *Annalen der Physik und Chemie*, 1896, **21**, 187–198; and on Planck's work, *e.g.* Martin Klein, 'Max Planck and the Beginnings of the Quantum Theory', *Archive for the History of the Exact Sciences*, 1962, **1**, 459–479.

15 Coblentz, *Investigations*, pp. 115 118.

16 *Ibid.*, pp. 11, 323.

17 Jones, 'Some Retrospective Thoughts', p. 5.

18 Coblentz, *Investigations*, p. 4.

19 William W. Coblentz, 'Early History of IR Spectroradiometry', *Scientific Monthly*, 1949, **68**, 102–107, on p. 107.

20 For an explanation of the principles of Raman spectroscopy and its use with the infrared technique see G. Eglinton, 'Infrared and Raman Spectroscopy', in *Physical Methods in Organic Chemistry*, J. C. P. Schwartz (ed.), Oliver and Boyd, London, 1964, pp. 35–125.

21 R. Bowling Barnes, 'IR-Spectra and Organic Chemistry', *Review of Scientific Instruments*, 1936, **7**, 265–271; R. Bowling Barnes and Lyman G. Bonner, 'A Survey of Infrared Spectroscopy', *J. Chem. Educ.*, 1937, **14**, 564–571, on p. 566; and Frederick K. Bell, 'The IR Absorption Spectra of Organic Carbonates', *Journal of the American Chemical Society*, 1928, **50**, 2940–2950.

22 See, *e.g.* Ta You Wu, *Vibrational Spectra and Structure of Polyatomic Molecules*, National University of Peking, Peking, 1939.

23 Frederick S. Brackett, 'Characteristic Differentiation in the Spectra of Saturated Hydrocarbons', *Proceedings of the National Academy of Sciences*, 1928, **14**, 857–864; and Urner Liddel and Charles Kasper, 'Spectral Differentiation of Pure Hydrocarbons: A Near Infrared Study', *National Bureau of Standards Journal of Research*, 1933, **11**, 599–618.

24 For a study of information flows related to the API project see Yakov M. Rabkin and Jean-Jacques Lafitte-Houssat, 'Cooperative Research in Petroleum Chemistry', *Scientometrics*, 1979, **1**, 327–338.

25 Edward W. Washburn, 'Fractionation of Petroleum into Its Constituent Hydrocarbons', *Industrial and Engineering Chemistry*, 1933, **25**, 891–894.

26 Frank W. Rose, 'Infrared Absorption of 19 Hydrocarbons, including 10 of High Molecular Weight', *NBS J. Res.*, 1937, **19**(2), 143–161; and Rose, 'Quantitative Analysis with Respect to the Component Structural Groups of the IR Molal Absorptive Indice of 55 Hydrocarbons', *NBS J. Res.*, 1938, **20**(2), 129–158.

27 On chemicalization see Yakov M. Rabkin, 'Chemicalization of Petroleum Refining in the United States: The Rôle of Cooperative Research, 1920–1950', *Social Science Information*, 1980, **19**, 833–850; and (a comparative analysis) Rabkin, 'La chimie et le petrole: Les debuts d'une liaison', *Revue d'Histoire des Sciences*, 1977, **30**, 303–336.

28 Kendall, *Applied Infrared Spectroscopy*, p. 2.

29 Norman Wright, 'Application of Infrared Spectroscopy to Industrial Research', *Industrial and Engineering Chemistry, Analytical Edition*, 1941, **13**, 1–7.

30 Harrison M. Randall, 'The Spectroscopy of the Far Infrared', *Review of Modern Physics*, 1938, **10**(1), 72–85; and Randall, 'Infrared Spectroscopy', *Science*, 1927, **65**, 167–173, on p. 168.

31 Coblentz, 'Early History', p. 107.

32 For an account of Sutherland's activities see Norman Sheppard, 'Sir Gordon Sutherland', *European Spectroscopy News*, 1981, **34**, 25–28.

33 Interview with Sir Harold Thompson, Oxford, Dec. 1981; and 'Professor Sir Harold

Thompson', *Eur. Spectr. News*, 1975, **1**(2), 13–18, on p. 14.

34 Interview with Sir Harold Thompson, Dec. 1981.

35 *Ibid.*

36 *Ibid.*

37 R. Bowling Barnes *et al.*, 'Infrared Spectroscopy: Industrial Applications', *Ind. Eng. Chem., Anal. Ed.*, 1943, **15**(11), 659–709.

38 Interview with Wilks, Nov. 1981.

39 Interviews with W. I. Kaye, Fullerton, California, and Robert Brattain (by telephone), Feb. 1982.

40 Barnes *et al.*, 'Infrared Spectroscopy'.

41 'Technical Advisory Committee of the Petroleum Industry War Council', an internal API report on the symposium 'New Physical Methods of Analysis as Applied to Petroleum Chemistry', held on 12 May 1944.

42 *Ibid.*; and Van Z. Williams, 'Infrared Instrumentation and Techniques', *Rev. Sci. Inst.*, 1948, **19**, 135–145, on p. 142.

43 Barnes and Bonner, 'A Survey of Infrared Spectroscopy'; also *J. Chem. Educ.*, 1938, **15**, 25–39.

44 Interview with Sir Harold Thompson, Dec. 1981.

45 Barnes *et al.*, 'Infrared Spectroscopy', p. 660.

46 *Ibid.*, pp. 661, 674.

47 Interview with Wilks, Nov. 1981.

48 'An ESN Talk with Professor Angelo Mangini', *Eur. Spectr. News*, 1979, **22**, 18–25, on p. 22.

49 R. Norman Jones, 'Vibrational Spectroscopy and Organic Chemistry: Thirty Years of Symbiosis', *Journal of the Spectroscopical Society of Japan*, 1981, **30**, 423–436, on p. 424.

50 'J. Lecomte', *Eur. Spectr. News*, 1977, **10**, 8–12, on p. 11. The publication Lecomte refers to must be Jean Lecomte, *Le spectre infrarouge*, Presses de l'Université de Paris, Paris, 1928; see also Jean Lecomte, 'Spectres dans l'infrarouge', in *Traité de chimie organique*, Victor Grignard (ed.), Masson & Cie, Paris, 1936.

51 Interview with Sir Harold Thompson, Dec. 1981.

52 *Ibid.*

53 See *e.g.* George L. Mosse (ed.), *Nazi Culture*, Simon and Schuster, New York, 1966; and Alan D. Beyerchen, *Scientists under Hitler*, Yale University Press, New Haven, CT, 1977.

54 Gurikov, 'K predystorii infrakrasnoi tekhniki', p. 56.

55 Beckman Instruments, 'The Age of Instrumentation and Beckman Instruments', 1963, p. 64.

56 Hugh Aitken's thoughtful account of the invention of the radio is an excellent example of this genre; see H. G. J. Aitken, 'Science, Technology, and Economics: The Invention of Radio as a Case Study', in *Dynamics of Science and Technology*, Krohn *et al.* (eds.), pp. 89–111.

57 Janich, 'Physics', p. 130.

58 Frank Greenaway, 'Instruments', in *A History of Technology*, Trevor I. Williams (ed.), Vol. VIII, Part 2, Clarendon, Oxford, 1978, p. 1204.

59 Janet Osteryoung, 'Developments in Electrochemical Instrumentation', *Science*, 1982, **218**, 261–265, on p. 265.

60 Greenaway, 'Instruments', pp. 1217–1219; and Donald S. L. Cardwell, *Turning Points in Western Technology*, Neale Watson, New York, 1972, p. ix.

Analytical Chemistry and the 'Big' Scientific Instrumentation Revolution*

DAVIS BAIRD

Department of Philosophy, University of South Carolina, Columbia, South Carolina 29208, USA, Email: bairdd@gwm.sc.edu

Introduction

Between the years 1920 and 1950 analytical chemistry underwent a transformation that has been described as 'the second chemical revolution'.[1] Prior to 1920 analytical chemists determined the chemical constitution of some unknown by treating it with a series of known compounds and observing the kind of reactions it underwent. After 1950, analytical chemists determined the chemical constitution of an unknown by using a variety of instruments which allow one to discriminate chemicals in terms of their physical properties.

This transformation did not involve changes in theory. Rather, it involved changes in the practice of analytical chemistry. It involved changes in the limits of possible analyses – with respect to the amount of sample required for an analysis, the time necessary for an analysis, and the precision with which trace quantities could be analysed. It involved the development of a new family of scientific instrument making companies, and a new level of capital expenditure necessary to do analytical chemistry. It involved the development of new means to disseminate information about scientific instruments.

Perhaps most important from my point of view, this revolution in analytical chemistry promoted a new kind of scientific knowledge – scientific instrumentation. It has been common to treat scientific instruments as instrumental goods which promote the final good of theory construction. In contrast, I urge that instruments also are final goods of science. I do not mean simply that technical

*Reprinted with permission from D. Baird, *Annals of Science*, 1993, **50**(3), 267–290, http://www.tandf.co.uk

innovation is a final good of science. Science provides knowledge. This is the final good of science which I am interested in, and when I say that instruments and theory are both final goods of science, I mean that both teach us what the world is like. Scientific instruments are elements of scientific knowledge.[2]

My claim invites a skeptical response from philosophers. Philosophers typically think of knowledge as a kind of belief. Instruments are not kinds of belief, and hence cannot be a kind of knowledge. This strikes me as philosophical prejudice. Philosophers express themselves in words, and so it is not surprising that they characterize scientific knowledge in the terms with which they are familiar. It is worth recalling Derek de Solla Price's lament, 'It is unfortunate that so many historians of science and virtually all of the philosophers of science are born-again theoreticians instead of bench scientists'.[3]

Historically scientists – natural philosophers – were born of philosophers. While there is much to Ian Hacking's idea that modern science is the result of a collaboration of doing and thinking, the doing has always been understood in a subservient or instrumental rôle. There have been occasional exceptions; Joseph Priestley was interested in building new instruments to create new phenomena and thereby to express the fullness of nature. But, Priestley is usually remembered as the stubborn backward scientist who resisted Lavoisier's brilliant theoretical insights concerning oxygen, phlogiston and combustion.[4]

The revolution in analytical chemistry is important because with it we have widespread recognition that building a new instrument can teach us about the world just as devising a new theory can. This revolution is not the kind discussed by Thomas Kuhn[5] and successive analyses. It marks a wholesale shift in our understanding of scientific knowledge. It is more akin to the 'probabilistic revolution,' where the possibility of probabilistic scientific knowledge emerged.[6] With this revolution, seen here in analytical chemistry, instrumentation emerges as a vehicle to carry knowledge.

Evidence of the Change in Analytical Chemistry

Virtually every history of analytical chemistry notes the revolutionary character of the changes in analytical chemistry. I quote one example, John K. Taylor:

> In 1985, it is hard for anyone to remember, and most analytical chemists have never known, an instrumentationless world. When one enters a modern analytical laboratory, one is surrounded by equipment so that the analyst may be dwarfed by the instruments at his or her command. Contrast this with the laboratory of the 1930s; the analyst was surrounded by chemical reagents and the most conspicuous installation was a fume hood. Several drawers contained the tools of the profession – beakers, filters, burets and pipets.[7]

Section 6 of his article is titled 'The Chemical Revolution'. In it he writes:

> Chemical analysis is undergoing a change of operational mode similar to the industrial revolution of a century ago.... The trend is from individual

Chemical laboratory, 1930s. Note the prominence of the microscope, now rarely found in chemical laboratories. Daily Herald Archive/NMPFT/Science and Society Picture Library

craftsmanship to mechanical outputs, using apparatus and equipment that is often poorly understood by the technical operator.[8]

Many other examples can be found.[9] None of the authors who discuss this transformation of analytical chemistry assert a detailed sense of 'revolution' with their use of the word. Their use, however, does signal dramatic changes in the field; developments in analytical chemistry during this period were not simply work as usual.

Textbooks provide nice summary pictures of analytical chemistry before and after these changes. Consider, for example, the elementary text by W. A. Noyes, *The Elements of Qualitative Analysis*.[10] The book has three main parts. The first part (16 pages) provides the theory of qualitative analysis; precipitation is qualitatively explained by ionization theory. The second, and longest, part of the book (82 pages) presents an empiricist's gold mine of descriptions of chemical reactions. Substances are classified according to their 'deportment toward various reagents'.[11] Reactions are discussed with the aim of separating and distinguishing the elements. The third part of the text (10 pages) presents an algorithm for determining the nature of an unknown drawing on the reactions studied in the first two parts of the book. In short, qualitative analysis works by running an unknown through a series of reactions designed to separate the various components and allow their identification from the kinds of reactions they participate in.

In contrast to qualitative analysis, quantitative analysis is concerned with determining the relative amounts of the different elements present in an unknown. Prior to 1920, quantitative analysis used two methods, gravimetric analysis and volumetric analysis. A brief example from George Smith's 1921 text

Volumetric analysis at the laboratories of the National Oil Ltd refinery at Llandarcy, Glamorgan, Wales, 1931. Photograph by James Jarché. A stop-clock is the most complex piece of apparatus in sight. Also note the absence of safety glasses, lab coat and gloves! Daily Herald Archive/NMPFT/Science and Society Picture Library

Quantitative Chemical Analysis explains both approaches:

[L]et us consider the determination of silver in a silver coin.

(a) Gravimetric Method. The weighed sample is dissolved in nitric acid, the solution diluted, and the silver separated from copper, by precipitation as insoluble silver chloride, with dilute hydrochloric acid. The precipitate is filtered off, washed, dried and weighed. From its weight, the weight of silver is calculated as follows:

Ag/AgCl × wt. of precipitate = wt. of silver....

(b) Volumetric Method. The weighed sample is dissolved in nitric acid, diluted as before, and the silver converted into the insoluble chloride by the gradual addition, from a burette, of a solution of sodium chloride of known concentration. As soon as, after stirring each time and allowing the precipitate to settle, the first drop is added which fails to induce further precipitation, the reaction is known to be complete; and the number of cubic centimeters required, multiplied by the silver equivalent of the sodium chloride solution per cubic centimeter, gives directly the weight of silver in the sample.[12]

Smith's work continues with detailed discussions of a large number of examples which illustrate both basic approaches.

A few other methods are mentioned in other more advanced texts. J. C. Olsen's 1916 text *Quantitative Chemical Analysis* is subtitled 'by Gravimetric, Electrolytic, Volumetric and Gasometric Methods'. Still gravimetric and volumetric methods make up the greatest proportion of the 555 page text. The book includes one (30 page) chapter on electrolytic methods. Here materials can be separated by dissolving them and passing a current through the solution depositing different substances on the different electrodes. It is clear from the discussion that electrolytic methods suffered from a lack of dependable sources of electricity. A single 23 page chapter is also devoted to gas analysis. Here the main difficulty lies in finding appropriate substances with which to absorb different gases. Once absorbed, a more standard gravimetric approach can be pursued.[13]

John Muter's 1906 *Short Manual of Analytical Chemistry* also is primarily devoted to gravimetric and volumetric methods. Muter includes one eight page chapter on 'alternative methods'. Here he discusses analysis by circular polarization with the saccharometer – two pages – spectrum analysis with a Bunsen burner and a prism spectroscope – one page – gas analysis with Hempel's 'gas-measuring apparatus which is reasonable in price, and yet is capable of measuring gas volumes with very fair accuracy' – four pages.[14] It is clear, however, given the short treatment, that these methods were not central to the practice of analytical chemistry.

Finally, it is worth mentioning William Lacey's 1924 text, *A Course of Instruction in Instrumental Methods of Chemical Analysis*. This is the earliest textbook with an explicit instrumental focus. The instruments, however, are fairly simple devices developed for special analytical needs. The polarimeter, which measures the angle through which polarized light is rotated when passing through a substance is particularly useful for determining the percentage of sugar in a sample, a commercially important piece of information. The book pays scant attention to the two areas which have become most important to instrumental analysis. Spectrographic analysis is covered in two pages. Electrochemical analysis is covered in one chapter where we find:

> Apparatus: Aside from the cell and burettes, the apparatus consists of some means of measuring the electromotive force produced. If a potentiometer with standard cell for comparison is used, the actual electromotive force may be determined.... [I]n most cases, the method used merely for the location of endpoints of titrations; and for this purpose it is only necessary to be able to follow relative changes in voltage without reference to the actual magnitudes.[15]

The book continues with a circuit diagram for constructing a simple means to measure these relative changes. For all its shortcomings, the book does suggest the importance of using instruments to make various determinations of physical properties. During the following decades this became the source of the radical transformation in analytical chemistry.

Consider a more recent text. Schenk, Hahn and Hartkopf's 1977 text, *Quantitative Analytical Chemistry*, can serve as an example.[16] The first part of the book, 'Fundamentals', provides a general introduction to chemical analysis. Here the student is presented with some of the important theoretical bases of chemistry (solution equilibria; acid–base reactions; oxidation–reduction reactions). The student is taught some of the important procedures of chemical analysis (data handling, *i.e.* statistics; sampling the material for which an analysis is sought; preparation of the sample). This is also where the classical – gravimetric and volumetric – methods of analysis are covered.

The second part of the text covers a variety of instrumental methods of analysis. Four chapters are devoted to optical methods of analysis. Here, for example, the student is taught about spectroscopy, where analysis relies on the fact that different atoms emit and absorb electromagnetic radiation at characteristic wavelengths. Analysts can also use electromagnetic radiation to identify chemicals in other ways: they can use the fluorescence of molecules and ions, where they absorb radiation at one wavelength and emit it at another wavelength; they can use the refractive properties of some materials, where radiation is 'bent' when passing through these materials.

Two chapters are devoted to electrochemical methods of analysis. Here analyses are accomplished by making measurements of various electrical properties. pH is routinely determined by measuring the electric potential across a special electrode. Electrogravimetry uses an electric current to cause deposits – which can be weighed – of ions on the electrodes in a solution. Coulometric analysis is accomplished by measuring the amount of current necessary to reduce or oxidize ions in solution. Polarography depends on the fact that different ions require different potentials to participate in a current; from a plot of applied voltage versus current – a 'polarograph' – an analyst can determine the concentrations of the different ions involved. All of these approaches rely on special optical and electrical instrumentation.

Schenk *et. al.*'s text is aimed at students who may use analytical methods in their work, but who are not pursuing a career in analytical chemistry. Douglas Skoog and Donald West have written a pair of texts, more thorough in treatment, aimed at students pursuing a career in chemistry.[17] Their first text, *Fundamentals of Analytical Chemistry*, does all that Schenk *et. al.*'s text does in greater theoretical detail. Ten chapters (225 pages of 765 pages) are devoted to volumetric methods. Nine chapters (207 pages) are devoted to instrumental methods – electrochemical (five chapters) and spectrochemical (four chapters). Gravimetric methods receive a treatment similar to Muter's treatment of 'alternative' methods – one (30 page) chapter with the following apology:

> Some chemists are inclined to discount the present day value of gravimetric methods on the grounds that they are inefficient and obsolete. We, on the other hand, believe that the gravimetric approach to an analytical problem – like all others – has strengths and weaknesses, and that ample situations exist where it represents the best possible choice for the resolution of an analytical problem.[18]

Skoog and West's second text, *Principles of Instrumental Analysis*, focuses exclusively on instrumental methods. They begin the preface noting, 'Instrumental methods of analysis have become the backbone of experimental chemistry'.[19] The book has detailed treatments of spectrochemical analysis (ten chapters), other electromagnetic–radiation-based methods (three chapters), mass spectroscopy (one chapter), radiochemical methods (one chapter), electrochemical analysis (six chapters), and chromatographic methods of separation and analysis (two chapters).

We can also get a sense of the changes in analytical chemistry by looking at the changes in the central American outlet for research in analytical chemistry: the journal, *Analytical Chemistry*. It was born in 1929 as an offshoot of the journal *Industrial and Engineering Chemistry*. From 1929 to 1948 it was published as the *Analytical Edition of Industrial and Engineering Chemistry*. After 1948, taking its current name, *Analytical Chemistry*, the journal became autonomous with its own subscription list and editorial policies. Volume numeration was kept continuous.

The journal has experienced dramatic growth. The first, 1929, volume contains 238 pages. The 1948 volume contains 1250 pages. The 1989 volume contains 2850 pages. Beside the considerable growth *per se*, there was a significant increase in the proportion of advertising pages. In 1929, 13% of the pages were devoted to advertising. In 1989, 33% of the pages were devoted to advertising. This increase represents the increasingly important rôle of analytical instrument makers to analytical chemistry.

In February 1943, editorial responsibilities passed from Harrison Howe to Walter Murphy. Murphy took a very active rôle in promoting the changes taking place in analytical chemistry. Several regular columns were introduced, the first on instrumentation. The scope of papers appropriate for publication in the *Analytical Edition* was broadened to include more theoretically oriented papers and papers focused on instrumentation. The editors were particularly insistent that one must take the 'chemistry' in analytical chemistry very liberally: 'The tools used may be chemical or physical, ... The physical chemistry, in many cases may approach pure physics'.[20]

Analytical Chemistry published three surveys of trends in the field. The first, in 1947, responded to the perception that the field was being taken over by instruments:

> Columns and editorials in *Analytical Chemistry* and the correspondence they quote show that many chemists are asking, 'Where is analytical chemistry heading, technically and professionally?' ...
>
> Though many teachers of quantitative analysis are adding as much material on the newer instrumental methods of analysis as time and equipment permit, there are many shades of opinion on what is significant enough to include and whether the new methods are important enough to be given at the expense of material on the classical methods.[21]

56% of the papers on analytical chemistry published in 1946 were on instrumental methods.[22]

Similar studies were done in 1955 and again in 1965.[23] By 1965, 40.5% of all papers concerned optical methods alone. Only 3.6% of papers concerned gravimetric methods, down from 10.7% in 1946.[24] Perhaps the most important point is that the 1965 analysis is not concerned with the percentage of instrumental as opposed to non-instrumental methods. That issue has been settled. Analytical chemistry has been won by the instrumental approach, although the early courses still present much basic chemistry.

From Separation and Manufacture to Identification and Control

Currently analytical methods are distinguished as either 'classical' or 'instrumental'. Typically, contemporary textbook authors play down differences between classical and instrumental methods. Everyone agrees that one mandate of the analytical chemist is to determine the constitution of the sample. On a cursory analysis, instrumental methods simply augment the arsenal of methods available to an analytical chemist. But, there are subtle differences between these two kinds of analyses which are important to notice.

In his 1969 *Instrumental Methods of Chemical Analysis* Galen Ewing writes:

> Historically, the development of analytical methods has followed closely the introduction of new measuring instruments. The first quantitative analyses were gravimetric, made possible by the invention of a precise balance. It was soon found that carefully calibrated glassware made possible considerable saving of time through the volumetric measurement of gravimetrically standardized solutions.
>
> In the closing decades of the nineteenth century, the invention of the spectroscope brought with it an analytical approach which proved to be extremely fruitful.[25]

Contrast this with the remarks of H. Laitinen and W. Harris, in their 'noninstrumental' text, *Chemical Analysis: An Advanced Text and Reference*:

> Classical methods of final measurement will long continue to be important. In the first place, they are inherently simple. For an occasional determination or standardization the use of a titration [volumetric] or gravimetric determination often will require the least time and effort and will involve no investment in expensive equipment. Second, classical methods are accurate. Many instrumental methods are designed for speed or sensitivity rather than accuracy, and often must be calibrated by classical methods....
>
> In summary, the thesis of this book is that knowledge of chemical reactions is important, first because it is needed for direct application to classical methods, and second because it is essential in instrumental methods where chemical reactions are involved in operations preceding the use of an instrument in the final measurement.[26]

There is no contradiction between Ewing's and Laitinen and Harris's remarks, but there is a difference in emphasis. For Ewing, new and better instruments are the bellwether of progress in analytical chemistry. One could imagine a Ewing-style analytical 'chemistry' which had little to do with chemistry *per se* and a lot to do with physics. Laitinen and Harris are not banking on such a course of events.

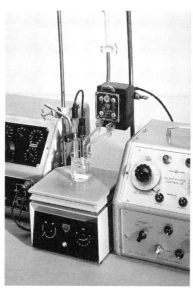

Pye 'Autotitration Controller', 1962. Science Museum collections, inventory number 1962–371. Science Museum/Science and Society Picture Library

Instrumental methods do not work the same way as classical methods. Contrast gravimetric methods with spectrographic methods. Put simply, atomic-emission-spectroscopy works because the wavelengths of light which an atom emits are characteristic of that kind of atom. When an analyst measures these wavelengths, the analyst is directly 'fingerprinting' the atom. By contrast, how much some precipitate weighs is not, in any general sense, characteristic of the composition of the precipitate. This measurement allows identification only within the context of the chemical reactions involved in the initial chemical separations.

This difference has ramifications for how analytical chemistry is conceived. The most careful theoretical development of analytical chemistry prior to 1920 is Wilhelm Ostwald's *Scientific Foundations of Analytical Chemistry*.[27] Ostwald starts at a high level of abstraction. Substances are distinguished by differences in their properties. Two substances are identical when they agree on all their properties. Here Ostwald is only concerned with *distinguishing* elements from each other. The hard part of analytical chemistry, however, comes from *separating* a complex mixture into its elements:

> From what has been said in the foregoing chapter it is apparent that the task of recognizing any given substance, ... is always more or less easy of accomplishment, ... But the problem becomes far more complicated when we have to deal, ... with a mixture; separation must here precede recognition, and the first-named operation is naturally much the more difficult of the two.[28]

It is with separating substances that the subtlety and craft of the analytical chemist was displayed.

Ostwald spends the remaining theoretical portion of the book discussing the various means available to separate different substances from each other. He discusses various *physical* means of separation – *e.g.*, using filters to separate liquids from solids – but by far the bulk of his text concerns *chemical* methods of separation. He focuses on the way in which ions behave in solution. The material is developed in considerably greater theoretical detail than in Noyes, or the other early texts mentioned above. But, from a practical point of view precipitation is the primary means to accomplish chemical separation. He does consider two other means of separation: the liberation of a gas from a solution, and the electrolytic method.

Ostwald spends one chapter on quantitative analysis. As with the other more elementary texts, Ostwald focuses on the two basic means to obtain quantitative data: gravimetric and volumetric. Ostwald distinguishes two kinds of quantitative problem: when the elemental components of a substance have already been separated out into their pure forms, and when the elemental forms are still combined. In the first case, one merely (!) needs to measure the quantity of the different pure forms involved. Unfortunately, substances are frequently not suitable for weighing in their natural state – they may too easily absorb water from the air, for instance. For this reason, substances are frequently weighed while in combination with some other elemental substance. Atomic weights and the law of constant mass proportions can then be used to determine the quantity required.

When pure substances have not been obtained analytic chemists can resort to other tricks to find the amounts of the components without separating them. For instance, in the case of two mixed liquids, one can determine the proportions of each in the mixture by comparing the specific gravity of the mixture with the known specific gravities of each substance in their pure states. Any such property, in which the two substances differ, and which can be measured on the mixture can be used in this way.

The second half of Ostwald's book concerns 'applications'. Here he discusses briefly the characteristic ways to identify and separate the various elements from each other. There is no attempt to present a systematic algorithm for doing so. Rather he discusses each element, one by one, mentioning the more important means available to separate and distinguish it from other elements.

Prior to the instrumental revolution in analytical chemistry, chemical separation, and not physical identification, was the central feature of this science. Noyes, in his elementary text, follows Ostwald:

> Qualitative analysis, with which we are here concerned, deals with the qualitative composition of bodies; *i.e.* with the separation (either free or in the form of characteristic compounds) and identification of the various elements present in them.[29]

Indeed, as late as 1929 the analytical chemist was defined as: 'a chemist who can quantitatively manufacture pure chemicals'.[30] Such manufacture is the result of separations performed by appropriate chemical reactions. With the introduction of instrumental methods of analysis, analytical chemists came to focus on the physics of elemental properties as a means of identification rather than the chemistry of reactions as a means of separation.

A Crisis of Identity

In March 1947, Walter Murphy published an editorial describing how the profession of analytical chemistry had changed. Previously analytical chemists were hired to do work that was 'largely repetitive, usually long drawn out, tedious, dull, uninteresting and therefore uninspiring to the truly professional'.[31] By 1947, because of advances in analytical instrumentation things were quite different:

> The widespread introduction of instrumentation has caused a sharp division in the analytical laboratory between those of professional and subprofessional training, experience and ability. Today thousands of analytical procedures are carried on readily by laboratory technicians. The true professional is expected to direct, to administer, and to pioneer research in analytical chemistry. He is therefore required to be an organic chemist, and may, at times, be expected to be a biochemist, a metallurgist, a specialist, if you will, in a dozen or more highly specialized fields. He most certainly must be somewhat of an expert in electronics – he must be almost as much a physicist as physicists themselves. In addition, he is usually expected to be specially skilled in some field within the profession of analytical chemistry.[32]

Unfortunately, Murphy noted, outside the field of analytical chemistry, the impression remained that the analytical chemist is only suitable for routine chemical determinations.

Murphy made some specific proposals to change the perception of his field. These included insisting on a sharp line between analytical chemists and technicians, updating curricula to come to terms with advances in analytical instrumentation, the establishment of an award for outstanding work in analytical chemistry, and the '[c]onsideration of ways and means of educating industry and particularly top-flight management on the true importance of analytical chemistry'.[33]

Murphy was remarkably successful in implementing his program. Distinctions between technicians and analytical chemists did develop. Curricula were revamped.[34] Awards were established to promote advanced research in analytical chemistry.[35] Even before the end of the year, Murphy commented on an article in *Fortune* discussing new methods of spectroscopic analysis: 'Modern chemical analysis has arrived!... Top management and executives are beginning to have their attention directed to the wonders of modern chemical analysis'.[36]

Perhaps the most important consequence of this editorial was the responses it provoked concerning the nature of the profession. In May 1947, Murphy reprinted at length the response of William Seaman of the Analytical Research

Laboratory at American Cyanamid. Seaman expanded on the difference between the professional analyst and a technician; the analyst is not 'a pharmacist filling a prescription', but more 'a doctor planning a course of treatment'.[37] The June editorial consisted of three pages of letters responding to Murphy's editorial.[38] In August a guest editorial by D. B. Keyes on the importance of analytical chemistry to industry was published.[39] Finally the November, December – 1947 – and January – 1948 – editorial pages consisted of invited responses to the question, 'What is Analysis?'

The responses point up the several ways in which analytical chemistry was in a state of confusion. B. L. Clarke of Merck & Co starts with the old-fashioned notion that analytical chemistry was the science of separating substances from each other, but he continues, recognizing the analyst's goal to determine the constitution of substances and recognizing the importance of instrumental methods of analysis:

> Thus, the analytical chemist is one who breaks substances down in order to find out of what they are made.
>
> Two points deserve emphasis. Because the analytical chemist is really a manufacturing chemist who works on a reduced scale, his basic training in the understanding of chemical reactions cannot be very different from that of the factory chemist.... Thus the analytical chemist is first and foremost a chemist....
>
> The other point is that modern analysis frequently avoids the actual physical destruction of the sample, by the use of instruments, like the spectrophotometer, that in effect extend the senses and allow the analyst to observe molecular structure without the crudity of picking the molecule apart. Not only are these instrumental methods more elegant; they are potentially more efficient, and are more and more used in industry where efficiency counts.
>
> Obviously, then, the curricula for the training of analytical chemists must give great emphasis to analytical instruments and to the physical basis underlying their operation.[40]

Clarke vacillated just on the point question which the development of the new instrumental methods raised: is analysis primarily a chemical process of separation and identification, or a physical process of direct identification. Everyone recognized the importance of these instrumental methods, yet few were able to embrace them wholeheartedly.

W. C. McCrone of the Armour Research Foundation took a more 'liberal' stand:

> The analytical chemist has been replaced not by a man having a different training but by a group of specialists in the determination of physical properties. In general, these specialists resent being referred to as analysts. They are instead physicists or chemists trained in the study of electron microscopy, tracer techniques, infrared spectrophotometry, X-ray diffraction, mass spectrometry, chemical microscopy, polarography, *etc.* A group

of people qualified in each of these phases of analytical work make up the modern analytical laboratory.[41]

This crisis of identity also produced problems in the administration of analytical laboratories. McCrone continues:

> The most appropriate name for such a group has not yet been found; the Armour Research Foundation has used Analytical Section, some groups prefer Chemical Physics, other possibilities are Instrumental Analysis, Analytical Physics, and Physical Analysis. It is desirable to have a new name, more dignified than 'analytical' alone, yet it is essential to retain the word, or at least the connotation, 'analytical'. Instrumental Analysis Laboratory is perhaps the best compromise.[42]

The confusion expressed itself in arguments about appropriate curricula for training in analytical chemistry. Most respondents urged that instrumental methods needed to have a greater portion of available course time. J. J. Lingane – noted for his work on electrochemical methods – disagreed:

> I venture the opinion that it is neither desirable nor feasible to attempt much serious instruction in 'instrumental analysis' in the undergraduate course in quantitative analysis. True, one can place potentiometers, spectrophotometers, pH meters, polarographs and the like in the laboratory and have the student 'make determinations' with them, carefully selecting the 'unknowns', of course, so no 'difficulties' are encountered. But since the undergraduate quantitative course is peopled chiefly by sophomores and juniors, who have not begun the study of physical chemistry, and whose background in physics and mathematics is meager, the educational value of such a scheme is questionable. Too much superficial 'modernization' of this kind tends to dilute the instruction in more fundamental aspects of analytical chemistry, which many will agree are still as essential to the education of an analytical chemist as they ever were.
>
> In our justifiable enthusiasm for the truly great accomplishments of 'instrumental analysis' it is easy to lose our sense of proportion and forget that the most important factor in a chemical analysis is the chemical experience of the analytical chemist rather than the final determinative techniques at his disposal.[43]

For Lingane, *chemical* experience was still central to analytical chemistry.

Lingane was not opposed to instrumental analysis. Nor did he doubt the need for instruction in these areas:

> No one will deny the increasing importance of physicochemical determinations in modern analytical practice, and the concomitant need for more systematic and more extensive education in these methods.[44]

He preferred to postpone such instruction to graduate-level courses. Lingane was inclined to agree with Ralph Müller's idea for special departments of instrumentation (see below). This would promote research in instrumentation and allow analytical chemistry to remain chemistry:

> But the science of instrumentation itself presents a larger problem. Month by month in this journal for two years our colleague Ralph Müller has been presenting convincing evidence that instrumentation in the broad sense has grown to such proportions that it merits recognition as a new branch of knowledge. Many others share the belief that haphazard instruction in this subject is no longer adequate if we wish to realize its potentialities fully. Perhaps graduate courses in instrumentation will suffice, although there are some who believe that special curricula will be required.[45]

These editorial remarks expose a crisis of identity for analytical chemists. Instrumental methods need not be chemical in nature. The theoretical underpinnings could come from physics, electrical engineering and instrumental design. The result made it unclear how analytical chemistry was chemistry. New chemical principles were not the only or primary goal for analytical research. A good portion of analytical chemical research focused on the development of new instruments.

By the 1960s the crisis had resolved. In his 1962 Fisher Award Address, H. A. Liebhafsky argued that 'Like it or not, the chemistry is going out of analytical chemistry'.[46] Instead, Liebhafsky saw 'modern analytical chemistry as the characterization and control of materials'.[47] Characterization involves ascertaining a material's composition, properties and qualities. Control involves using various kinds of sensors and feedback mechanisms to control the production and use of materials. Control is one new feature of post-revolutionary analytical chemistry made possible by the introduction of instrumentation. Even characterization, which bears the greatest similarity with pre-revolutionary analytical chemistry is different:

> Mellon [*Analytical Chemistry*, 1952, **24**, 924] described the older analytical chemistry as consisting of separation preceding determination: separation usually involving chemistry, determination being based on physics. If we adopt characterization as the essence of modern analytical chemistry, then separation needs replacing by the broader preparation, a change made necessary by the decreasing emphasis on ascertaining composition.
>
> Two trends are noticeable: toward less chemistry in the preparation, and toward less preparation prior to the determination,....[48]

Liebhafsky further notes that with the de-skilling brought about by the introduction of instrumental methods, analytical chemists are increasingly required to act as personnel managers directing the technicians running the instruments.

Ralph Müller's Science of Instrumentation

Not satisfied simply to solicit papers on the application and development of new instruments, the editors of *Analytical Chemistry* took an active rôle in providing information on instrumental methods of analysis. The entire October issues of the 1939, 1940 and 1941 volumes were devoted to instrumentation.[49] Ralph H.

Ralph Müller. Williams Haynes Portrait Collection, Chemists' Club Archives, Chemical Heritage Foundation Image Archives, Othmer Library of Chemical History, Philadelphia, Pennsylvania

Müller wrote the entire 1940 and 1941 issues. After the War, the editors decided on a more regular means to provide information on instrumentation in analysis. The first monthly column Murphy introduced to the *Analytical Edition* was Müller's 'Instrumentation in Analysis'. Müller's first contribution appeared in January, 1946, and he continued to write this column until the end of 1968, after which the column was written by invited authors, and had comments by Müller.

Müller did a variety of things in his column. He frequently discussed new instruments which would be of interest to analytical chemists. Initially, Müller discussed instruments in the research literature. In his first column, he described a paper by Lingane on measuring the amount of material deposited on an electrode by measuring the amount of current used.[50] In his second column, he described a paper by three Dow Chemical chemists – J. L. Saunderson, V. J. Caldecourt and E. W. Peterson – on a means to use photo multiplier tubes for the direct analysis of spectral intensity.[51] Eventually, many of the instruments Müller described were production models from commercial instrument makers. In April 1946, Müller described a new kind of vacuum gage sold by the National Research Corporation.[52] In July 1947, he described the Baird Associates–Dow Chemical direct reading spectrometer, a commercially made instrument based on the approach to direct reading spectrometers described in his February 1946 column.[53]

Müller also used his column to discuss the basic principles of the emerging 'science of instrumentation'. His third column discussed the 'three Rs of instrumentation, reading, 'riting, and 'rithmethic', which is to say 'indication, recording and computing'.[54] Of indication, he wrote:

The ideal instrumental indication is bold, legible, requires no interpolation, and reads the desired quantity directly....

We cannot evade the feeling that the laboratory will ultimately share with industry and commerce the equivalent convenience afforded by digit-type Telechron clocks, gasoline pumps which indicate gallons, price plus tax, and weighing machines which print the answer on a ticket.[55]

Müller noted that the recording of data has become particularly important because of the complexity of indications which newly developed instruments provide. Infrared absorption spectrometers – which note relatively how much infrared radiation a substance absorbs as a function of wavelength – would not be a useful analytical tool without recording absorption values as a function of wavelength; too much data is provided in too short a time. Müller's discussion of computations presages the dramatic developments in computing which have marked the past thirty years. He noted:

Lord Kelvin once said that 'the human mind is never performing its highest function when it is doing the work of a calculating machine'. The same may be said of the analyst and his chores. At present, under the compulsion of industry's pace, we are in a stage of extensive mechanization. That process cannot be stayed, however much the classical analyst bemoans the intrusion of the physicist and engineer upon our sacred domain. It is to be hoped, rather, that it will afford the analyst more time and better tools to investigate those obscure and neglected phenomena which, when developed, will be the analytical chemistry of tomorrow.[56]

When the three Rs are appropriately built into an instrument, we have a device which directly provides the desired information. This is how Müller defined an 'instrumental' method of analysis:

The true instrumental method of analysis requires no reduction of data to normal pressure and temperature, no corrections or computations, no reference to correction factors nor interpolation on nomographic charts. It indicates the desired information directly on a dial or counter and if it is desired to have the answer printed on paper – that can be had for the asking. It is strange and difficult to comprehend why the last few steps have not been taken by the analyst in bringing his instruments to this stage of perfection. They are minor details, the absence of which in his motor car, office equipment, or telephone he would not tolerate for a moment.[57]

While one may reasonably doubt how 'minor' these details are, Müller provides a clear notion of what the ideal instrument would be. One can also see from these remarks why John Taylor likens the instrumental revolution in chemistry to the industrial revolution. Both produce the same de-skilling of routine analytical work. Müller's fantasy, however, was that so relieved of routine analytical determinations, the analyst would be able to investigate 'obscure and neglected phenomena'.

Müller blamed the resistance of analytical chemists to perfecting their instruments on their overly provincial attitudes about the need for analytical chemistry to be chemistry. He used his column as a bully pulpit to urge a more cosmopolitan approach:

> Recently we were taken to task by one of America's distinguished chemists for emphasizing these distinctions [between a direct reading instrument and an instrument which provides data requiring further analysis]. 'All a matter of applied physical chemistry', he explained patiently, 'and therefore not particularly new'. We are obtuse enough to feel that physicochemical techniques bear the same relationship to instrumental analysis as the violent oxidation of hydrocarbons does to the modern motor car.[58]

There is more to building an instrument than knowing the basic principles behind the method of analysis. This something more made the difference between instrumentation and analytically unusuable theory.

Müller advocated a science of instrumentation with its own departments in universities. In November 1949, Müller wrote:

> The annual instruments issue of *Science* [1949, **110**, 2858] contains a number of articles which we believe, will interest the analyst. In the first of these, E. U. Condon, director of the National Bureau of Standards, raises a question to which we supplied an affirmative answer ten years ago: 'Is there a science of instrumentation?'[59]

Indeed in 1946 Müller had written:

> September 16 to 20 [1946] represents an important landmark in American instrumentation. The first National Instrument Conference and Exhibit was held in Pittsburgh with the theme 'Instrumentation for Tomorrow', ...
>
> ... This meeting has demonstrated that all the factors essential for a true profession [of instrumentation] are in evidence: a common interest, a well defined set of principles and practices, a wide assortment of special skills, and a well educated and trained body of experts completely dedicated to this field.[60]

The main issues for Müller were where the instrument-scientists would be trained and where the research into instrumentation would take place. He continued his comment on Condon's question about a science of instrumentation expressing the following concerns:

> We are in complete and enthusiastic agreement with, but wish to repeat that a profession cannot exist without adequate professional training. Consequently, we have been asking, by what type of academic osmosis are the prerequisites for this profession to be absorbed from our present curricula?[61]

In numerous columns Müller called for more instrumentation research in universities. Unfortunately, to his way of thinking, industry had taken the lead in training and developing instrumentation research:

> One of our academic friends expressed surprise and some resentment at the

preponderance of industrial and instrument company representatives at this conference [AAAS Chemical Research Conference, Colby College, 18–22 August, 1947]. By actual count we found that the universities were represented to the extent of about 15%.... This situation emphasizes the fact that the initiative and intelligent prosecution of instrumental research have long since passed to industrial research laboratories and a few instrument companies.[62]

The consequence was that general principles of and approaches to instrumentation would not be developed. Instead specific, commercially viable, instruments would be developed:

We have long insisted that research in analytical instrumentation of the 'useless' variety is urgently needed and that its proper place is the university. Not that this will be conceded in academic surroundings, because there one hears the constant complaint that there are already so many instruments that it is not possible to tell the students about them. This attitude cannot halt the march of progress, but it helps immeasurably.[63]

Müller had in mind research into instrumentation on the model of pure theoretical research.

While the universities may have missed Müller's boat, industry did not. The development of instrumental methods of analysis went hand in hand with the development of a new class of instrument makers. All of the instruments for instrumental analysis had to be developed, produced and marketed by instrument making companies. These companies had to hire personnel with an understanding of the physical principles underlying the operation of the instruments, an understanding of appropriate design for rugged reliable instrumentation, and an understanding of the manner in which these instruments would perform useful analytical tasks. New companies, such as Perkin–Elmer, Beckman, and Baird Associates sprang up to fill this need.[64] These new companies pursued research into instrumentation, although always with an eye toward commercial markets for their instruments.

The Instrumentation Transformation in Analytical Chemistry and Scientific Revolutions

There is little doubt that analytical chemistry underwent a radical change. The practice of the analyst, now dealing with large expensive equipment, is different than it was in 1930. Modern instrumental methods, by and large, are more sensitive, more accurate, have lower limits of detection, and require smaller samples; different kinds of analyses can be performed. Analytical chemistry is much less a science of chemical separations and much more a science of determining and deploying the physical properties of substances – which is not to say separations have disappeared from analytical chemistry, but they no longer are the centerpiece of the analyst's craft. Analytical chemistry is now a central part of much industrial research and control. Analytical chemistry is integrated into the business of making instruments both commercially and in the academy.

Given the extent of these changes it is surprising to note that the standard models for revolutionary scientific change do not fit this case. The revolutionary phase of Thomas Kuhn's *Structure of Scientific Revolutions* starts with a crisis: some problem which the established methods of normal science could not solve.[65] There was no such crisis in analytical chemistry. While one might imagine that analytical chemistry underwent a change of paradigm, there was no crisis which provoked this change. Pre-1930 analytical chemists did not bemoan the inability of their chemistry to solve certain problems. Instead new methods were developed which could solve established – solved – problems, but solve them better – more efficiently, with smaller samples and with greater sensitivity and lower limits of detection. These changes in analytical chemistry do not suffer from any kind of incommensurability: today one can easily enough understand what analytical chemists were doing in 1900 – although the idea that the analytical chemist is one who can quantitatively manufacture pure chemicals does jar a little.

Here is a central prejudice behind Kuhn's work. Scientific change is *theoretical* change. There was no Kuhnian crisis in analytical chemistry because new empirical findings did not threaten the theoretical edifice of analytical chemistry. Rather, new methods, judged better by all concerned, threatened the notion of analytical chemistry as a sub-discipline of chemistry. In the process these methods transformed the discipline.

I. B. Cohen provides a broader more historical framework for the discussion of scientific revolutions.[66] Cohen provides criteria for judging whether or not a given event in the history of science should be judged a revolution. A genuine revolution must:

1. Be identified as such in the testimony of scientists and/or non-scientists active at the time;
2. Have an impact on the treatises and textbooks written;
3. Be judged a revolution by competent historians of science;
4. Be judged a revolution according to the current opinion of scientists.[67]

The transformation in analytical chemistry passes all of Cohen's tests. The scientists active at the time were well aware of the radical – revolutionary – changes which were taking place in their field. The changes in analytical chemistry had a substantial impact on journal articles and textbooks. There has not been any discussion by professional historians of these changes in analytical chemistry. But, there have been several historical studies by chemists, and all of these studies note the dramatic changes in analytical chemistry.

Cohen also provides a general schema for the stages of a revolution in science. According to Cohen, a scientific revolution always starts with a private mental event.[68] Here a scientist conceives a radical means to solve some pressing problem. Already there is trouble fitting Cohen's model to the changes in analytical chemistry. There was no single 'purely intellectual exercise' which was the fountainhead for the succeeding changes in analytical chemistry.

One might look to Bunsen's 1860 invention of spectrochemical identification.

But, the revolution in analytical chemistry is not simply the result of the development of spectrochemical methods. The revolution in analytical chemistry is the result of the confluence of many events all of which showed how the introduction and development of physically based instrumental methods would improve the abilities of analytical chemists. One might say there were many small 'Cohen-revolutions', one in emission spectroscopy, one in pH meters, *etc*. But, this would miss the central feature of this revolution: it was the introduction of the instrumental-outlook, that transformed analytical chemistry, not the introduction of one or another particular instrumental methods.

As Kuhn does, Cohen sees revolutions primarily in terms of changes in scientific concepts and theories:

> Two popular models of scientific change are evolution and revolution.... One invokes the notion of a relatively slow and gradual process, consisting of a succession of small steps, perhaps occasionally punctuated by a greater one, while the other connotes a sudden change of a radical kind, a violent fragmentation *of a system of concepts and theories* that is followed by the introduction of something wholly new.[69]

Cohen recognizes the impact new instruments can have. But changes in instrumentation are not themselves revolutionary. In Cohen's view, a new instrument can radically alter the evidential base a theory must explain (*e.g.* the telescope), or a new instrument can radically alter the form theories can take (*e.g.* the computer). In either case, revolutions ultimately have to do with changes in theories.

Neither kind of impact appropriately characterizes the change in analytical chemistry. It is true that the introduction of instruments radically altered the evidential base of analytical chemistry. Still, it is not right to say that the evidential base was altered in such a way that analytical *theory* had to shift radically. Rather, these instruments radically altered the information which analytical chemistry could glean. This, not analytical theory *per se*, is the central goal of analytical chemistry. I am more comfortable with the idea that the introduction of instruments in analytical chemistry altered the form which analytical theories could take. While clearly, the instruments did not alter the form of theory itself, they did provide a new kind of outlet for analytical knowledge – instrumentation. It is the theoretician's bias – again – that results in analyses of scientific revolution not fitting the changes in analytical chemistry. The changes are real enough, they just are not changes in theory.

There is another model of scientific revolution which is more promising. This is the sense of revolution which talk of *the* scientific revolution calls on. *The* scientific revolution was a large-scale transformation in the nature of scientific knowledge itself. Kuhn has described the rise in importance of measurement in early nineteenth century physics, as the 'second scientific revolution'.[70] Given Kuhn's claimed ubiquity for '*Structure*-style' revolutions, this 'second' revolution must be a different sort of beast in Kuhn's mind. Ian Hacking has developed this idea further in his discussion of 'the probabilistic revolution'.[71]

Hacking calls these revolutions 'big revolutions'. Big revolutions have many characteristics, of which Hacking singles out four. First, they are interdisciplinary or, better, pre-disciplinary. Second, new social institutions appear with these revolutions. Third, part and parcel to these revolutions are dramatic social changes; societies in general organize themselves in different ways. Finally, these revolutions involve substantial changes in our attitudes toward the world; Hacking uses Herbert Butterfield's language: big revolutions are 'accompanied by a change in our sense of the "texture" of the world, in different "feel" for the world'.[72]

The Fourth Big Revolution

Hacking's first rule is 'Don't look for a big revolution until you find new kinds of institution that epitomize the new directions created by the revolution'.[73] The revolution in analytical chemistry involves two connected changes. First of all it involves instrumentation research, development, marketing and use. When scientific experiments were done with one-of-a-kind instruments, the research, development and use of an instrument was done by a single person or a single research lab. Marketing was not necessary. Now many instruments are bought 'off-the-shelf'. The research, development, marketing and use of instruments have become separate functions. This gives rise to a need for ways for the people involved in these separate functions to get together to coordinate their activities. New institutions have developed to fill this need.

Secondly, the research and development of new instruments requires money. Much has already been said about the National Science Foundation and the other institutions established to help fund science. I have nothing to add here. Relatively little, however, has been written about the new institutions which bring the researchers, developers, marketers and users of instruments together.

Spectroscopy was one of the first instrumental methods to have a big impact on analytical chemistry. But the scientists involved were scattered in many different academic disciplines, industrial laboratories and governmental laboratories. These scientists needed a means to get together to discuss their common interests. The MIT Summer Conferences on spectroscopy filled this need.

The first MIT Summer Conference on Spectroscopy was in July 1932. A Summer Spectroscopy Conference was held each summer thereafter until World War II interceded. The conferences brought together people with various professional affiliations. 88 Papers were published in the proceedings for the 1937, 1938 and 1939 conferences. Of these 41 (46%) were authored by employees of universities, 25 (28%) were authored by employees in industry and 17 (19%) were authored by government employees.[74]

Papers were presented by invitation, and George Harrison, who organized the conference, clearly had in mind to bring together the diversity of interests in spectroscopy. Some papers focused on new developments in spectrographic technique; R. A. Sawyer and H. B. Vincent of the University of Michigan report on 'Characteristics of Spectroscopic Light Sources' at the 1938 conference.[75] Some papers focused on new instruments; at the 1937 conference M. F. Hasler of Applied Research Laboratories described the first commercial grating spectrograph.[76] Some focused on applications; Joseph Walker, of the Massachusetts

State Police, reported on using the spectrograph to assist criminal investigations at the 1938 conference.[77] Some compared different kinds of instruments; G. R. Harrison and, separately, M. Slavin argued for grating as opposed to prism instruments.[78]

After the War, a variety of forums cropped up to provide for a means for instrument researchers, makers and users to get together. The Instrument Society of America put on the first National Instrument Conference and Exhibition, 16–20 September 1946. The conference also had exhibits of instruments by commercial instrument makers.[79] In 1947, 7000 persons attended this meeting and it offered 139 instrument exhibits.[80] This conference and exhibit became an annual event for September. In addition, there were other one-shot conferences and exhibits devoted to providing information about new instruments. Lawrence Hallett hailed the freer exchange of information between makers and users, 'it marks real progress and will result in faster development of this very important and fascinating part of applied science'.[81]

The most successful forum for the exchange of ideas between makers and users of instruments is the Pittsburgh Conference. The Conference was created from the marriage of the Society for Analytical Chemistry of Pittsburgh and the Spectroscopy Society of Pittsburgh. The Society for Analytical Chemistry formed in 1942 and it began putting on conferences in 1946. By 1949, 11 commercial instrument companies exhibited at the meeting. The Spectroscopy Society had held annual meetings since 1940. In 1949 the two societies decided to merge their meetings because of their common interests in analytical/optical instrumentation. Analytical chemists were very interested in learning about the possibilities of spectrochemical analysis and the spectroscopists were interested in closer contact with those who applied their techniques. The result was a great captive audience for spectrographic equipment makers. At the first, March 1950, joint meeting there were 56 papers and 14 exhibits by commercial instrument makers. The conference has been held every March since. In 1964 the 'Pittsburgh Conference' was incorporated and since 1968, because of its size, it has been held at various places other than Pittsburgh.[82]

The Pittsburgh Conference has grown to enormous proportions. The 1990 conference was held at the Jacob Javits Convention Center in New York. 12 500 hotel rooms were reserved in advance. The technical program included 25 symposia and over 1200 contributed papers. There were over 3000 instrument exhibits representing more than 800 different commercial instrument makers. This meeting is one key place where people in industry, government and the academy can meet, find out what each other is doing, share the results of their research, and negotiate plans for pursuing cooperative research.[83]

Besides new institutions, Hacking identifies three other central aspects to big revolutions. They are interdisciplinary, they are associated with dramatic social changes and they are associated with a change in the 'texture of the world'.

The interdisciplinary nature of the instrumentation revolution, should be a fairly obvious characteristic even if the bulk of this paper has concerned one particular discipline, analytical chemistry. A flock of new journals devoted to instrumentation were founded during this period. *The Review of Scientific Instru-*

ments first appeared in 1929. Its British counterpart, *The Journal of Scientific Instruments* first appeared in 1930. *Instruments: Industrial and Scientific* first appeared in 1928. *Instrument Abstracts* first appeared in 1945. These journals are not devoted to a single science, but cover the spectrum of sciences. Instrumentation is fully interdisciplinary, and if Ralph Müller's fantasies about a science of instrumentology are fully realized, these developments in instrumentation are appropriately pre-disciplinary.

Hacking finds dramatic social changes associated with big revolutions. Here I must be more circumspect. I would point to the rise of big government and the 'military–industrial complex'. Spurred by the demands of World War II, the federal government took the lead in paying for and promoting the dramatic build-up of the analytical-instrumentation industry. Many of the companies, originally founded to develop and supply analytical instrument would not have survived if it had not been for governmental contracts during the War.[84] The social implications of these developments are too vast and too ill-understood to bear much analysis here. The closing of the gap between unfettered capitalism and pure socialism is perhaps a product of the rise of big government and the associated government supported research.

Instruments are providing a different texture of the world. In a 1948 editorial in *Analytical Chemistry*, Walter Murphy describes an address by H. V. Churchill where modern *objective* methods of analysis are identified with *instrumental* methods of analysis.[85] This is a general phenomenon. Obstetricians used to use a variety of 'low-tech hands-on' means to follow the development of a foetus. Now, in many cases, an obstetrician is more likely to be touching the transducer of an ultra-sound imaging instrument than the belly of a pregnant woman. Ultrasound provides, as I have been told, the objective 'gold standard' with which to follow foetal development.[86] Whether or not objectivity is fully appropriated by instrumentation is not crucial. Instrumentation has become one important standard for objectivity, and in so doing it has become one important channel for the expression and development of scientific knowledge.

While this 'big' instrumentation revolution has not been discussed by other scholars, several have described changes which coincide with this revolution. I. B. Cohen is instructive.[87] While most of his *Revolution in Science* is devoted to smaller-scale revolutions, more akin to the kind of change Kuhn describes in *Structure*, Cohen does devote one chapter to big revolutions. He provisionally identifies four such revolutions and characterizes them in terms of institutional and conceptual changes.

The scientific revolution is institutionally characterized by the first organizations, such as the Royal Society, devoted to science. Conceptually, the scientific revolution is characterized by the emergence of the importance of experiment to knowledge. Cohen's second revolution is institutionally characterized in terms of the development of new scientific societies, such as the British Association for the Advancement of Science, for the burgeoning population of increasingly professional scientists. Conceptually, this second revolution is associated with the rise of the importance of measurement during the first half of the 1800s. This is Kuhn's 'second scientific revolution'. Cohen's third big revolution occurred around the end of the 1800s when scientific research centers and schools for the graduate training of scientists first appeared. Conceptually, it is associated with the rise of probability and statistics.

Finally, there is Cohen's fourth scientific revolution, 'one that has occurred during the decades since World War II'.[88] Given the time frame, Cohen's fourth big revolution is of particular interest. Cohen identifies this revolution with the expenditure of large sums of money on science and the necessary institutions to make this possible:

> In the United States these have included not only the specially created National Science Foundation (NSF) and the National Institutes of Health (NIH) but granting divisions in the armed forces, the National Aeronautics and Space Administration (NASA), and the Atomic Energy Commission.[89]

Curiously, Cohen comes up somewhat short in his description of the conceptual changes associated with this fourth big revolution:

> It is difficult to think of any ... single intellectual feature that marks the fourth Scientific Revolution. But of major significance is the fact that a considerable part (though by no means the whole) of the biological sciences can be construed as almost a branch of applied physics and chemistry. At the same time, in the world of physics, the most revolutionary general intellectual feature would be the abandonment of the vision of a world of simple elementary particles with only electrical interacting forces between them.[90]

Cohen has missed the major conceptual change here because of his focus on theory.

The major conceptual change associated with Cohen's fourth big revolution is not primarily involved with theory, it is the rise in the importance of scientific instrumentation. Among other things, this explains why science has become so expensive. Instruments cost money; theories are cheap. High energy physics is big science, not because of the abstract theories it involves, but because of the mammoth instruments which it develops and works with. Cohen notes the 'physicalization' of biology. He could have said the same for analytical chemistry – indeed chemistry in general. One way in which this physicalization took place is through the incorporation into instruments of physical approaches to measurement serving a chemical or biological end. Cohen's fourth big revolution is the instrumentation revolution in which analytical chemistry has played such a significant part.

I close with a quote from Ralph Müller:

> That the history of physical science is largely the history of instruments and their intelligent use is well known. The broad generalizations and theories which have arisen from time to time have stood or fallen on the basis of accurate measurement, and in several instances new instruments have had to be devised for the purpose. There is little evidence to show that the mind of modern man is superior to that of the ancients. His tools are incompar-

ably better. Indeed, the early philosophers disdained experiment and even common observation, if the results were contrary to sound logic. Although the modern scientist accepts and welcomes new instruments, he is less tolerant of instrumentation. He is likely to regard preoccupation with instruments and their design as 'gadgeteering' and distinctly inferior to the mere use of instruments in pure research. Thus, Lord Rutherford once said of Callendar, the father of recording potentiometers, 'He seems to be more interested in devising a new instrument than in discovering a fundamental truth'....

Fortunately, there is a great body of earnest workers, oblivious to these jibes, devoted to these pursuits, whose handiwork we may examine. They are providing means with which the 'Olympians' may continue to study nature.[91]

Notes and References

1 Jon Eklund, private communication.
2 See D. Baird, 'Five Theses on Instrumental Realism', in *PSA 1988*, A. Fine and J. Leplin (eds.), 1988, **1**, 165–73; T. Faust, 'Scientific Instruments, Scientific Progress and the Cyclotron', *British Journal for the Philosophy of Science*, 1990, **41**, 147–175; A. Nordmann, 'Facts-Well-Put', unpublished manuscript.
3 Derek de Solla Price, 'Philosophical Mechanism and Mechanical Philosophy: Some Notes Toward a Philosophy of Scientific Instruments', *Annali Dell'Instituto é Muséo di Storia Della Scienza di Firenze*, 1980, **5**, 75.
4 Ian Hacking, 'Representing and Intervening', (Cambridge, 1983); A. Nordmann, 'Engines of Spirit: The Word and the Air-Pump in the Chemical Revolution', unpublished manuscript.
5 T. Kuhn, *The Structure of Scientific Revolutions*, Chicago, 1970, (first ed. 1962).
6 L. Krüger, L. J. Daston and M. Heidelberger (eds.), *The Probabilistic Revolution*, MIT Press, Cambridge, MA, 1987.
7 J. K. Taylor, 'The Impact of Instrumentation on Analytical Chemistry', in *The History and Preservation of Chemical Instrumentation*, J. Stock and M. Orna (eds.), D. Reidel, Dordrecht, 1985, 1.
8 *Ibid.*, 8.
9 Sir Harry Melville, 'The Effect of Instrument Development on the Progress of Chemistry', *Transactions of the Society of Instrument Technology*, December 1962, **14**, 216–218; H. Laitinen and G. Ewing, (eds.), *A History of Analytical Chemistry*, The Division of Analytical Chemistry of the American Chemical Society, Washington DC., 1977, 109–110, 147; G. Ewing, 'Analytical Chemistry: The Past 100 Years', *Chemical and Engineering News*, 6 April 1976, 140; A. Ihde, *The Development of Modern Chemistry*, Dover, New York, 1984, (first ed. 1964), 559; I. Kolthoff, 'Development of Analytical Chemistry as a Science', *Analytical Chemistry*, 1973, **45**, 36A.
10 W. A. Noyes, *The Elements of Qualitative Analysis*, Henry Holt and Co, New York, 1911, sixth ed. revised in collaboration with G. McP. Smith; (first ed. 1887).
11 *Ibid.*, p. 17.
12 G. McP. Smith, *Quantitative Chemical Analysis*, Macmillan Co, New York, 1921, second ed. revised; (first ed. 1919), 2–3.

13 J. C. Olsen, *Quantitative Chemical Analysis*, D. Van Nostrand, New York, 1916, fifth ed. (first ed. 1904).

14 J. Muter, *A Short Manual of Analytical Chemistry*, P. Blakiston's Son, Philadelphia, 1906, (fourth American ed.), 232.

15 W. N. Lacey, *A Course of Instruction in Instrumental Methods of Chemical Analysis*, MacMillan, New York, 1924, 83.

16 G. H. Schenk, R. B. Hahn and A. V. Hartkopf, *Quantitative Analytical Chemistry: Principles and Life Science Applications*, Allyn and Bacon, Boston, 1977.

17 D. A. Skoog and D. M. West, *Fundamentals of Analytical Chemistry*, Holt, Rinehart and Winston, New York, 1976, third ed. (first ed. 1963); *Principles of Instrumental Analysis*, Holt, Rinehart and Winston, New York, 1971.

18 Skoog and West, *Fundamentals of Analytical Chemistry*, 135.

19 Skoog and West, *Principles of Elemental Analysis*, v.

20 L. T. Hallett, G. G. Gordon and S. Anderson, 'Editorial: Scope of the *Analytical Edition*', *Industrial and Engineering Chemistry, Analytical Edition*, 1946, **18**, 218.

21 F. C. Strong, 'Trends in Quantitative Analysis', *Industrial and Engineering Chemistry, Analytical Chemistry*, 1947, **19**(12), 968.

22 *Ibid.*, 969.

23 R. B. Fischer, 'Trends in Analytical Chemistry 1955', *Analytical Chemistry*, 1956, **27**(12), 9A–15A; 'Trends in Analytical Chemistry 1965', *Analytical Chemistry*, 1965, **37**(12), 27A–34A.

24 Fischer, 'Trends', 31A; Strong, 'Trends', 968.

25 G. W. Ewing, *Instrumental Methods of Chemical Analysis*, McGraw-Hill, New York, 1969, third ed. (first ed. 1954), 1.

26 H. Laitinen and W. Harris, *Chemical Analysis: An Advanced Text and Reference*, McGraw-Hill, New York, 1975, second ed. (first ed. 1960), 2–4.

27 W. Ostwald, *The Scientific Foundations of Analytical Chemistry*, Macmillan, London, 1895, translated by G. McGowan (first German ed. 1894).

28 *Ibid.*, 9.

29 Noyes, *The Elements of Quantitative Analysis*, 1.

30 C. Williams, 'Editorial: The Rôle of the Analyst', *Analytical Chemistry*, 1948, **20**(1), 2.

31 W. J. Murphy, 'Editorial: The Profession of Analytical Chemist', *Industrial and Engineering Chemistry, Analytical Edition*, 1947, **19**(3), 145.

32 *Ibid.*

33 *Ibid.*

34 J. J. Lingane, 'Editorial: The Rôle of the Analyst', *Analytical Chemistry*, 1948, **20**(1), 2–3.

35 W. J. Murphy, 'Editorial: Fisher Award', *Industrial and Engineering Chemistry, Analytical Edition*, 1947, **19**(10), 699; 'Editorial: The Merck Fellowship in Analytical Chemistry', *Analytical Chemistry*, 1948, **20**(10), 885.

36 W. J. Murphy, 'Editorial: We Have Arrived!', *Industrial and Engineering Chemistry, Analytical Edition*, 1947, **19**(12), 1131.

37 W. J. Murphy, 'Editorial: The Analytical Chemist: Dispenser of Analyses or Analytical Adviser?', *Industrial and Engineering Chemistry, Analytical Edition*, 1947, **19**(5), 289.

38 W. J. Murphy, 'Editorial: The Analytical Chemist', *Industrial and Engineering Chemistry, Analytical Edition*, 1947, **19**(6), 361–3.

39 D. B. Keyes, 'Editorial: The Importance of the Analytical Research Chemist in Industry', *Industrial and Engineering Chemistry, Analytical Edition*, 1947, **19**(8), 507.

40 B. L. Clarke, 'Editorial: What is Analysis?', *Industrial and Engineering Chemistry, Analytical Edition*, 1947, **19**(11), 822.

41 W. C. McCrone, 'Editorial: The Rôle of the Analyst', *Analytical Chemistry*, 1948, **20**(1), 2–3.

42 *Ibid.*, 4.

43 Lingane, 'Editorial', 2.

44 *Ibid.*, 1–2.

45 *Ibid.*, 2.

46 H. A. Liebhafsky, 'Modern Analytical Chemistry: A Subjective View', *Analytical Chemistry*, 1962, **34**(7), 23A.

47 *Ibid.*

48 *Ibid.*, 24A.

49 *Industrial and Engineering Chemistry, Analytical Edition*, 1939, **11**, 563–582; R. Müller, 'American Apparatus, Instruments, and Instrumentation', *Industrial and Engineering Chemistry, Analytical Edition*, 1940, **12**, 571–630; R. Müller, 'Instrumental Methods of Chemical Analysis', *Industrial and Engineering Chemistry, Analytical Edition*, 1941, **13**, 667–754.

50 R. Müller, 'Monthly Column: Instrumentation in Analysis', *Industrial and Engineering Chemistry, Analytical Edition*, 1946, **18**(1), 21A–22A.

51 R. Müller, 'Monthly Column: Instrumentation in Analysis', *Industrial and Engineering Chemistry, Analytical Edition*, 1946, **18**(2), 25A–26A.

52 R. Müller, 'Monthly Column: Instrumentation in Analysis', *Industrial and Engineering Chemistry, Analytical Edition*, 1946, **18**(3), 25A–26A.

53 R. Müller, 'Monthly Column: Instrumentation in Analysis', *Industrial and Engineering Chemistry, Analytical Edition*, 1947, **19**(4), 19A–20A.

54 R. Müller, 'Monthly Column: Instrumentation in Analysis', *Industrial and Engineering Chemistry, Analytical Edition*, 1946, **18**(5), 29A–30A.

55 *Ibid.*, 29A.

56 *Ibid.*, 30A.

57 R. Müller, 'Monthly Column: Instrumentation in Analysis', *Industrial and Engineering Chemistry, Analytical Edition*, 1947, **19**(1), 23A.

58 *Ibid.*

59 R. Müller, 'Monthly Column: Instrumentation in Analysis', *Analytical Chemistry*, 1949, **21**(6), 23A.

60 R. Müller, 'Monthly Column: Instrumentation in Analysis', *Industrial and Engineering Chemistry, Analytical Edition*, 1946, **18**(10), 25A.

61 R. Müller, 'Monthly Column: Instrumentation in Analysis', *Analytical Chemistry*, 1949, **21**(6), 23A

62 R. Müller, 'Monthly Column: Instrumentation in Analysis', *Industrial and Engineering Chemistry, Analytical Edition*, 1946, **19**(9), 26A.

63 R. Müller, 'Monthly Column: Instrumentation in Analysis', *Analytical Chemistry*, 1948, **20**(6), 21A.

64 D. Baird, 'Baird Associates's Commercial Three-Meter Grating Spectrograph and the Transformation of Analytical Chemistry', *Rittenhouse*, 1990, **5**(3), 65–80.

65 Kuhn, *The Structure of Scientific Revolutions*, Ch. 5.

66 I. B. Cohen, *Revolution in Science*, Harvard University Press, Cambridge, MA, 1985.

67 *Ibid.*, Ch. 3.

68 *Ibid.*

69 I. B. Cohen, 'Scientific Revolutions, Revolutions in Science, and a Probabilistic Revolution 1800–1930', in *The Probabilistic Revolution, Volume 1: Ideas in History*, L. Krüger, L. J. Daston and M. Heidelberger, (eds.), MIT Press, Cambridge, MA., 1987, 23, emphasis added.

70 T. Kuhn, 'The Function of Measurement in Modern Physical Science', in *The Essential Tension*, University of Chicago Press, Chicago, 1977, 178–224; originally published in *Isis*, 1961, **52**, 161.

71 I. Hacking, 'Was There a Probabilistic Revolution 1800–1930?', in *Probability Since 1800: Interdisciplinary Studies of Scientific Development*, M. Heidelberger, L. Krüger and R. Rheinwald (eds.), B. K. Verlag GmbH, Bielefeld, 1983; this essay was revised and condensed in *The Probabilistic Revolution, Volume 1: Ideas in History*, L. Krüger, L. J. Daston and M. Heidelberger (eds.), MIT Press, Cambridge, MA, 1987.

72 Hacking, *Probability Since 1800*, 51.

73 *Ibid.*, 49.

74 G. Harrison (ed.), *Proceedings of the Fifth Summer Conference on Spectroscopy and Its Applications*, John Wiley, New York, 1938; *Proceedings of the Sixth Summer Conference on Spectroscopy and Its Applications*, John Wiley, New York, 1939; *Proceedings of the Seventh Summer Conference on Spectroscopy and Its Applications*, John Wiley, New York, 1940.

75 R. A. Sawyer and H. B. Vincent, 'Characteristics of Spectroscopic Light Sources', in Harrison, *Proceedings of the Sixth Summer Conference*, 54–59.

76 M. F. Hasler, 'The Practice of Arc Spectrochemistry with a Grating Spectrograph', in Harrison, *Proceedings of the Fifth Summer Conference*, 43–46.

77 J. T. Walker, 'The Spectrograph as an Aid in Criminal Investigation', in Harrison, *Proceedings of the Sixth Summer Conference*, 1–5.

78 G. Harrison, 'A Comparison of Prism and Grating Instruments for Spectrographic Analysis of Materials', in Harrison, *Proceedings of the Fifth Summer Conference*, 31–37; M. Slavin, 'Prism Versus Grating for Spectrographic Analysis', in Harrison, *Proceedings of the Seventh Summer Conference*, 51–58.

79 R. Müller, 'Monthly Column: Instrumentation in Analysis', *Industrial and Engineering Chemistry, Analytical Edition*, 1946, **18**(10), 25A–26A.

80 L. T. Hallett, 'Monthly Column: The Analyst's Column', *Industrial and Engineering Chemistry, Analytical Edition*, 1947, **19**(10), 15A.

81 L. T. Hallett, 'Monthly Column: The Analyst's Column', *Analytical Chemistry*, 1948, **20**(10), 25A.

82 'Conference Announcement', *Applied Spectroscopy*, 1971, **25**, 123.

83 'Conference Announcement', *Applied Spectroscopy*, 1990, **44**.

84 D. Baird, 'Baird Associates's'.

85 W. Murphy, 'Editorial: Modern Objectivity in Analysis', *Analytical Chemistry*, 1948, **20**(3), 187.

86 A. Smyth, private conversation; see also, Z. Swijtink, 'The Objectification of Observation: Measurement and Statistical Methods in the Nineteenth Century', in *The Probabilistic Revolution, Volume 1: Ideas in History*, L. Krüger, L. J. Daston and M. Heidelberger, (eds.), MIT Press, Cambridge, MA, 1987, 261–286.

87 Cohen, *The Probabilistic Revolution*, Ch. 6; see also S. Brush, *The History of Modern Science: A Guide to the Second Scientific Revolution, 1800–1950*, Iowa State University Press, Ames, Iowa, 1988.

88 *Ibid.*, 93.

89 *Ibid.*, 94.

90 *Ibid.*, 96.

91 R. Müller, 'American Apparatus', 571–572.

The Rôle of Physical Instrumentation in Structural Organic Chemistry in the Twentieth Century*

PETER J. T. MORRIS[1] AND ANTHONY S. TRAVIS[2]

[1]Science Museum, London SW7 2DD, England, Email: p.morris@nmsi.ac.uk
[2]Sidney M. Edelstein Center for the History and Philosophy of Science, Technology and Medicine, The Hebrew University of Jerusalem, Edmond Safra Campus, Givat Ram, Jerusalem 91904, Israel, Email: travis@cc.huji.ac.il

Introduction

During the twentieth century, many changes have taken place in the organic chemist's laboratory, but none greater than the introduction of electronic instrumentation. The widespread adoption of instrumental methods has led to tremendous advances in the determination of complex structures. Between 1940 and 1970, structural studies were reduced from being life-long Nobel Prize-winning activities for leading professors to a day's work for graduate students and technicians. Doubtlessly, this shift has transformed the chemist's work and the nature of different jobs within chemical laboratories. Some organic chemists found more time to work on organic synthesis, others transferred their attention to biomolecular topics, and yet others were forced to find alternative careers outside chemistry. Even a leading pioneer of the new methods, Carl Djerassi, found reason to regret the passing of the former intellectual and creative challenge of structure determination:

> But if eliminating the need for 'wet chemistry' (the laboratory equivalent of 'Twenty Questions') saves a lot of time and material, it also makes structure

*Reprinted with permission from Peter J. T. Morris and Anthony S. Travis, 'The Role of Physical Instrumentation in Structural Organic Chemistry' in *Science in the Twentieth Century*, eds. J. Krige and D. Pestre, Harwood, Amsterdam, 1997, 715–740.

elucidation a more mechanical endeavor. Ironically, much of our own research into better flashlights [physical instrumentation] has made obsolete the traditional and often intellectually exciting ways of exploring dark rooms [organic chemical structures].[1]

Remarkably, the emergence of, and the present-day reliance upon, instrumental methods, has received little attention from historians, although it spans the history and sociology of science, the history of technology and even business history. The vastness, diversity and complexity of both the science and the technology of these developments explain the dearth of any significant historical literature. Furthermore, the displacement of classical methods by physical instrumentation was completed only in the last two decades. The story involves instrument manufacturers, chemists, physicists, government agencies and the chemical industry. Here, we will delineate the impact of instrumentation on the determination of the structures of organic compounds by focusing on the natural products that are the lifeblood of organic chemistry and the mainstays of the biomedical sciences.

During the second half of the nineteenth century, the structures of many important natural products had been established, with considerable degrees of accuracy, by lengthy and painstaking studies. By 1900, the chemical laboratories in which this work was carried out were characterised by reagent bottles and test tubes, the glass and porcelain apparatus employed in qualitative and quantitative analysis, and combustion furnaces necessary for routine, but extremely tedious, elemental analyses. Together, these made up the standard tools used to study the reactions and decompositions of natural products. They were often accommodated in cupboards and on shelves that overlooked long wooden, and highly polished, benches, with sinks installed at the ends. A few fume cupboards enabled the comparatively safe handling of dangerous chemicals. The laboratory techniques were what we now call 'wet and dry'. There were very few physical instruments to be found in such laboratories: a chemical balance, a refractometer, a microscope and perhaps a polarimeter, used to examine optically active compounds such as carbohydrates and proteins. By 1980, most of those methods had been superseded by physical techniques, which had not originally been developed for such studies.

Early Responses to the New Instrumental Technology

The most powerful, and ubiquitous, of the new tools used in structural elucidation is the nuclear magnetic resonance (NMR) spectrometer. In a 1985 interview Laurie Hall, the newly appointed professor of medicinal chemistry at the University of Cambridge, observed that 'NMR has developed in the past two and a half decades to the point where for many chemical scientists, it is now the dominant analytical and structural tool'. Twenty-six years earlier, Hall had 'quite by accident ... "discovered" NMR. What I didn't know at the time was that there were only half a dozen NMR machines in the whole of Britain; but I decided that I wanted to do NMR because even at that early date (1959), it was quite obviously going to become an important technique in organic chemistry'.[2]

A decade earlier, it was far more difficult to appreciate the potential offered by NMR. MIT physical organic chemist John D. Roberts first heard about NMR from Richard Ogg of Stanford University. 'One day in late 1949 or 1950, he [Ogg] was at MIT, and I invited him to lunch. He was really wound up and proceeded to tell me about the wonderful new magnetic resonance spectroscopy, with such promise for chemistry. I wish I could say that I could understand even 5% of what he told me, but I had too little knowledge of magnetism and absorption of radio frequency radiation – indeed, hardly any knowledge of other radiation... It was clear there were applications to chemistry, even if I didn't understand what they were'.[3]

These two responses to an emerging laboratory technology provide as good an insight as any into the ways in which organic chemists came to appreciate how, with the aid of instruments, physical properties could be related to structural features. Here we explore these developments during the middle of the twentieth century, requiring of the reader no special knowledge of the techniques, certainly no more than the level of initial understanding held by John D. Roberts in 1949. To place the story in perspective, however, it is necessary to review the classical, pre-instrumental methods used for determination of structure by chemical means.

Classical Methods

The modern representations of organic chemical structures originated with the concepts of the tetravalent carbon atom and the six-carbon benzene ring, as suggested by August Kekulé in 1858 and 1865, respectively. These enabled graphic representations of aliphatic and aromatic molecules. With the aid of chemical analysis that afforded empirical formulae, from which molecular formulae were derived, and prior knowledge of the atoms in functional groups, such as hydroxyl (OH) and carbonyl (C=O), good approximations of the structural formulae of relatively complex molecules could be drawn. A good early example is the structure of alizarin, a constituent of the madder root, and an important dyestuff. During 1868, using qualitative tests, Carl Graebe and Carl Liebermann showed that it was a quinone with two carbonyl groups, and that it probably contained two hydroxyl groups. However, it was the action of chemical 'brute force', in this case through zinc dust distillation, that provided the main clue to the constitution, and a partial structure. The reduction with zinc gave anthracene. By 1874, the total structure of alizarin was available, as were the structures of related hydroxyanthraquinones. Since these compounds were coloured, it was possible to identify them by spectral analysis, through their characteristic 'fingerprints'.

Other 'brute force' chemical methods included acid and alkali hydrolysis, and nitric and chromic acid oxidation. A series of specialised tests for particular functional groups was slowly built up, the most pungent of which was doubtlessly Wilhelm Hofmann's carbylamine test for primary amines. These methods permitted the structural elucidation by test-tube methods of indigo and purines.

Table 1 *Nobel Laureates in Structural Organic Chemistry*

1902	Emil Fischer	Sugars, purines
1905	Adolf von Baeyer	Indigo
1915	Richard Willstätter	Chlorophyll
1927	Heinrich Wieland	Bile acids, steroids
1928	Adolf Windaus	Steroids, vitamins
1930	Hans Fischer	Blood pigments, chlorophyll
1937	W. N. Haworth	Sugars, vitamin C
	Paul Karrer	Vitamins A and B_2
1938	Richard Kuhn	Carotenoids and vitamins
1939	Adolf Butenandt	Sex hormones, steroids
	Leopold Ruzicka	Polyenes, higher terpenes
1947	Robert Robinson	Strychnine, *etc.*
1958	Frederick Sanger	Insulin
1962	Max Perutz and John Kendrew	Myoglobin and haemoglobin
1964	Dorothy Hodgkin	Vitamin B_{12}
1965	R. B. Woodward	Mostly for his synthetic work

In later years it becomes harder to separate structural organic chemistry from molecular biology or biochemistry, for example, Alexander Todd's Nobel Prize (in 1957) for his work on nucleotides, or the medicine–physiology Nobel Prize awarded to James Watson, Francis Crick and Maurice Wilkins in 1962 for the determination of the structure of DNA.

The detection of molecular asymmetry through measurement of optical rotation with the polarimeter became an important tool in the investigation of the structures of carbohydrates and proteins by Emil Fischer, from 1891. Another important contribution to organic structural chemistry was ozonolysis, introduced by Carl Dietrich Harries during 1903–5. Because it revealed the position of carbon–carbon double bonds in molecules, ozonolysis was particularly important for the study of natural rubber by Harries between 1903 and 1916, and the investigation in the early 1930s of the vitamin A precursor α-carotene by Paul Karrer.

In general, parent molecules were reconstructed from known fragments, using intuitive assumptions, such as the 'isoprene rule' introduced by Leopold Ruzicka in 1922, and thinking by analogy. These 'wet and dry' methods were, however, time-consuming. Complex structures could take two to four decades to resolve. The work on chlorophyll, started by Richard Willstätter in the early 1900s, led to a 'clover leaf' structure in 1912, and the full structural determination in 1939 by Hans Fischer. This was confirmed (using physical methods) by Patrick Linstead in 1955–6. The magnitude of these tasks was such that they were often rewarded with the highest accolades, including the Nobel Prize (see Table 1). Certainly they added greatly to the stock of chemical knowledge and laid the foundations for the introduction of physical techniques in structural organic chemistry.

The golden era of classical determinations began around 1930 and lasted almost twenty years. It resulted from a 'critical mass' of accumulated data. In addition to chlorophyll, the triumphs included most of the vitamins (A, B_1, B_2, B_6, C, D, E), many steroids, quinine, and, above all, strychnine. The empirical formula of strychnine had been established as early as 1838; it took another sixty years to identify a benzene ring attached to a nitrogen atom in the molecule.

Between 1910 and 1932, William Henry Perkin, Jr. and Robert Robinson, made further contributions towards the elucidation of the structure of strychnine. In 1929, they showed that it was an indole derivative. The position of a second nitrogen atom could not be determined until 1948 when Robert Burns Woodward resolved several controversies and drew the first correct structural representation of strychnine. Arthur J. Birch summed up the chemist's approach during this time:

> Our classical natural products, typically extractable by organic solvents, made available interesting but not too complex molecular structures as exercises for chemical investigation and training. Many such substances were of interest in connection with human applications as dyestuffs, pharmaceuticals, tanning agents, psychedelic agents, and so forth. Classical structure determinations (as practiced until about the 1960s) are at a maximum difficulty with such initially totally unknown natural products... The principles of structure work then technically involved two aspects: to detect and interrelate functional groups by chemical means, and to obtain structural information on the atomic nature of the main nucleus. Until the mid-1970s* new substances could only be examined for their chemical transformations. When possible they were converted into previously known, or synthesizable, simpler compounds (*e.g.* by fission of unsaturation and by dehydrogenation into aromatic substances, for which synthetic methods were efficient). The interpretation of such work was facilitated by the efficient indexing of known compounds and their properties. For this and other reasons, organic chemistry is the best documented science (the order of a million substances on record).[4]

The indexing of information about the reactions and structures of organic molecules was, and remains, essential for both classical and instrumental approaches to structural determination. This necessity gave rise to the multi-edition, multi-volume and complex *Handbuch der organischen Chemie* founded by Friedrich Konrad Beilstein, who completed the first edition in 1882 after two decades' effort. The period between 1950 and 1970 saw the introduction of various 'atlases', compendia of the infrared, ultraviolet, NMR and mass spectra of numerous organic compounds.[5]

Evolution of Chemical Instrumentation

Before 1930, developments in physics had already contributed towards structural knowledge of simple molecules. The relative positions of atoms, and information about the nature of chemical bonds, had been made possible by numerical data provided by X-ray diffraction and dielectric behaviour. In particular, it was observed that certain organic functional groups gave characteristic absorption spectra. However, since the taking of measurements was an invariably lengthy process, requiring considerable skill, organic chemists were slow to

---

*Birch appears to be in two minds about the date of the changeover. We would date the changeover to the mid-1960s, in agreement with Birch's first estimate.

incorporate these techniques in their laboratory practice. The Second World War and the demands of the new petrochemical industry provided the stimulus, and resources, for the development of chemical instrumentation, and its increasing application to organic chemistry. A notable example was the use of infrared spectroscopy (and to a lesser extent ultraviolet spectroscopy) in the study of the structure of synthetic rubber by Paul Flory and John White at Esso Research in 1942; White moved to Perkin–Elmer in 1944.

The mediators in this endeavour were the instrument manufacturers, notably Arnold Beckman, who often established close links with the pioneers. Initially, the instrument companies tended to work with physicists, chemical physicists, and physical chemists interested in fundamental processes. When it became clear that companies, especially petroleum and petrochemical firms, rather than universities, would be their major customers, the instrument manufacturers collaborated with industrial researchers, most notably the link-up in infrared spectroscopy between Beckman and Robert Brattain at Shell Research with the encouragement of the US government's Petroleum Administration for War. For the most part, it was only after World War II that the developers of physical instrumentation established close links with organic chemists concerned with structural problems. This collaboration quickly deepened and once it was shown that the new instruments were of value in the elucidation of complex structures, particularly of natural products such as steroid hormones, funding for academic research became available from both government agencies and the chemical and pharmaceutical industries.

Although the use of instrumentation in routine organic analysis and structural determination became widespread only during the 1960s, its prior history can be traced back to the beginning of the twentieth century, notably with X-ray diffraction, an area not usually regarded as falling within physical instrumentation, but clearly a forerunner of later developments.

Development of X-ray Methods

In 1912, Max von Laue obtained the first X-ray diffraction pattern. His success encouraged the young Lawrence Bragg to investigate the structure of alkali halides, and Bragg's results were published in 1913. It soon became apparent that the method offered the possibility of identifying the location and nature of each atom in a molecule. There were of course problems, particularly the need for crystalline samples and the fact that organic molecules are much more difficult to analyse structurally than inorganic ionic lattices. Bragg later observed:

> The strong homopolar bonds between the atoms make the organic molecule a definite entity, which typically retains its individuality when the solid is melted or dissolved or even vaporized. In contrast, the forms of most inorganic structures only exist in the solid state … it [is] possible to investigate [inorganic] structures of quite high complexity … The organic molecule, on the other hand, is an entity which typically has an irregular shape and no symmetry … a correspondingly large number of parameters

Charles Supper precession X-ray camera, 1950s. Science Museum collections, inventory number 1977-57. This type of camera was often used to determine the structure of proteins. Science Museum/Science and Society Picture Library

must be determined to define the structure. It is not surprising that their analysis was for long regarded as an almost impossible task. They have, however, a compensating feature which helps analysis. The atoms are linked by homopolar bonds, and the lengths of these bonds and the angles between them can be established to a high degree of accuracy by the analysis of the simpler structures.[6]

The simpler structures were those of naphthalene and anthracene, the first organic molecules investigated by Bragg in 1921. Despite the initial difficulties, the structure of hexamethylenetetramine was established by Roscoe Dickinson and Albert Raymond in 1923, of polymers by Herman Mark in the mid-1920s, of hexamethylbenzene by Kathleen Lonsdale in 1928, and of ergosterol by John Desmond ('Sage') Bernal in 1932. The progress of X-ray crystallography was tied intimately to the development of computation, particularly the availability of the Patterson function (1934), Lipson–Beever strips (1936), and Patterson–Tunell strips (1942).

As in other physical techniques, X-ray structure determination depends to a considerable extent upon prior knowledge, in this case assigning likely positions of atoms in the unit cell. This led to the development of a trial and error technique that was particularly successful in the study of aromatic hydrocarbons, where structures were already known. In cases where there was no structural information available, the incorporation of a few atoms with atomic numbers much greater than those of other atoms in the molecule provided useful information. Once the locations of these 'heavy atoms' had been established, phase constants for these atoms were used to produce an electron density distribution. From this, peaks for light atoms, particularly carbon, nitrogen, and oxygen, were then established.

This 'heavy atom' method was particularly useful in the case of centrosymmetric structures in which the 'heavy' atom at the centre of symmetry is taken as the origin. It was first used by J. Monteath Robertson in 1935 on a molecule that came with the 'heavy atom' already in place, namely the first of the newly discovered synthetic dyes known as phthalocyanines, related to the natural pigments haeme and chlorophyll. Robertson and Ida Woodward (no relation to R. B. Woodward) published an electron-density map of platinum phthalocyanine in 1940. The 'heavy atom' method also enabled the full structural determination of a steroid, cholesteryl iodide, by Harry Carlisle and Dorothy Crowfoot Hodgkin, in 1945, which was in full agreement with structural information determined by chemical means in 1932 by Heinrich Wieland and Elisabeth Dane, and independently by Otto Rosenheim and Harold King in the same year. X-ray crystallography also resolved the debate over the structure of penicillin. Chemical studies during World War II had permitted two possible structures. Robinson favoured two separate rings, while R. B. Woodward preferred the fused β-lactam structure (which was supported by infrared studies at Shell Research). The definitive structure of penicillin and proof that it was a β-lactam was provided by the X-ray analysis of Crowfoot Hodgkin and Charles Bunn at Oxford, 'working in a state of much greater ignorance of the chemical nature of the compounds we have had to study than is usual in X-ray analysis'.[7] The X-ray analysis was simplified by the presence of a large sulfur atom. This study was completed in spring 1945, at about the same time that chemical work at Merck came up with the same structure. Similarly, Johannes Bijvoet determined the structure of strychnine in 1947–1949, although he was narrowly beaten, as we have seen, by Woodward's classical approach.

The structure of vitamin B_{12} was an outstanding application of X-ray diffraction to organic structure determination – Bragg described it as 'breaking the sound barrier … of telling the organic chemist something he did not already know'.[8] The history of vitamin B_{12} went back to 1855 when Thomas Addison reported pernicious anemia. However, the value of liver in its treatment was not realised until 1926. Partly because of World War II, another twenty-two years were to elapse before the active principle, vitamin B_{12}, was isolated by Karl Folker's group at Merck, and, independently, by Lester Smith, at Glaxo. Subsequently, the structure of this vitamin was extensively studied by Merck, Glaxo, British Drug Houses (BDH), and Alexander Todd's group at the University of Cambridge. The chemists clarified several important features of vitamin B_{12}, including the presence of pyrrole rings, but were unable to determine completely its extremely complex structure. Smith prepared a crystalline sample that Glaxo donated to Dorothy Crowfoot Hodgkin. She published the full structure of the cobalt-containing molecule in 1957, after eight years of study. It was the first time such a complex molecule had been almost entirely elucidated by physical methods. Even the molecular formula was deduced from the X-ray work!

The determination of the structure of vitamin B_{12} was significantly assisted by the growing power and availability of computers. Hodgkin later recalled:

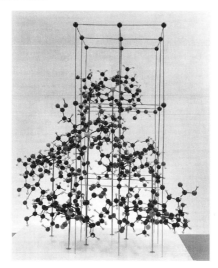

Model showing the crystal structure of vitamin B_{12}, made by Dorothy Crowfoot Hodgkin, c. 1956. Science Museum collections, inventory number 1959-135. Science Museum/Science and Society Picture Library

'And we were greatly helped by friends with computers: on a particularly happy day Kenneth Trueblood, on a casual summer visit to Oxford, walked into the laboratory and offered to carry out any additional calculations we needed on a fast computer in California, free and for nothing and with beautiful accuracy'.[9]

X-Ray crystallography was completely transformed by the arrival of electronic computers in the 1950s. They enabled routine determination of bond lengths, bond angles, and the spacing between non-bonded atoms. By the 1960s, three-dimensional electron-density distribution patterns, incorporating heavy atoms and isomorphous replacement, enabled the definitive solution to many structural problems. This breakthrough led to the determination of the structure of the complex biomolecules myoglobin in 1960 by John Kendrew, and haemoglobin by Max Perutz ten years later.

Ultraviolet and Infrared Spectroscopy

Spectroscopic methods based on absorption in the visible spectrum had been employed since the 1860s, mainly for 'fingerprinting'. Indeed this is how alizarin and other hydroxyanthraquinones were identified. By the early 1900s spectroscopy had found extensive application with coloured compounds such as chlorophyll, haemoglobin and dyestuffs, using simple spectrometers and colorimeters. Studies into the ultraviolet region had been pioneered by Walter Noel Hartley in the latter decades of the nineteenth century. Hartley was mostly interested in the use of ultraviolet spectroscopy in metallurgical analysis, but he also studied the ultraviolet spectra of organic compounds. For instance, in 1899 he used ultravio-

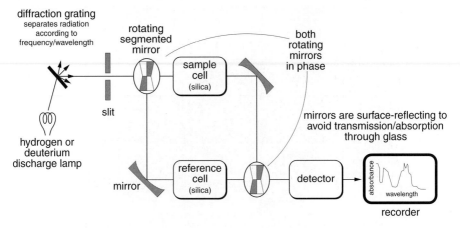

Ultraviolet spectrometer

let spectroscopy to study the vexed issue of tautomerism, and was able to show that isatin has a keto (lactam) rather than an enol (lactim) structure. Between 1907 and 1919, the French chemist Victor Henri studied numerous natural products including strychnine, chlorophyll and cholesterol. He showed that the ultraviolet spectrum of an organic compound was characteristic of certain bonds, rather than a function of the whole molecule.

Widespread use of ultraviolet spectroscopy relied on the introduction of quartz spectrographs in which spectra were recorded photographically. An early manufacturer of these instruments was Adam Hilger; its technical director Frank Twyman collaborated closely with Hartley. In 1906, Hilger introduced the first fixed adjustment quartz spectrograph, in 1910 the sector photometer, and in 1931 the photoelectric 'Spekker' photometer. Arthur C. Hardy at MIT designed an advanced, but expensive, photoelectric spectrophotometer which was commercialised by General Electric in 1933. Ultraviolet spectroscopy assisted the determination of the structure of thiamine (vitamin B_1) (VI) by Robert R. Williams in 1936. Several new photoelectric spectrophotometers were introduced during World War II. Beckman brought out the famous DU spectrophotometer in 1941, and it was followed by British-made instruments such as the Unicam SP 500 and Hilger Uvispeck. The spectra were obtained by measurement of point-by-point dial readings. Subsequently, recording spectrophotometers, with spectra recorded on paper charts, appeared on the market, although at three times the price of non-recording machines. Ultraviolet spectroscopy was particularly useful for identifying the presence of conjugated double bonds, and therefore applicable to structural work on carotenoids and steroids.

The principle behind infrared spectroscopy is similar. During 1881–82, William de Wiveleslie Abney and Edward Robert Festing were the first to relate infrared spectra to the structures of organic compounds. Much of the fundamental work was done by William Coblentz in the United States between 1903 and 1905, using a non-recording spectrometer. However, his efforts were not widely

Unicam SP 500 ultraviolet–visible spectrometer, c. 1949. Science Museum collections, inventory number 1976–419. Science Museum/Science and Society Picture Library

recognised. The first commercial infrared spectrometers were introduced in the early 1940s, notably by Beckman in 1942, and found widespread use in the wartime study of synthetic rubber. Infrared spectroscopy picks out functional groups, and is more useful as a backup to other methods, or for 'fingerprinting', rather than for direct determination of structures. It is used to establish branching in long chain hydrocarbon polymers whose properties are determined by the branched chain. Most functional groups in a pure compound or mixture can be determined within minutes. The method is non-destructive, and requires only a few milligrams of sample. About 1950, Van Zandt Williams of the American firm of Perkin–Elmer gave a talk about the new Mark 21 double-beam infrared

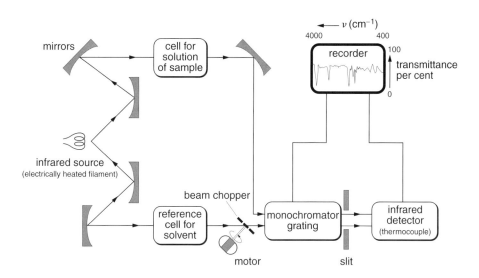

Infrared spectrometer

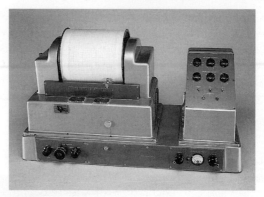

Perkin–Elmer D21 double-beam infrared spectrophotometer, 1950. Science Museum collections, inventory number 1970-205. This is the spectrophotometer which was shipped from Perkin– Elmer in the United States to Cambridge University, where it was used by Norman Sheppard. Science Museum/Science and Society Picture Library

spectrophotometer to Gordon Sutherland's infrared research group at Cambridge. Everyone there was very impressed by the value of infrared spectroscopy for structure elucidation, and Alexander Todd ordered a Perkin–Elmer 21 for the new university chemical laboratory. When the bench-top Perkin–Elmer 137 came out in 1957, Ralph Elsey, a technician at the Cambridge laboratory, remarked 'compared with the 21 the Model 137 is made of tin!'[10] But as it was much cheaper than earlier models, it brought infrared spectrophotometry within the reach of the ordinary chemistry laboratory and, in time, even the undergraduate teaching laboratory.

Woodward Rules

The availability of ultraviolet spectroscopic data encouraged several theoretical developments, of which the Woodward Rules for conjugated dienes and trienes in cyclic systems were outstanding. During 1941–42, the young R. B. Woodward made a significant contribution towards the establishment of structural knowledge from observed spectral data, in this case ultraviolet absorptions of steroids containing double bonds. Woodward undertook a careful numerical analysis of published spectral data for various steroidal ketones (containing the structures $C=C-C=O$) that gave intense absorption maxima around 230–250 nm. This enabled him to develop rules that were of general applicability, involving incremental additions to the wavelength of the original absorption maximum, depending on the arrangement of the double bonds and the substituents attached to them. With characteristic self-confidence, Woodward observed in the first of four papers on the subject: 'A few substances do not conform with the generalizations outlined above. The simplest and most probable explanation of this apparent anomaly is that the structures at present assigned to those compounds are incorrect'.[11] And, of course, he was right. Thus he showed that the accepted structure for a menthadiene derivative was incorrect.

This was the first systematic application of instrumentation, apart from the polarimeter, to a major area of natural product chemistry. There had been earlier attempts to establish numerical relationships between the ultraviolet absorption maximum in a steroid's spectrum and the number and type of conjugated double bonds and substituents that it contains, most notably by the German chemist Heinz Dannenberg in 1940, but they had little impact. By contrast, the Woodward rules, as extended by Louis Fieser in 1949, are still cited today.

Octant Rule

This was another remarkable achievement by R. B. Woodward that enabled a relationship to be drawn between structure and absorption. Again, it relied on prior studies, in particular Carl Djerassi's work with optical rotatory dispersion, or ORD. This rule is based on the fact that optically active compounds are associated with the rotation of the plane of vibration of light with which they interact, thereby providing clues to the stereochemistry. The conventional polarimeter, as introduced in the 1840s, uses the yellow D-line of sodium. The principle of ORD is based on the fact that optical rotation varies with wavelength. Jean Baptiste Biot established in 1813 that the rotation of plane polarised light of an optically active substance varies with wavelength. In 1895, Aimé Cotton discovered the effect named after him, *viz* that the sign of optical rotation changes markedly around an absorption maximum, typically in the far ultraviolet. It increases abnormally and falls off sharply close to a characteristic absorption band. However, optical rotatory dispersion languished for decades for lack of suitable instrumentation. Until 1950, nearly all measurements of optical rotation were at a single wavelength (usually the sodium D-line), which is of little value in the study of organic compounds.

In 1953, O. C. Rudolph & Sons introduced the first commercial spectropolarimeter. Carl Djersassi described it as 'the prince who awoke the sleeping beauty of rotatory dispersion'.[12] He measured this effect for steroids, asymmetric cyclic ketones (the same compounds behind the formulation of Woodward's rules), in the ultraviolet region: 'the laborious point-by-point wavelength measurements being taken by the wives of my graduate students'.[13] Woodward's involvement came six years later, when he introduced Djerassi to two theoretical chemists, William Moffitt and Albert Moscowitz, and they all developed the Octant Rule during a single brainstorming session in 1958. It was published three years later. Djerassi recalled that at the 1958 meeting, 'The octant rule explained virtually all of our published and unpublished results'.[14]

The octant rule relates the Cotton effect (positive or negative) to the substituents on the steroidal ketone according to their arrangements in space, thus clarifying the conformation as well as the gross structure. This is expressed graphically by placing the carbon of the ketone carbonyl group at the centre of an octant created out of three planes. Each segment of the octant carries a plus or minus sign. This can be demonstrated with cyclohexanone, as explained by Djerassi: 'In our original version ... the cyclohexanone model was divided into three planes corresponding to the nodal and symmetry planes of the 280 nm transition of the carbonyl chromophore. These three planes create eight octants, and the presence of substituents in each octant is given a qualitative rotational contribution'.[15] The sum of the differing contributions allowed the sign of the Cotton effect to be calculated. Conversely the sign of the Cotton effect could be used to choose between one of several possible structures. As with the Woodward rules, the derivation of the octant rules depended on empirical data, here the material painstakingly collected by Djerassi's team.

Mass Spectroscopy

Francis William Aston and Arthur Dempster independently made the first mass-spectrographs around 1919. The mass spectrometer produces positive ions from the sample, and uses a strong magnetic field to resolve them into a series of beams recorded on photographic plates or, more recently, by electronic detectors. They are then presented as a series of peaks representing mass/charge ratios. Alfred Nier developed the first high resolution mass-spectrometer at the University of Minnesota in the late 1930s. The Consolidated Engineering Company (CEC), an American firm founded by a member of President Hoover's family, moved into this field in the late 1930s with the aim of supplying the petroleum industry with instruments that could give rapid analyses of hydrocarbon mixtures. The first CEC instrument, model 21-101, was brought out in 1942, and Metropolitan Vickers produced the first British counterpart, the MS-2, eight years later. The CEC 21-103C instrument, introduced in the early 1950s, was widely used by organic chemists until it was displaced by double-focusing instruments, namely the British MS-7, introduced in 1958, and the CEC 21-110,

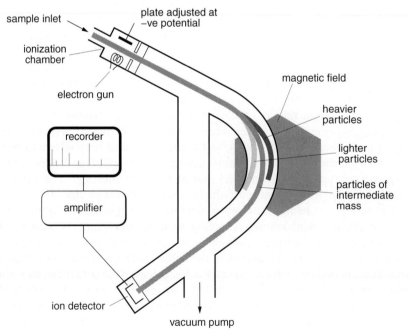

Mass spectrometer

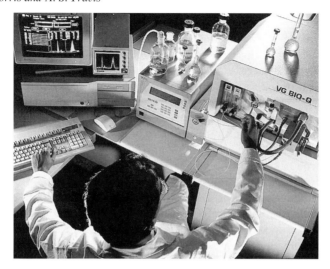

A mass spectrometer. Courtesy of Geoff Thompkinson/Science Photo Library

in 1963, which was widely used in the United States.

Until the mid-1950s, mass spectroscopy had found great value only in physics and physical chemistry. The key year for the application of mass spectroscopy to organic chemistry was 1956, when Fred McLafferty began his work on the fragmentation patterns formed when a complex organic compound breaks up in a mass spectrometer, and John Beynon used a high resolution mass spectrometer to work out the molecular formulae of organic compounds. Using fragmentation analysis, it gradually became possible to decipher the structure (or at least the partial structure) of an organic compound from the mass spectrum. An early pioneer was Rowland Reed at the University of Glasgow, who used fragmentation analysis in 1956 to determine the structure of the side-chains of various steroids. Computerised data-handling of mass spectra data also emerged in this period.

By 1959, Klaus Biemann, an Austrian chemist working at MIT, had entered this field and soon became 'a renaissance man in mass spectroscopy' according to his colleagues.[16] His first paper concerned the use of mass spectroscopy to determine the amino acid sequence of a peptide. Peptides and proteins have endured as one of his major interests. In the early 1960s, however, he mostly used mass spectroscopy to determine the structure of complex alkaloids, especially indole alkaloids. This work brought him into contact with R. B. Woodward, who used Biemann's expertise with the mass-spectrometer to identify intermediates in his famous organic syntheses. By early 1964, Biemann had acquired a CEC 21-110 that enabled the entire mass spectrum to be displayed on a single photographic plate. Using an IBM 7094, this data could be used to calculate the exact molecular mass of each fragment. Biemann presented this information in the form of an 'element map', a table with the fragments arranged in different columns according to their heteroatom* content. This was a very powerful

*A heteroatom is an atom of any element in an organic compound except carbon or hydrogen.

technique and could be used to determine the structure of compounds with only microscopic samples. For instance, in 1968, he determined the structure of the marine sex hormone, anthediriol from its high-resolution mass spectrum, in conjunction with ultraviolet and infrared spectra, and the NMR spectrum. Nothing was known about its structure beforehand and the mass spectrum was obtained from a sample of 'a few micrograms'.[17]

After the mid-1960s, Biemann concentrated on the linking of mass spectroscopy with gas chromatography. The ability of mass spectroscopy to give results with a sub-milligram sample makes it of great value with gas chromatography, in which very small amounts of pure substances can be separated from previously intractable mixtures. Although primitive gas chromatographs were developed during and after World War II, especially at Innsbruck (Biemann's original university), modern gas chromatography dates from the announcement of gas–liquid partition chromatography by Archer Martin and Tony James in 1951. This technique was rapidly taken up by academic chemistry and the petrochemical industry. In 1957, Roland Gohlke achieved the first hyphenation with mass spectroscopy (GC-MS) and Biemann (with J. Throck Watson) was the first to develop an effective method of removing most of the carrier gas before introducing the sample to the mass spectrometer. Biemann's work on GC-MS reached its literal high point in 1976 when, as part of NASA's Viking programme, he landed a GC-MS instrument on the surface of Mars to analyse any organic compounds that might be there. The GC-MS showed that there were no organic compounds on the Martian surface with a sensitivity limit of parts per billion.[18]

It is largely through the agency of GC-MS that it has been possible to study pheromones: chemicals that control the social behaviour, especially the sexual behaviour, of insects. The veteran steroid chemist Adolf Butenandt carried out the first major study of an insect pheromone, the attractant of the female silkworm moth (*Bombyx mori*) in 1959. However, he did not use GC-MS which was still in its infancy. Nearly all the early work on the structure of pheromones was surprisingly traditional: mainly chemical methods such as hydrolysis, ozonolysis, permanganate oxidation and infrared spectroscopy. However, as many pheromones are relatively simple compounds containing double bonds and only one or two other functional groups, these methods were adequate if enough material was available. It appears that Robert Silverstein, in his 1966 study of the male bark beetle (*Ips paraconfusus*), was the first to use preparative gas chromatography followed by an instrumental battery of infrared spectroscopy, mass spectroscopy and NMR. Thus he could show that a mixture of compounds was involved, thereby disproving Butenandt's idea of a single compound for a single rôle.

In December 1960, Carl Djerassi also entered the field of mass spectroscopy because of his interest in cacti alkaloids. His interest in these alkaloids stemmed from his early work at Syntex in Mexico City (and also gave rise to an infamous mescaline party at his home in 1954). Djerassi recalled:

As soon as I arrived [at Stanford University] in Palo Alto, I applied to the National Institutes of Health for financial support in buying a mass spectrometer in order to conduct a systematic study of the technique, using steroids as initial model substrates before applying it to the, structurally, much more diverse group of alkaloids. We wanted to determine whether special rules of fragmentation and reassembly could be developed ... which would make this method of more general utility. Using steroids as substrates, we set out to 'mark' certain portions of the molecule with stable, non-radioactive isotopes of hydrogen and carbon to facilitate the reassembly of the broken pieces. Eventually we used the marking technique, which on its own involved many man-hours of synthetic effort, to establish the rules of mass spectrometric decomposition for a wide variety of molecules, such as steroids, triterpenes and alkaloids.[19]

One interesting example of Djerassi's work was the discovery of gorgosterol. In 1969, Djerassi was sent a supposedly pure sample of a steroidal marine toxin by Paul Scheuer at the University of Hawaii. Djerassi takes up the story:

On subjecting his sample to mass spectroscopic analysis, we found it to consist of at least three sterols: two conventional ones of the cholesterol type, and a third one with a seemingly unprecedented number of carbon atoms. I encouraged Scheuer to isolate more of that 'impurity', which we then subjected at Stanford to ... nuclear magnetic resonance and mass spectroscopy. In a joint communication with Scheuer, we published the structure of 'gorgosterol', which has the same tetracyclic steroid nucleus as cholesterol, but an extremely unusual 'side chain' – an assembly of eleven carbon atoms [containing a cyclopropane ring] attached to position 17 of the steroid skeleton. We determined the complete structure of gorgosterol only by means of X-ray crystallography ... and promptly got hooked on a research line [marine sterols] my group pursued for the next twenty years.[20]

In the mid-1960s, through his Stanford colleague Joshua Lederberg, Djerassi also became involved with the early stages of the project to find out if there was life on Mars, specifically to develop a program that could be used to analyse the data sent back by the Mars-based GC-MS. Lederberg commented that 'We are trying to teach a computer how Djerassi thinks about mass spectrometry'.[21]

Nuclear Magnetic Resonance

The NMR effect was first reported in 1946 by Felix Bloch at Stanford and, independently, by Edward Purcell at Harvard University. Subsequently, Stanford became a leading centre for research into NMR. In the simplest version of NMR, a sample of a compound is bathed in a very strong constant magnetic field (nowadays generated by superconducting magnets) and bombarded by radio waves of a single frequency. The magnetic interaction of the radio waves by protons (or other suitable atoms such ^{13}C or ^{19}F) over a narrow range of radio frequencies is recorded and studied. The dependence of the resonance frequency of a proton on its chemical environment, as measured by the 'chemical shift', was

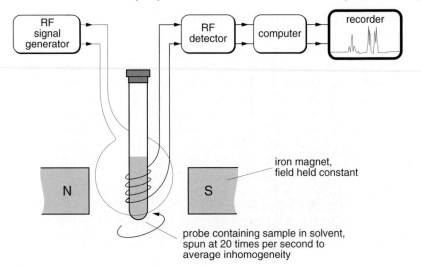

NMR spectrometer

noted by several researchers in 1949–50. Shortly thereafter, in 1951, the first NMR spectra were obtained by James Arnold, a postgraduate at Stanford. These showed separate resonances for protons located at different positions in the molecule: the field-dependent chemical shifts. The field-independent separation between peaks, spin–spin coupling, which represents the interaction between protons, was discovered by Warren Proctor and Fu Chun Yu at Stanford, and by Herbert Gutowsky and David McCall at Illinois during 1952–53. Herb Gutowsky was one of the first to introduce NMR into organic chemistry.

Varian launched the first commercial NMR spectrometer in 1952, which operated at 30 MHz. It found immediate application in industry, especially at Du Pont, whose NMR expert, William Phillips, described to his visitor John D. Roberts the proton spectra of *N,N*-dimethylformamide, in which two methyl signals coalesced at elevated temperature due to increased rates of rotation about the carbon–nitrogen amide bond. That was in 1954, when at $26 000 such an instrument was beyond university budgets. With the help of Linus Pauling, Roberts convinced Caltech's Board of Trustees of NMR's value for structural studies, and received a grant equal to the purchase price. This was used to acquire a 40 MHz instrument, introduced in 1955, which, according to Roberts, 'was to be the first commercial NMR spectrometer to be sited in a university. If it was not the first piece of such equipment, I'm sure it was the first to be put under the jurisdiction of an organic chemist'.[22] The Varian 40 MHz instrument was also the first to be introduced into Britain, at Cambridge University.[23]

Leaders in NMR structural organic chemistry in the late 1950s were: Jim Shoolery, of Varian Associates Inc in Palo Alto, California, who worked on steroids and in 1955 introduced spin decoupling; Raymond Lemieux, University of Ottawa, noted for carbohydrate research; Basil Weedon, Imperial College, London, who worked on polyenes; and Karl Folkers of Merck who studied ubiquinones. Ray Lemieux's experience of the initial problems in 1955 was not

In 1953, Varian's first NMR applications laboratory included an early 30MHz NMR system, pictured here with Jim Shoolery and Virginia Royden. Jim was a physical-organic chemist who was one of the first to see the true potential of NMR spectroscopy in solving the structure of organic molecules. Photograph courtesy of Varian Inc.

untypical. After managing, with difficulty, to set up a suitable homogenous magnetic field, every effort had to be made to maintain that field. However:

> It often did not last through a run (about 20 minutes) because of a sudden change in line voltage, a passing truck, or simply the temperature change caused by someone opening the laboratory door. Many of the problems could be minimised by working between 1 and 5 am, when the city line voltage was more constant and the street traffic was relatively quiet. Under these conditions, often, the main worry was, 'Would the recorder pen work throughout the experiment?'

Some aspects of NMR were less readily apparent. Thus Lemieux observed 'I well remember when we learned that our spectra would be much improved by spinning the sample tube'.[24] Bloch had this bright idea in 1954 while he was stirring a cup of tea. Once the teething problems were overcome and improvements in resolution were achieved, there were rapid developments in conformational studies on six-membered ring compounds.

An early example of the use of NMR in structural work was the determination of the structure of the alkaloid aspidospermine by Harold Conroy at Brandeis University between 1957 and 1959. Initially, the NMR spectrum suggested the presence of a *N*-methyl group, but this was ruled out by classical chemical methods and it was then possible to arrive at the correct structure. The X-ray determination of the structure of aspidospermine was published simultaneously by John Mills and Stanley C. ('Scan') Nyburg at University College of North Staffordshire (now Keele University) in 1959. Biemann subsequently studied the aspidospermine series of alkaloids in the early 1960s using mass spectrometry.

The interpretation of NMR spectra was greatly improved by the growing

The Varian HR-220 NMR spectrometer, introduced in 1964, featured the first superconducting NMR magnet. Photograph courtesy of Varian Inc.

availability of stronger magnetic fields and hence the use of higher radio frequencies. Higher field strengths produced proportionally greater chemical shift separations, which allowed chemists to distinguish between peaks created by spin–spin coupling and peaks from wholly different protons. Spin–spin decoupling could also be used to confirm such interpretations. In 1961, Varian brought out the first frequency-locked NMR machine, which operated at 60 MHz and surprisingly, Varian marketed the first commercial instrument to use a superconducting magnet (220 MHz) in the same year. Bruker introduced the first commercial Fourier transform NMR spectrometer in 1969 (90 MHz), which was followed a year later by a superconducting Fourier transform model operating at 270 MHz.

During the early 1960s, organic chemical shift data became available in tabulated form and in spectral atlases. With advances in resolution and sensitivity, chemists were able to correlate spin–spin coupling constants with physical features of molecules. The publication of the Karplus equation in 1963 provided the basis for relating the spin–spin coupling constant to the dihedral angle between bonds. The Nuclear Overhauser Effect (NOE), discovered in 1953, was introduced into chemical NMR by Frank Anet and Tony Bourn in 1965. It is useful in the study of conformations, because it provides information about the positions of protons in space, rather than the formal structure. For instance, Lemieux discovered in the late 1970s that NOE could be used to study the electron-withdrawing effect of an oxygen atom in a fucose ring on a proton on another ring which was nonetheless close to it in space.[25]

As with other instrumental methods, NMR has been widely used to determine chemical structure, often requiring 'hit or miss' assumptions similar to those employed in classical structural determination. This has the danger that chemists studying the 'entrails' of a complex NMR spectrum often find the structure they want, not necessarily the correct one. Hence the widespread use of NMR (and also mass spectroscopy) has led to a degree of uncertainty creeping back into the published chemical structures.

Modern NMR spectroscopy uses a combination of ^{1}H (proton) and ^{13}C NMR for structural work and to a lesser extent, nitrogen, fluorine and phosphorus NMR. There has been an explosion of techniques since 1979, notably in one and two dimensional NMR spectral measurements by modern pulse techniques. Some of these have produced strange acronyms, such as FOCSY, NOESY, COSY, SECSY, HOHAHA, INEPT, INADEQUATE, WALTZ, DANTE, IN-FERNO, SIMPLE and SPOTS.[26]

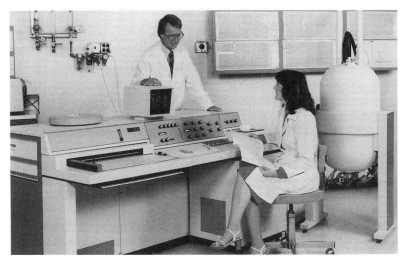

In 1978, Varian introduced the XL-200, a superconducting FT (Fourier-transform) NMR spectrometer that included many advanced software features. It could be set up to probe several atomic nuclei, including C-13 and P-31. Photograph courtesy of Varian Inc.

Promotion of Physical Methods Through Publications

The diffusion of instrumental methods into mainstream organic chemistry was a slow process. The new methods had to gain acceptance from chemists who had got where they were using well-established chemical techniques. Very often, organic chemists only had a broad understanding of physical chemistry and an even weaker grasp of quantum physics and electronics. To win them (and perhaps more importantly their postgraduate students) over to the new physical instrumental methods, it was necessary to use propaganda in the form of monographs and textbooks. The late 1950s was an exciting period to be a chemist working with these new techniques and it was also a golden age for publishers of chemical monographs, most notably the New York firm of McGraw-Hill. Roberts has recalled the thrill of publishing his monograph *Nuclear Magnetic Resonance* in 1959:

One day William A. Benjamin, a young editor from McGraw-Hill, dropped in and asked if I would consider writing a book about NMR. Bill was enthusiastic about the prospects for sales. He was also enthusiastic about modeling the art work after my slides, even wishing to see if the illustrations could be done in color. The latter surely would be a first at that level of monographs, and Bill ... liked to be first.... To my knowledge, this book was the first to include the now ubiquitous spectroscopic problems that, in the simple way I used them, had only a proton spectrum and a molecular formula.... Bill ... was eager to get my book out because it would be complementary to Pople, Schneider and Bernstein, which was also being published by McGraw-Hill.... The book turned out to be about 125 pages long and, despite a rather high price of $7.50, sold about 8000 copies.[27]

Characteristically, Djerassi was both innovative and astute in his dealings with McGraw-Hill over the publication of his *Optical Rotatory Dispersion: Applications to Organic Chemistry* in the same year:

One aspect of my publishing contract with McGraw-Hill was unusual. When the publisher invited me to prepare this first monograph dealing with the organic chemical applications of optical rotatory dispersion, I insisted on a penalty clause, whereby my royalties would escalate by one percent for each week the book's appearance might be delayed beyond my requested publication date. To everyone's surprise, the McGraw-Hill lawyers accepted my proposal, provided I agreed to return the corrected page proofs from Mexico City [where Djerassi was research vice-president of Syntex] within twenty-four hours of their receipt. This was supposed to prevent a horror scenario, whereby my royalties might escalate to unprecedented heights were I simply to sit on the page proofs. As a final compromise, the publisher set each chapter in print as it was received, rather than waiting for the entire manuscript. I managed to finish the book in time by sticking to a rigid Monday–Wednesday–Friday writing schedule, and McGraw-Hill was equally diligent. The royalties from the book eventually paid for a swimming pool at my new house in California, whose steps were set in Mexican tiles reading 'built by optical rotatory dispersion'.[28]

1959 and 1960 were key years for publications on the new methods and, together with what was happening in the laboratories, must be considered a major watershed. In addition to Roberts' book, Lloyd Jackman, *Applications of Nuclear Magnetic Resonance to Organic Chemistry*, and John Pople, William Schneider and Harold Bernstein, *High Resolution Nuclear Magnetic Spectroscopy* also helped to introduce NMR to organic chemists. John Beynon, *Mass Spectrometry and its Applications to Organic Chemistry* (1960), and Klaus Biemann, *Mass Spectroscopy* (1962, also McGraw-Hill) played a similar rôle for mass spectroscopy. It must be debatable, however, how many organic chemists fully understood the very technical monograph by Pople, Schneider and Bernstein, which may explain the failure of Roberts' second foray into this field.

I told Bill Benjamin that ... I would write *An Introduction to the Analysis of*

Spin–Spin Splitting in High-Resolution Nuclear Magnetic Resonance Spectra [published in 1961]. It was a long title for what was to be a short book. Indeed, it was not a lot more than a fuller explanation of a relatively few pages in Pople, Schneider and Bernstein.... Bill Benjamin wanted this book to be a showpiece to convince potential authors that he could do a very high-quality publishing job in half or less of the time normally required. The heat was on, and this was a text with a lot of mathematical equations, graphics and spectra.... The book sold several thousand copies, but I had remarkably little feedback on its utility. Perhaps there weren't that many people who were that interested in understanding the basis of what is involved in spin–spin splitting.[29]

Surveys of the entire field also played an important rôle in the assimilation process. The first volume of *Determination of Organic Structures by Physical Methods* (edited by Ernest Braude, and Frederick Nachod) appeared in 1955, and had an immediate impact; the second volume (edited by Nachod and William Phillips) followed in 1961. *Chemical Applications of Spectroscopy*, edited by William West as part of the widely circulated series on *Technique in Organic Chemistry* edited by Arnold Weissberger, brought the latest developments in infrared and ultraviolet spectroscopy to a wider audience when it was published in 1956. Another important overview – *Elucidation of Structures by Physical and Chemical Methods*, edited by Kenneth Bentley – was published in 1963. The task of winning organic chemists over to the new methods was assisted by the availability of undergraduate texts that emphasised the rôle of physical methods in organic chemistry, most notably John D. Roberts and Marjorie Caserio's *Basic Principles of Organic Chemistry*, which was published by Bill Benjamin who had just set up on his own as W. A. Benjamin. Roberts recalled:

The first edition was, I think, a landmark book with many features which are standard today. It was also a shocker to many, especially through the very early introduction of spectroscopy; the many follow-up spectroscopic problems; the inclusion of problems, as appropriate, right within the text; and perhaps more than anything, the unusual length for an elementary book. When I asked my organic chemistry colleagues, who complained about the length, for a list of things they would cut out, I never got much help on that score. The usual response was that perhaps we might 'somewhat enlarge' the sections that covered the suggestor's own field of interest.[30]

However, Roberts and Caserio was too basic for the more specialised British degree, and Oliver and Boyd published *Physical Methods in Organic Chemistry*, edited by Peter Schwarz, in the same year. McGraw-Hill countered in 1966 with *Spectroscopic Methods in Organic Chemistry* by Dudley Williams (who had worked with Djerassi) and Ian Fleming. Its reading lists helped to sell other McGraw-Hill books, and unlike the other two books, it has endured, with a fifth edition in 1995.

Conclusion

In summary, infrared and ultraviolet spectroscopy are used as 'fingerprinting' techniques for known compounds, and for identification of functional groups. Mass-spectroscopy is also a 'fingerprinting' technique; the fragmentation pattern can be used to determine both molecular mass and structure. NMR is the most powerful and widely employed technique and is particularly useful for conformational analysis. While NMR and mass spectroscopy can give excellent results for relatively small molecules (such as the pheromones), they can yield debatable structures, at least in the case of the larger more complex natural products.

While X-ray crystallography can give an unambiguous result, not every compound is studied by X-ray crystallography, which is usually performed away from the organic chemistry laboratory by different staff (who often have different priorities) and thus takes extra time and money. Furthermore, as Djerassi has emphasised,[31] X-ray crystallography requires a crystalline sample and not all organic compounds can be crystallised in a suitable form or obtained in amounts large enough to crystallise. Taken together, however, these techniques enable us to determine chemical structures rapidly and accurately with very small amounts of material.

Thomas S. Kuhn has described the paradigm as a conceptual model used to provide direction to a scientific activity by 'implicitly [defining its] legitimate problems and methods'.[32] Scientific revolutions occur when the ruling paradigm is overthrown by a new one. The introduction of instruments into the methods of structural determination, particularly in organic chemistry, is an illustration of this paradigm shift. The development of the new paradigm started in the early 1940s, began to seriously displace the old paradigm by the late 1950s, and the process was complete by the end of the 1960s. The routine of one type in chemistry was transformed into routine of another type, with major implications for the field of chemistry in general and the chemists that carry out such work. The introduction of electronic instrumentation after 1940 was nothing less than a scientific and technological revolution. It has led to the near-total displacement of classical 'wet and dry' methods in organic structure elucidation.

Further Reading

William H. Brock, *The Fontana History of Chemistry* (Fontana Press, London, 1992). A comprehensive history of chemistry in a readable style with good coverage of modern developments.

Theodor Benfey and Peter Morris (eds.), *Robert Burns Woodward: Architect and Artist in the World of Molecules* (Chemical Heritage Foundation, Philadelphia, 2001). This book provides accessible historical and scientific introductions to several topics covered in this chapter including the Woodward rules, the octant rule, chlorophyll, strychnine and vitamin B_{12}. It also reproduces Woodward's own papers on these topics.

Mary Ellen Bowden and Theodor Benfey, *Robert Burns Woodward and the Art of Organic Synthesis* (Beckman Center for the History of Chemistry, Philadelphia, 1992). A brief but user-friendly introduction to organic chemistry and physical instrumentation during the period covered by this chapter.

Robert Bud and Deborah Warner (eds.), *Instruments of Science: An Historical Encyclopedia* (Garland Publishing, New York and London, 1998). See, in particular, the entries on the mass spectrometer by Keith Nier (son of Alfred Nier), and the spectrophotometer by Jon Eklund and Peter Morris.

C. B. Faust, *Modern Chemical Techniques* (Royal Society of Chemistry, London, 1992). A first-rate introduction to the physical techniques covered by this chapter (with the exception of X-ray crystallography) and chromatography. It is an example of that rare phenomenon, a really accessible book, which can be understood without any prior knowledge of the techniques (or even chemistry) with the aid of very clear diagrams.

James Feeney, 'Development of high resolution NMR spectroscopy as a structural tool' in Robert Bud and Susan E. Cozzens (eds.), *Invisible Connections: Instruments, Institutions, and Science* (SPIE Optical Engineering Press, Bellingham, Washington, 1992). An excellent account of the development of NMR in structural chemistry, which is complemented by J. W. Emsley and J. Feeney, 'Milestones in the first fifty years of NMR', in *Progress in Nuclear Magnetic Resonance Spectroscopy*, 1995, **28**, pp. 1–9. [Also see David M. Grant and Robin K. Harris (eds.), *Encyclopedia of Nuclear Magnetic Resonance*, Volume 1, *Historical Perspectives* (John Wiley & Sons, Chichester, 1996).]

I. L. Finar, *Organic Chemistry*, Volume 2, *Stereochemistry and the Chemistry of Natural Products*, 4th edition (Longmans, London, 1968). The second volume of a standard undergraduate textbook, but one of the very few books to give a clear overview of the determination of the structure of numerous natural products. The 5th and final edition (1975) is not so good in this respect.

Jenny P. Glusker, 'Brief History of Chemical Crystallography II: Organic Compounds' in J. Lima-de Faria, *Historical Atlas of Crystallography* (International Union of Crystallography, Dordrecht, 1990). A straightforward and readable account, with extensive references, but very brief; ignore the 'atlas' in the book title, it has no bearing on this contribution. Also see the works by Bragg and Ewald referenced in the footnotes.

Herbert A. Laitinen and Galen W. Ewing (eds.), *A History of Analytical Chemistry* (Division of Analytical Chemistry, American Chemical Society, Washington DC, 1977). Accounts of the history of various aspects of analytical chemistry (including mass spectroscopy, UV and IR spectroscopy, and NMR) by experts in the fields concerned. Relatively narrow in focus but reliable and one of the few secondary sources available.

Peter J. T. Morris, 'From Basle to Austin: A Brief History of Ozonolysis' in A. R. Bandy (ed.), *The Chemistry of the Atmosphere – Oxidants and Oxidation in the Earth's Atmosphere*, RSC Special Publication no. 170, Royal Society of Chemistry, Cambridge, 1995, pp. 170–190. The only modern attempt to cover the history of classical methods of structure determination. [Also see Leo Slater, 'Woodward, Robinson and Strychnine: Chemical Structure and Chemists' Challenge'. *Ambix*, 2001, **48**, 161–189.]

Yakov M. Rabkin, 'Technological Innovation in Science: The Adoption of

Infrared Spectroscopy by Chemists,' *Isis* 78 (1988) pp 31–54. An academically rigorous but readable account of the history of infrared spectroscopy up to the 1950s. One of the few scholarly publications which touches on the subject of this chapter.

J. C. P. Schwarz (ed.), *Physical Methods in Organic Chemistry* (Oliver and Boyd, Edinburgh, 1964). An authoritative and very accessible introduction to the use of physical instrumentation in organic chemistry in the 1960s.

James N. Shoolery, 'The development of experimental and analytical high resolution NMR' *Progress in Nuclear Magnetic Resonance Spectroscopy* 28 (1995) pp 37–52. A fairly technical but readable overview of the development of NMR by one of the pioneers.

'The Instrumental Revolution, 1930–1955' in Dean Stanley Tarbell and Ann Tracy Tarbell, *Essays on the History of Organic Chemistry in the United States, 1875–1955* (Folio Publishers, Nashville, 1986), pp 335–352. An overview of the same topic as this chapter, but also covering other techniques such as Raman scattering and chromatography; it also contains useful references.

Dudley H. Williams and Ian Fleming, *Spectroscopic Methods in Organic Chemistry* 5th edition (McGraw-Hill, London, 1995). A good undergraduate-level introduction to the current state-of-the-art in UV and IR spectroscopy, NMR and mass spectroscopy in organic chemistry with a special chapter on structure determination.

'Profiles, Pathways and Dreams' series, edited by Jeffrey I. Seeman and published by the American Chemical Society, Washington D.C.

Arthur J. Birch, *To See the Obvious* (1995); Carl Djerassi, *Steroids Made It Possible* (1990); Raymond U. Lemieux, *Explorations with Sugars: How Sweet It Was* (1990); John D. Roberts, *The Right Place at the Right Time* (1990); chatty autobiographies by some of the leading chemists in the field. The volumes by Djerassi and Roberts are particularly worth reading.

Also see Djerassi's 'lay' biography: Carl Djerassi, *The Pill, Pygmy Chimps, and Degas' Horse: The Autobiography of Carl Djerassi* (Basic Books, New York, 1992).

Notes and References

1 Carl Djerassi, *The Pill, Pygmy Chimps, and Degas' Horse: The Autobiography of Carl Djerassi*, Basic Books, New York, 1992, p. 104.
2 'Laurie Hall: Opening up a Pandora's box', *Chemistry in Britain*, December 1985, **21**, 1057.
3 John D. Roberts, *The Right Place at the Right Time*, American Chemical Society, Washington DC, 1990, p. 151.
4 Arthur J. Birch, *To See the Obvious*, American Chemical Society, Washington DC, 1995, pp. 56–57.
5 The pioneering effort was the American Institute of Petroleum Research Project 44 which put out atlases of infrared, ultraviolet and mass spectra (mainly hydrocarbons) between 1947 and 1959. Other atlases of infrared and ultraviolet spectra were published by Stadler Research Laboratories of Philadelphia and the *Documentation of Molecular Spectroscopy* by Butterworths in collaboration with Verlag Chemie. R.

Mecke and F. Langenbucher, *Infrared Spectra of Selected Chemical Compounds*, 1965, Heyden and Sons, London and *The Aldrich Library of Infrared Spectra*, 1970, by C. J. Pouchert followed somewhat later. H. M. Hershenson compiled indexes (but did not reproduce the spectra) for infrared and ultraviolet spectra for the period 1930–1959 (UV) and 1945–1962 (IR), which were published by Academic Press. The earliest NMR atlas was N. S. Bhacca, D. P. Hollis, L. F. Johnson and E. A. Pier, *Varian High Resolution NMR Spectra Catalog*, two volumes, Varian Associates, Palo Alto, 1962 and 1963. The equivalent publication for mass spectra was F. W. McLafferty, E. Stenhagen and S. Abrahamsson, *Atlas of Mass Spectral Data*, three volumes, Interscience Publishers, New York, 1969.

 6 Sir Lawrence Bragg, in *The Development of X-ray Analysis*, D. C. Phillips and H. Lipson (eds.), G. Bell & Sons, London, 1975, pp. 176–177.
 7 Bragg, *The Development of X-ray Analysis*, p. 189, quoting Hodgkin and Bunn's original paper.
 8 W. L. Bragg, 'The Growing Power of X-ray Analysis' in *Fifty Years of X-Ray Diffraction* P. P. Ewald (ed.), International Union of Crystallography, Utrecht, 1962, pp. 130–131.
 9 Dorothy Crowfoot Hodgkin, 'The X-ray Analysis of Complicated Molecules', *Science*, 1965, **150**, 979–988; quote on p. 983.
10 Norman Sheppard, 'The UK's Contributions to IR Spectroscopic Instrumentation: From Wartime Fuel Research to a Major Technique for Chemical Analysis', *Analytical Chemistry*, 1992, **64**, 881A.
11 R. B. Woodward, 'Structure and the Absorption Spectra of α,β-Unsaturated Ketones', *Journal of the American Chemical Society*, 1941, **63**, 1125.
12 C. Djerassi, *Optical Rotatory Dispersion: Applications to Organic Chemistry*, McGraw-Hill, New York, 1960, p. 19.
13 C. Djerassi, *Steroids Made It Possible*, American Chemical Society, Washington DC, 1990, p. 54.
14 Djerassi, *Steroids Made It Possible*, p. 59.
15 Djerassi, *Steroids Made It Possible*, pp. 59–61.
16 *Life Search*, Time-Life Books, Alexandria, Virginia, 1989, p. 25.
17 G. P. Arsenault, K. Biemann, Alma W. Barksdale and T. C. McMorris, 'The Structure of Antheridiol. A Sex Hormone in *Achlya bisexualis*', *Journal of the American Chemical Society*, 1968, **90**, 5635–5636.
18 *Life Search*, pp. 15–28.
19 Djerassi, *The Pill, Pygmy Chimps, and Degas' Horse*, p. 101.
20 Djerassi, *The Pill, Pygmy Chimps, and Degas' Horse*, pp. 103–104. Also see Djerassi, *Steroids Made It Possible*, pp. 115–116.
21 Djerassi, *The Pill, Pygmy Chimps, and Degas' Horse*, p. 102.
22 Roberts, *The Right Place at the Right Time*, p. 154.
23 Unfortunately neither this instrument (which was transferred to the Colchester Institute in the 1960s) nor the second 40 MHz machine in Britain (formerly at Liverpool University) have survived.
24 Raymond U. Lemieux, *Explorations with Sugars: How Sweet It Was*, American Chemical Society, Washington DC, 1990, pp. 30–31.
25 Lemieux, *Explorations with Sugars: How Sweet It Was*, pp. 105–106.
26 Alex Nickon and Ernst F. Silversmith, *Organic Chemistry: The Name Game. Modern Coined Terms and Their Origins*, 1987, Pergamon Press, New York, pp. 183–185.
27 Roberts, *The Right Place at the Right Time*, p. 170 and p. 172. This quotation was from the version printed in *Science in the Twentieth Century*.

28 Djerassi, *The Pill, Pygmy Chimps, and Degas' Horse*, pp. 97–98.
29 Roberts, *The Right Place at the Right Time*, pp. 173–174.
30 Roberts, *The Right Place at the Right Time*, pp. 227–228.
31 Djerassi, *The Pill, Pygmy Chimps, and Degas' Horse*, p. 84.
32 T. S. Kuhn, *The Structure of Scientific Revolutions*, second ed., University of Chicago Press, Chicago, 1970, p. 10.

Instrument Development in Social, Economic and Political Context

Then . . . and Now

DAVID KNIGHT

Department of Philosophy, University of Durham, 50 Old Elvet, Durham
DH1 3HN, England, Email: d.m.knight@durham.ac.uk

One of my Durham colleagues was a student in Rome in the 1940s and says that
he received a seventeenth-century philosophical education, duly delivered in
Latin. The chemistry that I learned in school and at university in the 1950s was
essentially nineteenth-century. The school laboratory was built at the end of that
century, and the wooden benches, stools, bottle racks and sinks had worn well.
We squirted each other with wash bottles, dissolved pennies illicitly in concentrated nitric acid, learned to treat the concentrated sulfuric acid with respect, and
found that our fingers got gradually inured to holding hot test-tubes or boiling-
tubes. We used fragments of porous pot to stop things sploshing. We became
familiar with Kipps' apparatus, learned how to fold filter papers in whichever of
the two permitted ways was appropriate, and rolled corks beneath our feet
before boring them so that glass tubes (which we learned to cut and bend) could
be put through and apparatus fixed together. Glassware (connected through
corks, or occasionally rubber bungs, with glass and rubber tubing) was carefully
clamped to massive iron retort-stands; evaporating basins or round-bottomed
flasks were supported on wire-gauze and asbestos over tripods beneath which
Bunsen burners roared. Our Liebig condensers (connected so that water came in
from below) gurgled. We poured fluids carefully down glass rods, especially when
it was necessary to end with two liquid layers. Our burettes mostly had a rubber
tube and clamp to control the flow; taps were felt to be rather trendy, effete, and
liable to jam. We inadvertently drank nasty solutions through pipettes. We were
taught to work our way systematically through the analysis tables 'in the wet
way' carefully eliminating possibilities; and to be scrupulous about avoiding
short cuts involving flame tests or even blowpipes and borax beads on charcoal
blocks, which should be used only for confirmation after all the stages involving
hydrogen sulfide in acid and alkaline solution had been gone through. Weighing
was a great ritual, involving first weights and then finally a rider slid along the
balance beam with the case shut so that draughts would not affect the result.
Books described the various gases as having characteristic odours, which was

not very helpful: an important part of a chemical education was to take a whiff of this or that. Chlorine and sulfur dioxide once smelt were never forgotten; and teachers prided themselves (like wine tasters) on having a well-educated and sensitive nose. Chemical smells are still amazingly evocative; and we were even encouraged, though not so much as the earlier generation had been, to taste things; again, hard to describe but also hard to forget. Colours also had to be seen and made familiar. Even noises (popping and bumping) could be diagnostic: chemistry was the science of the secondary qualities.

From school I went into the Army, and then on to Oxford University in 1957. Oxford chemistry had an excellent reputation, with Robert Robinson and Cyril Hinshelwood as its leaders. But we undergraduates were learning a craft and a tradition as much as a science. My tutor, George Parkes, remembered as a student meeting William Odling, who had been a delegate at the Karlsruhe Conference of 1860 and subsequently Professor at Oxford (where he published only on Greek verse). In my college, Keble, there were a dozen chemists in my year (some 200 in the whole University, making it one of the biggest 'schools'). My chemistry teacher had been a pupil of Parkes, and I found after a bit that much the same was true of all the others. The college was not an elitist one, but clearly there was a strong feeling of a chemical family being established. Parkes was a great admirer of William Pope, the eminent stereochemist, and our required reading thus included some publications half a century old: on the other hand, he was said not to believe in electrons, which was a libel but, like the best libels, half true. Some of our other tutors were much younger and more up-to-date, though we seem to have heard rather little about R. B. Woodward; but Hinshelwood's lectures on chemical kinetics were focused on the work he had done in the 1920s, and it was said that once when he lost his place he was prompted by a student whose father had passed on his notes. C. A. Coulson gave very interesting lectures on molecular orbitals: but as perhaps might have been expected when Oxford was a centre of Logical Positivism, he told us that the wise man uses theories, but does not believe them. This was just what had been said about chemical atomic theory a hundred years before, when the Positivism was Auguste Comte's. Chemistry remained empirical, Baconian and experimental in its ethos: and chemists were scornful of physicists' attempts to explain the facts of chemistry 'in principle' – explanation in detail was the only thing that counted.

Laboratory work was an essential and time-consuming part of the course, and the labs were open all day rather than just for a few sessions. Health and Safety regulations hardly existed; I never remember a fire drill, and we would have thought anybody a wimp who supposed that benzene should only be handled in a fume cupboard, or mercury not handled at all. High technology in the physical chemistry lab meant a mechanical calculator, but it was slow and laborious and we were recommended to use slide rules for calculations, and tables of logarithms when more accuracy was required. Progressing from bending glass tubes, we had all to make a T-piece, which was very tricky, and the result generally wonky if watertight. In organic chemistry, our white crystals had (after their melting point had been determined) to be sealed into a little tube and preserved. We even had six-hour practical examinations at the end of the course, brewing up some coffee

Analytical laboratory, 1957. Painting by Wall Cousins. Science Museum/Science and Society Picture Library

in a beaker as necessary: during these, those who had done badly in the written examinations were summoned for a *viva*, and we all trembled lest we should be called away for interrogation. Nevertheless the organic chemistry practical examination involved an ether distillation, and having sixty or seventy of these going simultaneously under exam conditions with few invigilators would give a modern Safety Officer a fit. We had also to pass an examination in German, another archaic provision by 1957; rapidly and roughly translating (with the help of a dictionary) a chunk of chemical prose. My conversational German is to this day mostly concerned with Bunsen burners and hydrogen, and thus not very apt for most situations. All this was a very character-building and bonding experience, like basic training in the army, or like a craft apprenticeship; and intended to develop chemical intuition.

Oxford chemistry was already a four-year honours degree, the final year being spent on a project: most of these were experimental, and no doubt in 1960 my contemporaries then met in the post-graduate labs the expensive toys from which undergraduates were excluded. We had indeed heard of such measurements in lectures, and seen rather puzzling slides of the output of complex machinery: but our hands-on experience had been with test-tubes. I chose to do a historical project, and found myself working on Humphry Davy's chemistry; and this duly propelled me into subsequently writing a doctoral thesis on nineteenth century atomism. To someone with my training, the history of chemistry in its golden age (when it seemed the fundamental science and when there might be a black market in tickets for a chemistry lecture) was accessible. It was no surprise that Jacob Berzelius should have written a whole book about using the blow-pipe, or Michael Faraday a stout volume on *Chemical Manipulation* (still full of useful tips to my generation, on weighing, getting ground-glass stoppers out of

Franz Schmidt and Haensch polarimeter, c. 1900. Science Museum collections, inventory number 1963-278. In the nineteenth and early twentieth century, scientific apparatus was usually crafted in brass, lacquered metal, glass and wood. They were admired for the quality of their craftsmanship and their expensive refinements. Science Museum/Science and Society Picture Library

bottles, and distilling); or that William Ramsay prided himself on his glass-blowing. Eminent physicists since the early nineteenth century have sometimes been ham-handed; but for chemists that was impossible. Physicists might look upon them as upgraded cooks; but chemists knew that they had learned a craft the hard way. They did not work with black boxes but with the transparency of glassware. Buying in apparatus was time-saving but not essential, and the really good chemist could be his own technician. Faraday was also a great believer in recycling glassware: in accordance with the Victorian maxim, 'waste not, want not', it seemed clear that chemistry should be done as inexpensively as possible. Chemists also perceived the danger that an expensive toy, like the great batteries of the early nineteenth century, will be played with in time that, with more thought and less gadgetry, might be used for real discovery.

Like handloom weavers, drivers of steam-engines and cavalrymen, chemists felt the threat of being deskilled, of finding that in the new world of physical methods all that they had learned, the very essence of being a chemist was becoming irrelevant. Sir Harold Hartley, who had been a Fellow of Balliol College, Oxford, in the 1890s and was a great supporter of historians of science told me how strongly physical methods had been resisted; and that is something which is brought out by Charlotte Bigg in her discussion of spectroscopes. Chemists did not understand the basis of spectroscopy, while they were familiar with the diagnostic reactions underlying the analysis tables; spectroscopes had to be made foolproof before they would find their way generally into chemistry laboratories, like equipment being adapted for the army – and we learn something about military connections of science in Terry Shinn's and Davis Baird's

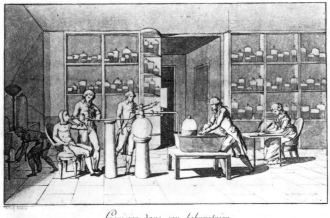

Lavoisier dans son laboratoire
Expériences sur la respiration de l'homme au repos

Fac-simile réduit d'un dessin de M.me Lavoisier

Antoine Laurent Lavoisier in his laboratory, carrying out experiments on human
respiration, c. 1789. Armand Seguin, a rich army contractor, was the experimental subject.
From a drawing by Mme Lavoisier, seen seated at a desk on the right. Science
Museum/Science and Society Picture Library

chapters. But that is only part of the story. Chemists were also from the early
twentieth century conscious that their science had been, or seemed to have been,
reduced to physics. Its underlying assumptions were derived from physicists'
atomic models. My contemporaries were strongly tempted to go either into
theoretical physics, where it seemed that fundamental explanations of chemical
phenomena might be sought; or into biochemistry, where the recent unravelling
of the DNA structure promised to chemists a deep understanding of the phenomena of life – something hoped for since the days of A. L. Lavoisier. Nowadays
structural chemists claim that their science is no more reduced to physics than
engineering or architecture are; but that is a new confidence. The arrogance of
physicists in the mid-century was daunting.

The theme of this book is that there was a revolution in chemistry in just about
those years, the 1950s and 1960s, when physical methods and automatic machinery replaced the traditional methods in both chemistry and chemical industry.
That seems from my experience and that of my contemporaries to be undoubtedly true. Johannes Kepler claimed that the coming of logarithms had doubled the
life of astronomers by halving the time it took them to make calculations. To old
fogies no doubt it is a matter of sadness that old practices have become obsolete,
and that studying chemistry at school and university these days must be less
physically exciting, and a less good preparation for cooking meals. But to those
involved, a great deal of drudgery has been taken over by new equipment – the
chemical equivalent of washing machines and vacuum cleaners. And moreover
there are now questions that can be asked, and also answered, given the new
apparatus, which could never have arisen in the days of craft chemistry. Science
after all is not fundamentally about techniques, important though they are, but

understanding: Faraday distinguished 'professional' activity, such as analyses, from his real science involved with the nature of electricity. Perhaps 'chemical heritage centres' could be set up to keep alive such rewarding techniques as cork-boring and heating things on water- and sand-baths.

When I moved into the history of science, a previous generation had devoted themselves to the Scientific Revolution of the seventeenth century: the age of Galileo, Descartes, Boyle and Newton. It seemed that then a decisive step towards the modern world had been taken, and a new method of getting reliable knowledge worked out; that was the period to which one should be devoted, and intellectual history was its key. We Young Turks bit the hand that fed us, and turned to the time when science began to deliver the goods, about 1800. This was especially easy for historians of chemistry, for clearly the late eighteenth and early nineteenth centuries were a time of tremendous transformation: Lavoisier himself used the word 'revolution'. We could see specialisation, the emergence of professional science, the coming of purpose-built laboratories and new institutions of all kinds, and the evolution of chemistry teaching: and associated with all this was chemical industry. There were popular books, textbooks, journals of various kinds, shops selling apparatus and chemicals, and popular enthusiasm, notably for things that frighten us today like explosives and fertilisers. We came to reject the distinction between 'internal' and 'external' history, looking instead for full context: and to deplore the Whig history (often found, and perhaps appropriate there, in brief historical sections of chemistry textbooks) which judged the past by its approximation to the present, and in the sciences led to a few great men being remembered as chemical grandfathers or secular saints with constants, units, reactions or institutions named after them.

J. T. Merz wrote his great *History of European Thought in the Nineteenth Century* a hundred years ago, having retired from his job as an electrical engineer: his theme was essentially intellectual history, but he would have been much interested in Stuart Bennett's discussion of control devices and their development. Merz reckoned that because of people he had known as well as what he had experienced, he had first- or second-hand knowledge of the whole century; and that it was thus possible to write its history. The earliest histories were usually of the historian's own times; and many scientists have been good historians of their science. But to someone who has been comfortable writing about the nineteenth century and its coda up to the middle of the twentieth, there seem to be problems about writing very recent history. Of course, it is possible that these are just what a now-Old Turk sees when looking at the current Young Turks moving on from where he happens to have worked. But the problems seem to be associated with technicality and with perspective.

Technicality is hard to avoid, and perhaps improper to avoid, in writing the history of science or other technical activities, such as law. But if we want to reach readers who have not got recent degrees in the science we are describing, then we must avoid the forbidding jargon (as it seems to them) of modern names for substances, apparatus and reactions. I managed to write a history of chemistry up to 1914, which included only two equations, both leading to water. That would be much more difficult to carry into the late twentieth century, and it would be weird to describe a machine based upon nuclear magnetic resonance without making it at all clear what the phenomenon is. Knowing what to leave out is one of the great academic gifts. With almost contemporary science, the historian neither wants to become a promoter or populariser of science, nor to do the kind of review of recent literature better done by a practitioner. For such internal history, and for obituaries, it is probably impossible to beat what scientists write for each other; but then what is the historian doing that is special? Perhaps the answer might be, looking for patterns that participants do not perceive, taking a longer view than those involved, being somewhat detached, keeping 'then' and 'now' apart.

That brings us to perspective. My colleagues in Durham's History Department (which is particularly strong on medieval history) sometimes grumble that students coming to university have only studied the twentieth century seriously. The result of this concentration upon the 'relevant' is that they do not appreciate that the past is a foreign country, where they did and saw things differently. They lack disciplined historical imagination, the real objective in studying the subject at all. The more remote past then becomes 'heritage', a matter of actors in machine-washable medieval garments playing rôles in beautifully-maintained ruined castles or great houses, now fitted with efficient plumbing, restaurants and gift shops. The distance that one feels in dealing with the real past is valuable, giving that shock of surprise that always accompanies discovery. Historians become international commuters, moving between that foreign past and our present, where the things taken for granted are different. They also come to appreciate the contingency of human affairs, which can be balanced against the determinism of science: things need not have turned out the way they did, other pathways were or seemed open, previous generations would experience that shock of surprise on seeing us and our world.

This unease about recent history was largely dissipated during the conference, for as this book shows participants did wrestle with questions with which chemists themselves might well not trouble, as well as illuminating the development of the instruments and their effect on chemical science. We find discussions of context – the Depression, the Cold War – of chemical professions and societies and their transformations, of law and litigation, of economics, of standardisation. The claim is and was often made that chemistry is all-pervading in modern life, and here we find chapter and verse. We may associate with wartime the maxim 'waste not, want not' which I had applied to Faraday; but as we learn from Terry Shinn and Davis Baird, war is also a great promoter of expenditure in science. Research, and the making of expensive instruments, can go ahead without the usual uncertainties of how the bills will be paid; and after World War II government investment in science continued at a high level, making large-scale chemical instrumentation possible. Similarly, control devices and servo-mechanisms have become part of everybody's life, as mechanical flappers and other Heath-Robinson arrangements give way to digital electronic devices: Stuart Bennett illustrates this in the chemical industry. Within all the discussions of technological evolution perhaps we find less about ethical issues than might be expected, for applied ethics is going to be the growth-industry of the early twenty-first century. It is safe to predict that ethical questions in the whole field of chemistry will assume ever-greater importance; and will help us in engaging with the wider world by coming from life to the lab rather than the other way about.

This book is based upon what was a coherent but varied conference, extremely stimulating and making those of us who had thought that shiny instruments were a little dull think again. Coming at modern science through its apparatus – 'ideas made brass', as earlier instruments have been called – proved highly profitable. In modern times as earlier, signs of the great scientist are knowing when to leave a field because instruments are not available to take one further, and when to seize upon new apparatus and adapt it to one's needs in spite of tradition. Chemistry clearly went through a revolutionary transformation in the second half of the twentieth century. Scientific revolutions can evidently be set off not only by intellectual leaps, like those of Lavoisier or Charles Darwin; but also by new instruments, like Galileo's telescope and then the spectroscope, the ultracentrifuge, the thermostat and their successors described here. However they happen, there will be those raised in the old paradigm who deplore and resist the new one: science is a conservative as well as a radical activity. This second chemical revolution is clearly something that should be more known about: it led to an excellent international and interdisciplinary conference, and now to a valuable book delineating the changes between 'then' and 'now'. Read it, and enjoy it.

Research-technology Instrumentation: The Place of Chemistry

TERRY SHINN

GEMAS, La Maison des Sciences de l'homme, 54 Blvd Raspail, F-5006 Paris, France, Email: shinn@msh-paris.fr

Introduction

Relations between science and technology are usually discussed with reference to five possible configurations. Science drives technology. Technology drives science. They develop independently from one another. Science and technology constantly push and pull at each other. Lastly, there is no distinction between science and technology; they form an unbroken continuum – technoscience.

The instrumentation perspective that I propose here offers another interpretation of the relations between technology and science.[1] I suggest that one category of instrumentation, what I call 'research-technology', stands between science and technology. Research-technologies belong neither to science nor to technology, but instead operate in a space between them. From this vantage-point, research-technology instrumentation serves and services academia, industry, metrology, state technical agencies, the military and so forth. Research-technologies fuel science and technology, and moreover, they provide a common language (a sort of *lingua-franca*) that gives rise to a technical pragmatic-universality.

But precisely what is research-technology? What are its constituent parts and how does it function? What are the material, intellectual and institutional mechanisms that permit some categories of scientific instruments to travel outward from their point of origin into numerous and diversified fields of application? What does such transverse instrument mobility imply for the relations between scientific specialities, and more generally, for the relations between science and technology?

My historical and sociological research on instrumentation has primarily focused on physics-based apparatus, such as nineteenth century automatic registering and switching devices; and in the twentieth century, devices like high intensity magnets, the ultracentrifuge and the rumbatron. Does research-technology occur more generally? It is important to ascertain to what extent, and in what sub-field and contexts research-technologies arise in chemistry, and what special rôle, if any, chemistry instruments, research and chemistry-related applications play in the workings of research-technology.

In this paper I will first present an example of research-technology and then discuss research-technology's historical appearance and evolution in Germany and America. Next I will outline the key characteristics of the category of instrumentation and will specify why it is important. Finally, I will open a discussion of research-technologies in chemistry, and will suggest that the discipline constitutes a particularly significant landscape for the transverse trajectories of this crucial category of instrumentation.[2]

The Ultracentrifuge

In the 1930s, 40s and 50s, the American Jesse Beams (1898–1977) developed the modern ultracentrifuge. The device and the man do not fit neatly into any standard institutional, professional, or intellectual mould. Long-time chairman of the University of Virginia physics department, Beams also sponsored two firms, acted as a key consultant to four additional companies, participated in the Manhattan Project, worked for the military during the 1940s and 50s, and contributed to numerous NSF science programmes. Beams was not the classical academic, engineer, entrepreneur, nor technical consultant. Although often located at or near the University of Virginia, his principal connection to that academic institution was the huge and well-equipped workshops that he developed there during decades of arduous endeavour.

Beam's ultracentrifuge had a parallel life. The ultracentrifuge was a by-product of his 1924 doctoral dissertation which focused on rapidly rotating mechanical systems. Assigned by his thesis director to investigate the speed of quantum absorption events, Beams developed a high-speed rotating technique for the accurate measurement of very short intervals of time. This device, and not the study of physical phenomena, was the centrepiece of his successful dissertation. An interest in multi-purpose, multi-audience technical apparatus rather than a focus on the stuff of the physical world emerged as Beam's guiding logic. Yet this focus did not make Beams an engineer or technologist in the usual sense of the term.

His initial devices employed air-driven turbines. However, their performance was limited by mechanical factors as well as by air friction. He first augmented speed by introducing a flexible drive-shaft which allowed for adjustments in the centre of gravity, thereby multiplying rotating capacity. He next placed the rotating vessel inside a vacuum, thereby eliminating air friction but nonetheless shaft mechanics continued to restrict performance. To solve this, Beams employed magnets to spin his vessel. The vessel was suspended inside a vacuum,

Ernest O. Lawrence and Jesse W. Beams at Yale. Photograph courtesy of the Lawrence Berkeley National Laboratory

Jesse Beams in his laboratory at the University of Virginia. From Jesse Beams Papers (RG-21/80.861), University Archives, The Albert and and Shirley Small Special Collections Library, University of Virginia Library

thanks to a magnet-based servo-mechanism. This constituted his consummate ultracentrifuge which rotated at previously unheard-of rates.

The ultracentrifuge became an important element in bio-medical research on bacteria and viruses, and they soon figured centrally in medical diagnosis and treatment. Beams engineered devices for radioactive isotope separation in the late 1930s which were effectively tested in the Manhattan project and became commercially viable in the 1950s and 60s. The Beams ultracentrifuge served in early ram jet propulsion research, and it was also used to do physics and

Jesse Beams. Spiral notebook. ca. 1940–41. On this page, Beams summarises three types of ultracentrifuging, including the 'straight centrifuging method in which the materials enter and leave the machine in a manner similar to that of the cream separator.' From Jesse Beams Papers (RG-21/80.861), University Archives, The Albert and and Shirley Small Special Collections Library, University of Virginia Library

engineering research on the strength of thin films. A Beams device rotating at over three million revolutions per second was used by physicists to measure light pressure. A somewhat different instrument enabled enhanced precision in the measurement of the gravitational constant.

As an author Beams published abundantly, sometimes in disciplinary periodicals, but much more of his written output appeared in instrumentation journals, such as the American *Review of Scientific Instruments*. A high proportion of his writings took the form of unpublished technical reports and he co-sponsored half a dozen patents. Beams' written productions were equally divided between the public and private spheres: between articles and patents on the one side (public), and confidential reports and consultancy on the other (private). Concurrent with these publications, he continued to build influential artefacts.

Beams and his devices crossed innumerable boundaries, circulating in and out of institutions, and shifting from employer to employer. He belonged to many organisations, movements and interests. He was neither a-institutional nor anti-institutional, but rather multi-institutional. He had no single home; his home lay everywhere. He explored and exploited the laws of nature as embedded in instruments and like Beams himself, his ultracentrifuges also crossed a multitude of boundaries. They were open-ended, general-purpose devices, which came to perform a host of functions, and found their way into a variety of non-academic publications and applications.

A special vocabulary and way of seeing events developed in conjunction with the Beams device. Light pressure and gravitation, isotope separation and thin films, microbes and viruses came to be spoken about in terms of rotational speeds and centrifugal pressures. 'Rotation' emerged as a *lingua-franca* for a disparate spread of fields and functions, extending from academia and research to industrial production and medical services. The rotational vocabulary and imagery of Beams' instrument percolated outward. Beams' approach and his artefacts thereby helped coalesce dispersed technical, professional and institutional worlds in physics related domains. It will be shown below that similar or parallel types of devices occur in chemistry.

Origins and Evolution of Research-technology

The historical roots of research-technology lay in Germany, yet other countries quickly followed, among them France, England, the US, and somewhat later Japan and the USSR. While each nation pursued a particular path, there were nevertheless elements shared in common. War played a central rôle as did nationalism, rapid industrial growth, and the simulation of newly emerging technologies and scientific disciplines.

German Origins

The rise of research-technology in Germany is attributable to military, industrial and nationalistic forces, as well as to characteristics of Germany's professional science community. The wars of Prussian expansion and German unification of the 1860s and 1870s alerted officers in the Prussian military to the potential importance of instrumentation. Officers perceived telegraphy, electricity, optics, mechanics and chemistry as central to military projects, and all these spheres entailed the use of high precision apparatus. In 1868 the army commissioned a report on the state of German precision instrumentation. The report concluded that Prussia stood a poor third behind France and Britain, and that the country badly required up-grading in the field. A Crown Commission was immediately set up, and it recommended that a special instrumentation section be established inside the Berlin Academy of Science to promote the development of scientific instrumentation.

The idea was immediately quashed by the Academy however. Academicians argued that science's task was the investigation of nature and not the design, construction and diffusion of instrumentation. The Academy insisted that instrumentation was carried out by instrument-makers who worked together with scientists, but instrument-makers must not become confused with scientists.

Nevertheless, instrument awareness in the military and imperial court gained strength, as instrumentation was increasingly viewed as synonymous with national interest. Some Berlin instrument-makers also began to see their traditional rôle in broader terms. Why should instrumentation *per se* not become an autonomous pole of activity they asked? Why must high precision continue to be a sub-routine inside existing areas of specialisation rather than emerging as an independent referent? Perhaps it could be transformed into a transverse domain with links to science, industry, metrology, state technical services, the military *etc.*, but at the same time remain semi-autonomous from all!

In 1878 a group of Berlin instrument artisans (R. Fuess, C. Bamberg, H. Haensch) created an instrument association, the Deutsche Gesellschaft für Mechanik und Optik, whose goal was to promote multi-purpose, general instrument systems having applications in many technical spheres.[3] The Society organised instrument discussion groups and workshops in Berlin and later elsewhere. It invited representatives from industry, public metrology services and academia to contribute ideas and to take stock of the new devices being proposed. The Society also founded an instrument school.

It embarked on three projects designed to transform the technical, cognitive and institutional landscape of instrument building. First, the Deutsche Gesellschaft für Mechanik und Optik organised instrument fairs according to a new format. In the past, instrument exhibitions had been structured around specific artefact or product themes (instruments for telegraphy, astronomy, surveying, hydrology, electricity, chemistry and so forth). Instrument exhibits were thus parcelled out according to their applications and thus fragmented. In the future, however, Germany's new breed of instrument men wanted all instrument exhibitions to be grouped and centralised. New categories of devices were being designed and produced that had a bearing on a broad range of technical arenas. There consequently existed a corresponding underlying instrument logic which transcended the former specificity of narrow applications. In view of this it made good sense to present instruments in terms of an autonomous referent. It was the job of future users to visit the central instrument stand of fairs and to take away new thoughts on how to tailor general devices to their narrow niche uses. This multi-purpose, generalist instrument logic governed the organisation of many instrument fairs in Berlin, Brussels, Paris, Chicago and Saint-Louis between 1880 and 1914.

Second, the Deutsche Gesellschaft für Mechanik und Optik founded a monthly journal specialising in instrumentation, the *Zeitschrift für Instrumentenkunde*. The journal was published from 1881 to 1944 and again between 1955 and 1967. Two categories of articles predominated. Approximately two thirds of the texts reported on highly specialised devices which were designed for a single application. Authors were characteristically academics or industrial engineers. However, about one third of the *Zeitschrift für Instrumentenkunde* texts focused on instrument theory and basic principles of instrumentation. Examples include texts that discussed principles of automatic switching or automatic registering. Instruments of this category had applications in biology, medicine, astronomy, spectroscopy, colorimetry, meteorology and so forth.

Based on an examination of over one hundred people who published this kind of work in the *Zeitschrift*, authors tended to report their artifact output in a far broader range of arenas than most academics and engineers involved with narrow-niche single application apparatus. Research-technologists published in as many as forty journals – in academia, industry, professional engineering reviews, trade publications and so forth. Moreover, they often wrote technical reports associated with private consulting, took out patents and wrote confidential reports for government, the military or industrial concerns. In effect, their standard arena of practice and impact stretched far beyond most other groups of instrument-makers and technical people.

Third, the membership of the Deutsche Gesellschaft für Mechanik und Optik insisted on operating on the fringe of existing groups and interests. One example of this was the project to force the establishment of an instrument section in the German Association of Scientists and Physicians, the Versammlung Deutscher Naturforscher und Ärzte, an organisation that at the end of the nineteenth century boasted over 5000 members. There existed special sections for astronomy, geology, optics, electricity, chemistry, architecture, biology and medicine – in all some forty sections. In the past, instrumentation had constituted a sub-activity inside every one of the separate sections. However, the Deutsche Gesellschaft für Mechanik und Optik struggled for the opening of a new section devoted only to high precision. The idea was not to become separated from the already established sections, but instead to stand at the crossroads between all existing technical sections. From this vantage-point, German research-technologists judged that their general purpose apparatus could circulate outward in all directions. In 1892 a special instrument section was at last founded. As will be shown, this *in between* stance is crucial to research-technologies.

Research-technology in the US

American research-technology only fully came into its own during the 1950s with the creation of the Scientific Instruments Society of America and the *Journal of the Scientific Instrument Society of America*.[4] The journey toward success was slow and often perilous.

A society entitled The Scientific Apparatus-makers of America was set up at the turn of the century. Its purview was mainly the circulation of information about east coast instrument-making firms – product descriptions, prices and the applications of devices. The organisation did not sponsor instrument research nor did it seek to promote an instrumentation mentality or instrument community. This situation remained static up to World War I.

Consciousness about the strategic importance of instrumentation developed in the US between 1914 and 1918. Scientists, particularly physicists, were recruited by the army and employed by industry in order to design and manufacture precision devices. According to one of the leading historians of twentieth century American physics, Daniel Kevles, scientific instruments became exceptionally important at this time and this situation was accordingly acknowledged in the scientific community. In 1917 the Optical Society of America was founded, along with its journal. The journal featured so many articles related to instrumentation that in 1920 the Society decided to add a special instrument section to the Journal. Some academics, particularly theoreticians, expressed fears that instruments were receiving too much attention. Moreover, some of the instruments reported on in the *Journal of the Optical Society of America* had little bearing on university research, instead being oriented to engineering or industry.

In response to this situation, in 1930 a new review was created, devoted

Advertisement for the Thwing pH meter, 1934. Reproduced from Instruments, *1934, 7, 266*

specifically to instrumentation – the *Review of Scientific Instruments*. Indeed by the 1930s the quantity of instrument articles in many US journals was staggering. The American Chemical Society abstracting service listed over 1000 articles for 1934 and again for 1935. The content of the *Review of Scientific Instruments* differed from the optics journal in two ways. First, technical spheres outside of optics appeared regularly. Chemistry texts appeared routinely, and during the 1930s, 40s and 50s articles in electricity, magnetism, electronics and nucleonics became predominant. There were also a few articles concerned with medicine and biology. Second, the review published articles from many perspectives. There were texts by academics who presented instrument designs for specific experiments, pieces by industrial engineers who had solved a restricted but crucial problem, and not least of all, texts by men in academia, metrology or industry concerned with basic metrology and instrument theory. Jesse Beams, whose research-technology was described above, published several key instrument theory texts in the *Review of Scientific Instruments* in the 1930s.

Despite this, during most of the 1930s and 1940s general instrument research remained marginal in the US. While other instrument journals prospered, like *Instrumentation* and the *Instruments* magazine, they emphasised single purpose, narrow-niche devices. Indeed, it is fair to say that there existed resistance in many instrument circles to the development of a generalist instrument current.

In 1935 the Instrument Publishing House was set up by a small group of east coast academics, instrument-men and industrialists to advance the cause of general, multi-purpose instrumentation. The idea was that certain apparatus, as in the area of electronics (the oscilloscope), had to be studied and understood from a more theoretical standpoint. Some apparatus had the potential to find applications in academia, metrology, industry, the army, *etc.*, and this kind of device was not reducible to narrow-niche instrumentation. The Instrument Publishing House thus published generalist instrument studies and sometimes organised research-technology regional workshops. In the late 1930s the Carnegie Institute too began to promote the research-technology perspective. It hosted generalist instrument conventions where research-technologists discussed their research and suggested specific applications for devices. This approach took on additional momentum during the war, and it was further stimulated by the immense instrument outlets associated with magnetism, nucleonics and low voltage electricity and electronics. (As reported in the book edited by B. Joerges and T. Shinn, the rumbatron, liquid scintillation counter and Fourier transform spectroscopy are three instances of how nucleonic, electronics, and later informatics research-technology spread outward.)

In 1954 the Scientific Instrument Society of America was created along with its journal. While much of the activity of the new Instrument Society involved single purpose, narrow-niche devices, there was also a call for research-technologists, and both currents found a home inside the nascent organisation. For example, in the early 1950s the National Bureau of Standards and the Air Force advertised for 'instrument theoreticians' and 'instrument generalists'. In 1954 and 1955 the *Journal of the Scientific Instrument Society of America* officially voiced this demand. It regularly carried editorials that stressed the need for research-technology as a component of the nation's instrument community. Technical areas where generalist devices had a key place were specified in journal articles, and the most advantageous organisational and functional relations between instrument generalists and narrow-niche specialists were discussed. The official stance of the Society was that both research-technologists and restricted application specialisation were complementary instrument modes that had to be promoted and expanded.

Components of Research-technology

Research-technology involves three key features.[5] 'Genericity' refers to the general-purpose and open-ended design quality of devices which can later be tailored for application in multiple local environments. 'Interstitiality' characterises the in-between and multi-institutional professional position and status of practitioners who are free to communicate with sundry user groups and assist them. 'Metrology' is crucial, for through the inclusion of basic metrological principles devices can percolate outward to many applications.

Genericity

Devices like mechanical and electrical-mechanical automatic systems for registering and switching, the ultracentrifuge, servo-mechanisms, cybernetic controls, lasers, and so forth are template apparatus. All of these material or methodological instruments enshrine general principles, and it is this fact that makes the devices open-ended and multi-purpose. Research-technologies are thus hub apparatus, base-line equipment, which carry the seeds for a myriad more narrowly intended specific machines or procedures.

Generic instruments may arise inside academia as well as inside industry, state metrology laboratories, the military, *etc*. Whatever their place of origin, genericity makes it possible to transfer across boundaries, be them disciplinary, economic sectors and so forth. Due to this transverse feature, generic instruments are thereby capable of fuelling activities in a multitude of spheres. It is precisely in conjunction with this that emerges research-technology's potential to structure and manage the science/technology divide in an unusual way.

Interstitiality

Research-technologists do not belong to academia, a scientific or engineering discipline, enterprise, the state service or the military. Nevertheless, they are habitually the employees of one of these institutions – or more frequently a combination or succession of them. Practitioners tend to move *between* employers and organisations. They are less a-professional than they are non-professional.

The interstitial stance and status are fundamental for two reasons. First, by working in between groups and organisations, practitioners are suitably positioned to glean new ideas and information for new generic devices from innumerable quarters. Second, once the generic design is completed and built, research-technologists thereby have the chance to promote their apparatus in a range of technical niches.

The interstitial stance enables research-technologists to structure divisions of labour in a dynamic and creative fashion. When there is a need for access to heterogeneous audiences professional cleavages and boundaries are traversed. The barriers entailed in professional divisions of labour are lowered. Conversely, during that phase of their work when research-technologists require abundant time and tranquillity (freedom from short-term technical pressures of niche audiences) to design a generic instrument, the barriers entailed in the division of labour are raised. Just as in the case of genericity, interstitiality plays a key rôle in regulating transverse science/technology relations in which fluidity occurs and also where each continent of the knowledge/artefact community retains its specific institutions, norms and goals.

Metrology

Generic instrumentation involves metrologies in the form of new units of measurement, norms and standards. Metrologies may alternatively take the form of new visualisations, representations, or perhaps even paradigms: CAT scan (images of structure densities), NMR (imaging through molecular structures), cybernetics (reiterating feed-back loops), *etc*.

The metrologies of research-technologies open the way to a kind of pragmatic universality. When specific features of a generic device are tested and then incorporated in a large number of diversified audience niches, the common denominator technical capabilities shared by all audiences (resulting from experience with a common generic research-technology device) becomes an individually experienced yet commonly held reservoir of practices, solutions and truths. The terminology associated with a research-technology becomes a technical *lingua franca* and research-technologies emerge as a sort of pan validation.

The metrological, generic and interstitial traits of research-technology make it a highly transverse force – indeed it may even be characterised as essentially a *transverse* science and technology culture. Its transverse mode of action and organisation are distinguishable from the more uni-institutional and uni-objective operations of the disciplinary science and technology culture, the utilitarian science and technology culture and the more complex transitory science and technology culture more familiar to historians and sociologists of science and technology. Research-technology manages to stretch across all of the other three cultures, providing them with a measure of convergence, if not unity. Such technical convergence is functionally useful in a cognitive, institutional and professional world often marked by acute specialisation, and sometimes fragmentation.

What About Chemistry Instrumentation?

Several of the episodes in the history of chemistry instrumentation that are presented in this book can be interpreted in the research-technology perspective. The trajectories of Frank Twyman, the constant deviation wavelength spectrometer and the Adam Hilger instrument company exhibit features of genericity, interstitiality and metrology. Charlotte Bigg (in this volume) shows that components of the technologies involved in the constant deviation wavelength spectrometer developed at Hilger in the 1920s contained the seeds for extending the device's sphere of application into many domains. The degree to which genericity constituted in this instance a premeditated instrument strategy requires additional exploration. Whatever the case, one of the spectrometer's architects, Twyman, grasped the instrument's potential for the development and extension of metrology. This metrological concern had a two-fold effect: it reinforced the generic potential of the spectrometer, becoming a mechanism by which the device's underlying theoretical instrument principles could be made visible and further extended, and also, through metrology, the possible utility of the general, open-ended apparatus could be demonstrated to instrument users in a range of specialities extending from academia to industrial chemistry and metallurgy. As depicted by Bigg, Twyman occupied an interstitial arena between the Hilger instrument company and the heterogeneous groups of users. Twyman's success in re-embedding the spectrometer was achieved through documenting the connection between the device's generic features and user-specific technical potentials and through educational programmes adapted to the demands of particular niches.

The persistent primacy of basic research on instrumentation, as distinguished from designing devices for a particular purpose and using them, is a key feature of research-technology. Peter Morris's fine-grained description of the work of James Lovelock (in this volume) shows how some chemists are first and above all tied to such basic research on apparatus. In his preliminary laboratory work on what would become the electron capture detector, Lovelock, the self-styled instrument-man, repeatedly expressed his wish to continue to probe the underlying operations of ion capture devices. He was himself disinclined to use them for the amassing of biomedical data. Once Lovelock fully developed the principles of his detector, however, he became eager to see it used in a number of areas, and he himself pioneered the use of the detector in the study of perfluorocarbon and other trace compounds diffused in the atmosphere.

Of course, not all instrument designers and builders and instrument companies belong to research technology. It is safe to say that it is only a small minority. The distinction between research-technologists and narrow-niche instrument-men is sometimes a fine one, but nevertheless an extremely important one. The relationship between practitioners of the research-technology bent and specialised instrument-makers emerges in Davis Baird's chapter on the instrument company, Baird Associates (in this volume). The fulcrum of intellectual and material activities of W. S. Baird, the principal founder and long-time head of the instrument company that bore his name, lay at the intersection of many communities including his own manufacturing company, academia, and industrial, government and military related clients that purchased his innovative high-performance spectrometers. Baird was above all an instrument researcher and this fact often generated problems. Since he had little interest in manufacturing and economics he and his firm sometimes found their survival problematic. Baird was only interested in instrument use and clients to the extent that they nourished basic instrument research. He thus saw himself as functioning in a space between numerous poles of reference, and Baird's main concern was consistently to garner those resources necessary to continue to engage in interesting broad scoped instrument inquiry.[6]

The Case of Interactive Molecular Graphics

Interesting aspects of research-technology dynamics come to light in the recent study of interactive molecular graphics carried out by the sociologist of chemistry Eric Francoeur in collaboration with Jérôme Segal.[7] The molecular graphics instrument is used today in molecular chemistry, biochemistry and physical chemistry. It involves computer-based three dimensional representations of molecular components, structures, their relations and the relations between molecules. Their visualisations require sophisticated representation programming as well as advanced analytic models of molecular features and dynamics. In interactive molecular graphics, simulated dynamic relations in and between molecules can be modified by a researcher in real time, thereby allowing at a rapid glance information about how changes in one molecular parameter may affect a variety of other molecular parameters and molecular activities.

The first interactive molecular graphics techniques, and first research to incorporate this techniques, arose in the mid-1960s in Cyrus Levinthal's bio-

Linus Pauling and Robert Corey with a molecular model of their protein α-helix, c. 1953. In 1953, Pauling and Corey developed a system of wooden molecular models using a scale of 1 inch per angstrom and plastic models with a scale of 0.5 inch per angstrom. Science Museum/Science and Society Picture Library

chemical laboratory at MIT. The study of macromolecular structure of proteins was among the foremost concerns of Levinthal and his group. At the time, MIT had become the host of Project MAC, which provided the local research community with a time-shared mainframe computer, the first of its type in an academic environment. Most of the terminals connected to this computer were paper terminals. However, there was also a prototype graphic display terminal developed by MIT's Electronic System Laboratory. Levinthal was introduced to this graphics terminal in late 1964 and rapidly began to see ways to use computer graphics in the investigation of macromolecular structures.

Why this adoption of a new instrument? Since the 1930s, and particularly during the 1940s and 50s, molecular chemists in many fields had increasingly used physical models in the course of their research. Physical models helped scientists picture molecular components and to think about possible relations. They were also used to try out how a change of elasticity, tension, distance and so forth of molecular components or between molecules might affect dynamics. Not least of all, physical models played a rôle in chemistry teaching.

Physical models proved central to Linus Pauling's research on protein molecular structure, as well as to the investigations of Francis Crick and James Watson on the geometry of DNA. Physical models were an integral part of contemporary chemistry, and although sometimes expensive, there existed an extensive demand for them, to the point that several firms profitably manufactured them.

In some respects, this prevalence and standard use of physical models in chemistry, which began with late nineteenth and early twentieth century stereochemistry, paralleled their use in physics. The French physicist and philos-

*Reconstruction of the double helix model of DNA built by Francis Crick and James
Watson in 1953 using some of the original metal plates. Science Museum inventory number
1977–310. Science Museum/Science and Society Picture Library*

opher Pierre Duhem characterised English physics in terms of its dependence on
physical models built of wheels and string, and we know that scientists like
Kelvin in fact sometimes turned to such devices in their research and teaching. In
mid-twentieth century chemistry two kinds of physical models were common.
Space-filling models, of the kind employed by Pauling, represented the distances
and geometric configurations between macromolecules. Skeletal models were
used by chemists to depict and investigate molecular structure, for example, the
crucial characteristics of elasticity, tension and angles of rotation.

According to Francoeur's study, chemists often found physical models difficult
to manipulate. Many models were complicated, containing up to six hundred
separate pieces. Although components were commercially available, researchers
arranged them according to their experimental needs. If one or a few components
were removed or exchanged to visualise the consequence for the whole, this
frequently involved an immense amount of labour for a modest result.

Physical models occasionally measured several feet across and in height. This
was the case with some of the models used by Levinthal at MIT in the 1960s. His
models of protein macromolecules were indeed so big that their components
often collapsed under their own weight. The smallest mishap in inserting an
additional element could also lead to disaster. Levinthal and his group briefly
considered building their giant models in a large swimming pool where the
effects of gravitation would no longer pose a problem. In short, even though
physical models were helpful to research, they were also very problematic!

The prospect of computer generated models thus offered a possible solution to
an increasingly acute difficulty. Working with the MIT Project MAC, members
of the macromolecule laboratory developed programs that depicted the molecular components of interest to them that had previously figured in skeletal and

space-filling physical models. The new molecular graphics models gradually improved in quality. Chemists in the laboratory were soon able to represent and to easily modify and manipulate the targeted elements, and were thereby able to visualise molecular configurations and how configuration changes might impact molecular structure and dynamics. As computational capacity grew, it became possible to generate increasingly complex and complete simulations, and changes could be introduced in real time. This meant a tremendous simplification of some research work and meant a speed up in the pace of investigations.

Little by little, interactive molecular graphics became a fairly common tool in macromolecular research in a number of chemistry-related fields. Has the new instrument changed chemistry in the sense that it allows discoveries that were perhaps previously unattainable? This is by no means certain. Is interactive molecular graphics modelling an extension of physical modelling? In some ways no. Since interactive graphics modelling is direct and hands on, it permits a multi-dimensional approach that was never realised with physical models. It is suggested by Francoeur and Segal that by offering a new quality of image and new manipulative possibilities, it may also open new epistemological horizons.

Computers most definitely belong to the category of devices that I have called research-technology, and computer graphics is in itself perhaps a research-technology. Computer graphics arose independently of interactive molecular graphics, initially arising in engineering and technology. Moreover, interactive molecular graphics did not develop fully inside chemistry. For a decade beginning in the mid-1960s, Levinthal and his group focused their activity on the use and development of computer techniques for research purposes. And in this capacity, they occupied a niche *between* the chemical scientist and the computer scientist.

It is clear that the discipline of chemistry offers, and has traditionally offered, a field of predilection for the diffusion of research-technologies and research-technology-like instruments. Interactive molecular graphics spread rapidly in many sub-disciplines of chemistry: molecular chemistry, biochemistry and physical chemistry. The instrument's suitability and effectiveness were thus tested, verified and extended in important ways within a host of audiences.

Conclusion

In summary, since chemistry is the discipline which trains more students, mobilises more laboratories and employs more scientists than any other field, it constitutes a potentially exceptional terrain for the creation, adaptation, absorption and validation of generic apparatus. Moreover, one can hypothesise that chemistry will be basic to the trajectory of research-technology both because of the variety and volume of generic devices which the field has historically absorbed, and because of its capacity to generate generic apparatus which is re-embedded outside the discipline, for example in biology and medicine.

In particular, research-technologies may be seen as playing a role in the linkage of chemical technologies and academic chemistry. In some respects, the history of chemistry-related control engineering instrumentation (see Stuart

Bennett in this volume) and chemical engineering[8] represent a transverse coupling mechanism between science and technology. It is a task for future historical studies to discover additional examples of such linkage in the republic of chemistry, as well as to identify additional rôles of research-technology.

Notes and References

1 B. Joerges and T. Shinn (eds.), *Instrumentation Between Science, State and Industry*, Kluwer Academic Publishers, Dordrecht, 2001.

2 For studies of instrumentation relevant to the history of fundamental and applied chemistry see: D. Baird, 'Analytical Chemistry and the "Big" Scientific Instrumentation Revolution' *Annals of Science*, 1993, **50**(3), 267–290; F. L. Holmes and T. H. Levere (eds.), *Instruments and Experimentation in the History of Chemistry*, MIT Press, Cambridge MA, 2000; H. A. Laitinen and G. W. Ewing (eds.), *A History of Analytical Chemistry*, American Chemical Society, 1977; Y. Rabkin, 'Technological Innovation in Science: The Adoption of Infrared Spectroscopy by Chemists' *Isis*, 1987, **78**, 31–54; J. T. Stock and M. V. Orna (eds.), *The History and Preservation of Chemical Instrumentation*, Dordrecht, 1986; A. Travis, 'Surrogate Instruments: Industrial Chemical Reactors and Organic Chemistry' in *Instrument – Experiment*; *Historische Studien*, Ch. Meinel (ed.), Verlag für Geschichte der Naturwissenschaften und der Technik – Diepholz, Berlin, 2000, pp. 201–216.

3 T. Shinn, 'The Research-Technology Matrix: German Origins, 1860–1900' in *Instrumentation Between Science, State and Industry*, B. Joerges and T. Shinn (eds.), Kluwer Academic Publishers, Dordrecht, 2001, pp. 29–45.

4 T. Shinn, 'Strange Cooperations: The US Research-Technology Perspective, 1900–1955' in *Instrumentation Between Science, State and Industry*, B. Joerges and T. Shinn (eds.), Kluwer Academic Publishers, Dordrecht, 2001, pp. 69–96.

5 B. Joerges and T. Shinn, 'A Fresh Look at Instrumentation: An Introduction' in *Instrumentation Between Science, State and Industry*, B. Joerges and T. Shinn (eds.), Kluwer Academic Publishers, Dordrecht, 2001, pp. 1–13; B. Joerges and T. Shinn, 'Research-Technology in Historical Perspective: An Attempt at Reconstruction' in *Instrumentation Between Science, State and Industry*, B. Joerges and T. Shinn (eds.), Kluwer Academic Publishers, Dordrecht, 2001, pp. 241–248.

6 For more on this subject see: D. Baird, *Thing Knowledge: A Philosophy of Scientific Instruments*, University of California Press, Berkeley, forthcoming.

7 E. Francoeur and J. Segal, 'From Physical to Virtual Models: Macromolecular Structures and the Origins of Interactive Molecular Graphics' in *Displaying the Third Dimension: Models in the Sciences, Technology and Medicine*, S. de Chadarevian and N. Hopwood (eds.), Stanford University Press, Stanford, forthcoming.

8 C. Divall, 'Education for Design and Production: Professional Organization, Employers and the Study of Chemical Engineering in British Universities, 1922–1976', *Technology and Culture*, 1994, **35**, pp. 258–288.

Adam Hilger, Ltd and the Development of Spectrochemical Analysis

CHARLOTTE BIGG

Max Planck Institut für Wissenschaftsgeschichte, Wilhelmstrasse, 44, 10117 Berlin, Germany, Email: bigg@mpiwg-berlin.mpg.de

Introduction

The instrumental revolution, the radical transformation of the material practices of chemistry in the middle decades of the twentieth century, has been well documented. Chemists and historians have pointed to the dramatic substitution in universities and industry of traditional 'wet' chemical procedures with a new battery of instrumental methods based on the physical properties of atoms and molecules.[1] While it was not the first time that chemists adopted a technique associated with the physical sciences (witness the debates surrounding Lavoisier's introduction of the balance in the eighteenth century[2]), these recent developments have transformed chemistry in its material, methodological and conceptual aspects to an unprecedented extent.

   Much of the existing literature on the instrumental revolution has focused on the transformation of organic chemistry in the 1950s and 1960s by such techniques as absorption spectrometry, mass spectrometry or nuclear magnetic resonance, for instance Reinhardt and Slater in this volume. This paper examines the earlier instance of visual and ultraviolet emission spectroscopy for analysis, which appeared in the mid-nineteenth century but only came into its own in the 1930s. Emission spectroscopy of the sort examined here arguably acted as a trail-blazer of the instrumental revolution, and as such provided a precedent and a starting point for the introduction into chemistry of the other techniques mentioned above. Its history also reveals the considerable initial difficulties involved in the creation and popularisation of these instrumental methods.

This case is illustrated by a study of the development of the spectroanalysis of metals, concentrating especially on the contribution of one of the earliest instrument-making firms to become involved, Adam Hilger, Ltd and its manager Frank Twyman. A theme common to the different accounts of the instrumental revolution (including Davis Baird's history of Baird Associates in this volume) is the central rôle played by instrument makers, who undertook much of the laborious work of making spectroscopy relevant to chemistry and, crucially, of convincing academic and industrial chemists of its usefulness. The particular status of instrument makers in this process as channels of transmission and translators of practices between different communities makes them appear in the light of Shinn's contribution as typical research technologists.

Spectrum Analysis Before the First World War

> But what became of Kirchhoff and Bunsen's beautiful conclusions? They were held to be mistaken! Spectral analysis no longer possessed the sharpness or the infallibility ascribed to it by its founders. From an analytical perspective the good old methods of pure chemistry remained unsurpassed, and spectral analysis was a complicated science which was better left to the specialists.[3]

Writing in 1913, Georges Urbain, professor of chemistry at the University of Paris and himself a keen spectroscopist, thus deplored the lack of consideration for spectral analysis in his discipline. Notwithstanding the fact that it was, famously, the outcome of a collaboration between a chemist and a physicist, there remained by then few traces of the early cross-disciplinarity of spectral analysis. Chemists' routine use of spectroscopy was circumscribed to a few isolated applications, such as the flame analysis of alkali metals and alkali earth metals after the example of Gustav Kirchhoff and Robert Bunsen's original investigations in the 1860s, or Julius Plücker's method for the identification of gases with Geissler tubes. In industrial settings, a few early attempts to introduce emission spectral analysis mostly floundered, including the attempt of Bunsen's student and collaborator Henry Enfield Roscoe to monitor the Bessemer process spectroscopically.[4] The optimism of the savants was not shared by the metallurgists and few spectroscopes were to be seen in metallurgical or chemical works before the Great War.[5] Despite the efforts of propagandists such as Urbain, who attributed such a neglect to ignorance and lack of training, chemists showed little interest in spectral analysis.

The reluctance of chemists to embrace the technique could not entirely be put down to apathy, however. As it stood then, spectral analysis was a rather complicated affair. For a start it was soon realised, *contra* Bunsen and Kirchhoff, that the emission spectra produced by substances varied not only with the experimental conditions under which they were studied (arc, spark, flame, tube), but also according to their concentration and physical state. No single method could be applied indiscriminately to all elements, because most only brought up spectra under specific conditions. It took a long time before electric currents became stable enough to produce controlled sparks and arcs. Further, the extreme sensitivity of the spectroscope to *some* elements and the recurrent presence of the air and sodium spectra delayed the establishment in the first place of reliable spectra for each element. When reference spectra were eventually obtained at great cost of time and labour, they were often so complicated, so rich in lines that no characteristic patterns could be recognised at a glance to identify elements. It was then necessary to measure spectral lines and to reduce the results by calculation, creating more complications. Twyman thus described the operations necessary before spectrum analysis could be carried out routinely:

> Before [the metallurgist] could apply the instruments to the daily routine, he had to find the best lines for identifying the various constituents, to assure himself that lines of other possible constituents would not be a source of confusion, and to decide how to estimate the quantities present and whether the accuracy of determination was sufficient for his purpose. Again he had to decide whether he should use the arc or spark, if the latter, what was the best self-induction and capacity to employ. He had also to discover whether the procedure appropriate for one alloy was also suitable for another. Even when he had found, by experience, the answer to these questions, the technique so developed was not always reliable in the hands of another observer who, though apparently doing the same thing, was actually departing from the original procedure in some way not then thought to be of importance.[6]

The practice of spectral analysis, in short, necessitated considerable amounts of experience and skill on the part of the experimenter, to manipulate the apparatus, interpret spectra, distinguish between evidence and noise and to produce consistent results.

This was an expertise which most chemists did not possess, the reason being that spectral analysis had become associated since the 1880s with physics, and later also with astrophysics, rather than with chemistry. Spectroscopes were only actively used in the research laboratories of a few pioneering physical chemists while they rapidly populated astrophysical observatories and physical laboratories. Physicists used them pedagogically, to teach the principles of optics to their students and to convey disciplines of precision measurement. In their research, they applied the spectroscope to investigate optics, electromagnetic radiation or the properties of matter under different states and conditions of pressure, under magnetic or electric fields. Astrophysicists, for their part, undertook vast programs to constitute spectral catalogues of stars and substances or monitored the sun spectroscopically.

For these physical scientists, spectroscopes belonged to a group of optical devices which also included photometers, refractometers and goniometers. Learning to work with them was not an exceptionally arduous task since the expertise gained on one of these instruments could be more-or-less readily transferred to the others. They all featured telescopes, systems of lenses and/or prisms, and often identical means of adjusting focus and of measuring angles and distances. Most physicists would have had to practice with several of these optical instruments as students. On the other hand, for chemists used to working with the traditional furnaces and glass and porcelain apparatus, spectroscopes represented a somewhat alien breed of instrument and were not so easy to manipulate and appropriate. Learning spectroscopy signified for them acquiring a physical technique, since it was usually not taught in chemistry courses, while it featured largely in physics lectures and textbooks.[7] The physical scientists' apparent monopoly over spectroscopy was reinforced by the fact that its instrumentation was more often than not listed in instrument makers' catalogues in the sections devoted to physical instrumentation.[8]

The active involvement of physical scientists with spectroscopy meant in addition that it developed along lines congenial to them, making them often unattractive to their chemist colleagues. An instance was the largely qualitative aspect of spectral analysis as developed by astrophysicists. It was very good to be able to identify the components of a substance, but chemists more often than not also needed to know in what proportions they were present. And in this respect spectroscopy did not deliver satisfactorily, despite some research in this direction by a few physical chemists in the first decades of the twentieth century. Similarly, the reference tools elaborated by physical scientists, the spectral tables on which all other measurements were to be indexed, were considered by chemists of little practical value. The product of the physical scientists' pursuit of extreme precision in the measurement of spectral lines in the 1880s and 1890s, grew to be, by chemical standards, exceedingly precise and unnecessarily complex, such that, as a German instrument maker put it,

> [The chemist] does not know where to start with the hundreds of pages filled with tables of the emission lines of the elements, the product of the diligence of two generations of physically-educated spectroscopists; for the chemist, they constitute an overwhelming amount.... This prevents the chemist from exploiting the physicists' work on emission spectroscopy. Spectral analysis cannot so far be described as a branch of physical chemistry, which would otherwise be self-evident.[9]

This is not to say that chemistry was not a precise science, of course. But the kind of precision pursued by physicists was not that valued by chemists, and the latter by and large were disinclined to take up a technique increasingly identified with physics, and as such of little use to them.

Adam Hilger, Ltd

Throughout the late nineteenth and early twentieth centuries, however, a small number of physical chemists took interest in spectroscopes, contributed to develop a chemically oriented practice for them and widely advertised its benefits, part of the wider movement which led to the formation of a distinct field of physical chemistry. These scientists were joined from the turn of the century in their advocacy of instrumental methods in chemistry by a group of instrument makers, starting with the London-based firm Adam Hilger, Ltd. The workshop had been founded by Adam Hilger, who after several years of training in Munich,

FRANK TWYMAN, F.R.S., F.Inst.P.
PRESIDENT OF THE OPTICAL SOCIETY
1929-31

Frank Twyman. Reproduced from Transactions of the Optical Society, *1929–30*, **31**, *facing p. 8*

Paris and London had set up his own workshop for the production of optical instruments in 1875, and was soon joined by his brother Otto. The company produced custom-made scientific instruments for a mostly academic and international public. The Hilgers acquired early on a solid reputation amongst scientists at home and abroad especially for their spectroscopic apparatus.

Shortly after Adam Hilger's untimely death in 1897, Frank Twyman, an electrical engineer, was recruited by Otto Hilger.[10] At Hilger, Twyman initially worked at the bench, testing optical systems, cutting spectroscope slits, adjusting instruments, and helping Otto Hilger with optical calculations. Twyman was soon in charge of 'all the testing of the optical work and the adjustments of all the instruments ... and until 1910 designed and made complete drawings for all the instruments',[11] as well as superintending their manufacture. With Otto Hilger's death in 1902, Frank Twyman's rapid rise within the company culminated with his accession, at the age of twenty-six, to the general managership of the firm and he remained its head executive in different capacities until his retirement in 1952.[12]

At the turn of the century the company had twelve employees and was housed in a two-storey building next to the Hilger family's home in Stanhope Street. In the workshop '[t]here were no power-driven machines, the lathes being worked by treadle and a small shaper by hand. There was no typewriter and no telephone. The instruments were then made almost entirely out of tube and sheet with very few castings'.[13] The workshop typified then the nineteenth century instrument-making firm: small, family based, working to order and catering to a niche market.[14] In the first couple of decades of the twentieth century, a growing proportion of the production was devoted to making lenses for non-scientific, usually military instruments, for instance range finder lenses for Barr & Stroud, which made up half their income in 1913. The connection with the British infantry was reinforced during the war, when the Hilger workshop was expanded to house the production of 'thousands of sets of binocular lenses, at one time ... up to 300 a week, that is three thousand lenses'.[15]

Like other optical instrument makers, notably Carl Zeiss and C. A. Steinheil Söhne in Germany, Hilger oriented its activities in the early twentieth century towards the mass production of lenses for commercial and military usage. And like them, Hilger drew most of its income from this routine activity, while its reputation was maintained by the production of small run, high expertise scientific instruments. After the war their general policy veered away from this pattern, however. The German instrument-making firms continued growing along with the market for camera lenses. Hilger, in contrast, completely ceased to supply the military in the mid-1920s, and turned instead to the production of instrumentation for routine chemical analysis and for the testing and control of industrial processes, hoping to open up new and profitable markets.

Black-boxing the Spectroscope

Twyman was inspired to design a spectroscope specifically for chemists by the work of Herbert Jackson, the professor of chemistry at King's College, London and an old customer of the firm. Jackson, according to Twyman, 'was a meticulously accurate inorganic chemist and had constant recourse to a spectrometer of the simple table form'.[16] He employed this instrument as a complement to wet analysis to ascertain the completion of reactions. Twyman became convinced that the value of spectroscopy for chemistry warranted its much larger use in the field and he set out in about 1900 to manufacture an instrument which would be appealing specifically to chemists, with the intention of tapping a wide public of analysts.

Twyman thereby espoused the view of the physical chemists advocating spectroscopy that part of the reason for chemists' widespread disinterest in the technique was their perception of the spectroscope as a physicist's instrument. He later recalled that:

> Thus although spectroscopy developed, it was mainly under the auspices of the physicist and the astronomer. For these it rapidly became an instrument of supreme importance, but in the academic laboratories it was chiefly used, if at all, as a means of teaching optics, and instruments were deliberately made with as many adjustments as possible in order that the students might learn the principles of the instrument.[17]

Examining the instruments produced by his firm, Twyman found that they embodied the physical scientists' cultures of precision measurement. The divided circles and numerous screws for modification, adjustment and measurement which had increasingly been built into spectroscopes represented a considerable hindrance for chemists. What they needed, the manufacturer reflected, was

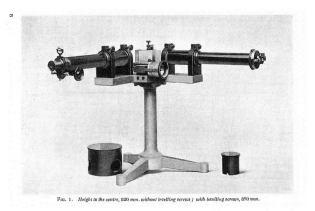

FIG. 1.   *Height to the centre, 320 mm. without levelling screws ; with levelling screws, 370 mm.*

Hilger CDW spectrometer, 1919. Reproduced from The Hilger Wavelength
Spectrometer... (*Adam Hilger, Ltd, February 1919*), *p. 8*

Detail of the drum of the Hilger CDW spectrometer, 1919. Reproduced from The Hilger
Wavelength Spectrometer... (*Adam Hilger, Ltd, February 1919*), *p. 10*

simple effective instrumentation for analysis rather than intricate set-ups for
reflexive experimentation.[18]

The instrument maker's strategy thereby differed from the tactics of the
physical chemists, who contended that chemists should learn more physics, who
wrote textbooks to popularise spectroscopic methods and worked to convince
universities to adopt them in chemical curricula. Twyman sought rather to adapt
spectroscopy to the specific skills and needs of chemists. In keeping with his
expertise and his interests, he focused on developing a material culture of
chemical spectroscopy, arguing that 'perhaps the most potent reason why the
chemist avoided spectroscopy was the lack of a suitable apparatus'.[19] Twyman
believed the problem lay not with the attitudes of chemists but with the instrumentation which he judged made readings cumbersome and time consuming. On
prism spectroscopes, by far the instruments most commonly used in visual
spectroscopy until the 1930s, spectral lines were usually measured on an arbit-

rary scale and translated into wavelengths by means of standard formulas which took into account the type of glass used for the prism, a complicated process which involved enough optical theory to be unappealing to chemists unfamiliar with physics. Twyman imagined that 'if one could read the wavelengths of the distinctive lines on a simple scale, it would remove the chief difficulty which stood in the way of such [chemical] use'.[20]

His first move was to create the 'constant deviation wavelength spectrometer' in 1903, basing it on a prism arrangement devised by Philibert Pellin and André Broca in 1899.[21] In this set-up, the telescopes and collimator were fixed, with all the wavelengths appearing successively in focus when the prism was rotated. This enabled the suppression of the divided circle, thus reducing considerably the number of preliminary adjustments. Measurements were taken by means of a fine micrometer screw tangent to the prism, soon replaced by Twyman's innovative 'wavelength drum', which gave readings in wavelengths directly, thus dispensing with calculation altogether.[22] These modifications amounted to a black-boxing of the apparatus, the transformation of an instrument into a reliable, robust machine meant to perform well under any circumstance. It was not a tool for investigating spectra, optics, or to develop spectroscopic methods. It was a pre-adjusted instrument ready to use for routinely analysing substances. The result, as a subsequent catalogue put it, was 'a construction ... which is at once extremely convenient and mechanically sound'.[23]

Many spectroscopists were taken by the new device. At the 1905 Optical Convention, a member of the executive committee praised 'the great purity of the spectrum produced by these instruments, as made by Messrs Hilger & Co, and their extreme simplicity in use'.[24] The spectrometer was very successful, this type of set-up soon becoming virtually synonymous with the company's name.

But to Twyman's (only partial) disappointment, these spectrometers were not bought by chemists: 'It was quite a success but not in the direction in which I had anticipated. It was purchased by physicists for academic use all over the world'.[25] And indeed, as a physicist wrote, by 1925 'constant deviation dispersing devices have become of such common use that an experimentalist accustomed to them struggles to do without them'.[26] The instrument was popular in particular with physicists investigating the structure of fine spectral lines and the Zeeman and Stark effects, the splitting of lines under magnetic and electric fields, respectively. The institutionalisation of astrophysics as well as Niels Bohr's derivation of the Balmer formula for the spectrum of hydrogen from his quantum atomic model in 1913 also helped broaden the field of application of spectroscopy in physics. Accordingly, 'the model was not discarded but its scope for carrying out various experiments of academic interest was extended'.[27] For this public Twyman devised versions of the spectrometer which allowed a number of extensions, including interferometric apparatus, photometers, thermopiles, *etc.*[28] He wrote that '[t]he basic instrument can be added to by clamping to the bar a range of accessories for carrying out laboratory experiments in practically every branch of spectroscopy'.[29] For these experiments the black-boxing of the dispersive apparatus was an advantage which let physicists focus on the manipulation of the more problematic extensions.

Quartz Spectrographs E 4

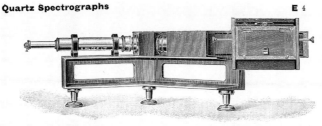

Fig. E 3.

Quartz Spectrograph, size (*c*) (*see* Fig. E 3).—Lenses of 24 inches (610 mm.) focus, the instrument giving a spectrum from W.L. 2100 to W.L. 8000 of about 200 mm. long. Prism, 41 mm. high × 65 mm. long face. Size of plate, 10 × 4 inches. The slit is our No. 2 (*see* Spectroscope slits, Section F). The dispersing system consists of one Cornu prism.

Hilger quartz prism spectrometer, 1912. Reproduced from a Hilger catalogue, dated September 1912, p. E4

Meanwhile, Twyman continued producing versions of the spectrometer specifically for chemical use. He devised a spectrograph, based on the same principle, but adapted for the ultraviolet, a region of the spectrum especially important for the investigation of metallic spectra. The quartz spectrograph, marketed from 1906, was modelled on an instrument built for the Dublin chemist Walter Noel Hartley's researches on the quantitative analysis of metals.[30] Twyman also developed the first commercial infra-red spectrometer in 1913, with 'the scale graduated in wavelengths and thus for the first time it became possible for chemists to study infra-red spectra'.[31] In later models of the chemical spectrometers, the instruments were literally black-boxed, the whole optical system being encased in a cast aluminium shroud, where 'the axis can be adjusted once and for all'.[32] This instrumentation was primarily bought by research chemists carrying out investigations in spectroscopy.

The constant deviation wavelength spectrometer thus rapidly evolved into a family of instruments catering to a wide and diverse public, and the Hilger literature was careful not to exclude any potential clients:

Very considerable modifications have been made in the design of the Hilger Constant Deviation Wavelength Spectrometer. These are mostly such as to make the instrument suitable for chemical laboratories where conditions render desirable the utmost freedom from corrosion, damage to optical work, alteration of adjustments, *etc*. The advantages of the new type will be appreciated by many physicists, although the older type, in which the removal of the telescope and collimator is easy, will still be found useful in many physical laboratories where it has been customary to use the separate parts of this instrument in a great variety of investigations.[33]

Faced with the failure to capture a large public of analytical chemists, Hilger was led instead to market their spectroscopes to a differentiated body of researchers. This is a reminder of the importance of the user in instrument design, and this in different ways. The instrument maker behaved as a user of scientists' researches when he adapted prototypes constructed to order into catalogue

Cover of a Hilger catalogue, The Hilger Wavelength Spectrometer ... (*Adam Hilger, Ltd, February 1919*)

types aimed at a wide public; and the scientists were equally active users when they bought instruments and modified them for specific investigations, leading in turn to the production of new variants by the manufacturer.[34] The modular system introduced by Hilger, with a standard spectrometer core to be customised with a large selection of appendices was their answer to the problem of simultaneously accommodating a broad range of specific requirements while still retaining a reasonably efficient production routine, a mode of production intermediate between the manufacture of individual custom-made objects and the industrial production of standardised artefacts.

Inventing the Routine

The lack of success of the spectrometer with chemists in the first decades of the century did not diminish Twyman's commitment to popularise spectrochemical analysis, and he continued devising new tactics to this end. Just after the First World War, Twyman's business strategy shifted towards the production of instrumentation for use in industrial contexts, especially metallurgical firms. Hilger's advocacy of routine spectroscopy made sense economically. Routines meant standard instruments, which were easier and cheaper to produce than individualised objects. Standard instruments were easy to master, so that relatively untrained people could use them, and this increased the potential market for spectroscopes. Twyman's strategy was to turn scientific instrumentation into a commercial product, that is to convince a large number of people to use spectroscopes for non-research purposes, reflected in his wish to make the spectroscope as common as the balance, it being the scientific, yet commercial instrument *par excellence.*

In this he was prompted by the first-ever sale in 1913 of a quartz spectrograph for commercial use, purchased by the American Brass Company, the largest producer of brass in the United States. He was also encouraged by the war-time pioneering application of the spectral analysis of metals by Arnaud de Gramont in France and William Meggers and Keivin Burns at the National Bureau of Standards in the United States. These researchers subsequently wrote articles detailing their methods for the routine quantitative analysis of metallic samples. Basing their procedures on methods originally developed by Hartley and de Gramont, they claimed that they had achieved promising results, and encouraged chemists to consider the spectroscope as an alternative to wet methods of analysis.[35]

For what he called 'works chemists', Twyman elaborated an innovative approach, with the development of an appropriate instrumentation and especially of an after-sale service in the form of a set of reliable routines of spectrum analysis. No longer seeking to break the association of spectroscopy with physics and to promote its chemical uses, and relying on the war-time experiences of the French and American investigators, Twyman now worked to show that spectral analysis could be practised by non-specialists. Perhaps realising the very real practical difficulties involved in performing especially quantitative spectral analysis, he concentrated on making its practice less daunting.

In the first place, this meant taking over a large part of the preliminary calibrating operations necessary to set up quantitative routines. While Hilger catalogues of the 1900s had simply recommended to the buyers of their spectroscopes the reading of the papers of de Gramont and Hartley, they increasingly mediated between the work of these researchers and prospective spectroscopists, by giving advice as to how users should prepare their experiment and make their own reference tables. But Twyman soon felt this was overly time consuming and left too much uncertainty in the hands of the users. After 1918, Hilger proposed to its commercial clients to perform these preliminary studies for them:

> On the purchase of an instrument for any stated purpose, we can in most cases send with the instrument tables for use with them.... Where some new application is in question, we are in most cases prepared to carry through for an agreed fee the experimental work necessary for the establishment of a convenient method. In such cases we supply free of charge a preliminary report on the prospects of the application being successful at the same time quoting a fee for the work.[36]

Further, Hilger offered to help firms choose the right instrumental arrangement for their particular needs. A booklet entitled *Hilger Instruments in Industry* listed different kinds of industries and the best instruments for each of them.[37] The instrument manufacturer continued devising new tools for assisting metallurgical chemists in their choice and use of spectrographs. Hilger's *Spectrographic outfits for metallurgical analyses* listed 'set menus' for various kinds of users. The first outfit 'comprising everything needed for the application of spectrographic analysis to metallurgy and suitable for any materials except those with very complex spectra' listed no less than 17 items for the total price of £324.7s, including a suitable spectrograph, condenser, arc outfit, spark outfit, high purity electrodes, photographic plates, *etc.*[38]

Together with these outfits came two booklets prepared by Twyman, his *Wavelength Tables for Spectrum Analysis* and *Practice of Spectrum Analysis with Hilger Instruments.*[39] The latter was a textbook detailing the procedures left to the operator once the pre-adjusted instrument had been set up and the preliminary studies delivered by Hilger. The *Tables* were a compilation of wavelength tables adapted from Hartley and de Gramont for use as a reference in the laboratory. Twyman's differed considerably from the physicists' comprehensive tables, in that they did not aim to list wavelengths exhaustively but featured only a small number of characteristic lines of the elements necessary for their identification.

In the early 1920s, Twyman also introduced the R[aies] U[ltimes] Powder, a careful mixture of powdered metals whose spectrum was to serve as a reference in the identification of impurities in metals. The powder was supplied with a spectrograph on which the lines were identified, to help the novice teach himself to produce spectra, subsequently to speed up analyses. From 1923 Hilger supplied, on the suggestion of their in-house researchers, the H[ilger] S[pectroscopically standard] Substances, pure substances which gave spectra free from lines due to impurities, also to be used as reference in metallurgical laboratories.[40]

All this information detailing the methods of spectrochemical analysis was conveyed by a stream of articles and books on the subject (most of them authored by Twyman) issued by the Hilger publishing department, amounting altogether to more than a hundred contributions between the two world wars. This literature advertised Hilger products and surrounded customers with a web of technical information and methodological advice, all meant to help the budding spectral analyst.

Twyman and others at Hilger thus devoted much work to transform the analytical methods developed by researchers into quick and reliable routine-like procedures which could be rapidly mastered. Just as he black-boxed the spectroscope, Twyman sought to black-box its manipulation. He was proud to announce that, with Hilger instruments and methods, spectral analysis could be performed by untrained individuals:

> Hilger spectrographs, by the suppression of all but essential adjustments and by careful design, are pre-eminently suitable for carrying out modern spectrum analysis. They have been developed over two decades to be of use to the chemist and metallurgist, who do not want to be hampered in their work by the necessity of acquiring a new and elaborate technique. The work of taking a spectrogram with the Hilger spectrograph can safely be left to an intelligent laboratory assistant. It is a simple, certain and very speedy matter with a Hilger spectrographic outfit to detect the whole of the metallic constituents of any metal or alloy.[41]

In producing and distributing this paraphernalia to accompany their chemical spectroscopes, Twyman was in effect hoping to create a culture of spectrochemistry which could be embraced by industrial chemists, a culture which contrasted with the dominant physical spectroscopy. Indeed, while astrophysicists and physicists came together into well-defined communities to codify and stabilise spectroscopic practices, which were then conveyed by education, the instrument maker looked for ways of sparing analytical chemists the need to come together to solve common problems, or to become too closely involved in spectroscopic experimentation.

Hilger reference material, instruments and literature indeed all contributed to narrow the user's practice of spectral analysis into one pre-determined method which could be rapidly learnt, but did not leave much scope for its modification or elaboration, and made inevitable the reliance of the spectroanalyst on the instrument maker. The instrument maker thereby proposed himself as the provider of tools, material and methodological, of a set of isolated users, an indispensable intermediary translating the work of researchers for analysts dependant upon his know-how. Twyman thus announced the era in which, as Terry Shinn and others have shown, instrument makers became increasingly important components of scientific activity, helping determine the research agendas of scientists keen to justify the investment in an expensive piece of instrumentation.[42] In the context of the metallurgical industries, this signified in the first place the substitution of spectral analysis for wet methods, to be performed equally routinely by the same works analysts, after a period of re-training. Twyman had to show to works owners hesitant to invest in the new (and expensive) instrumentation that spectroscopic methods were at least as good and rapid as traditional ones, and could be rapidly learnt by their employees. Ultimately, though, helped by the increasing automation of spectrographic equipment from the 1940s, the instrument manufacturer opened up the possibility of removing the analysts entirely, of replacing those sometimes referred to as 'mere analytical machine[s]'[43] in view of the routine-like low-status nature of their work, with literal machines. In the somewhat more optimistic words of a metallurgist, spectrochemical analysis would enable the 'freeing [of] hundreds of youths from the boredom, drudgery and monotony of chemical analytical work'.[44]

Routines Adopted

What may be described as the first major public campaign in favour of spectrochemical analysis eventually paid off, and Twyman was satisfied that '[t]he spade work which had been done since 1904 at length began to bear fruit in 1926'.[45] But the widespread adoption of spectral analysis was not all Hilger's doing. It is necessary, to understand Twyman's strategy and the reasons for its success, to take into account the broader context of the inter-war relations between science, government and industry.

Indeed, the First World War was a significant turning point in the history of the British Government's attitude towards science and industry. While the State had previously cultivated an explicit *laissez-faire* approach, as early as 1915 a number of industries were taken over by the Government with the aim of organising, coordinating and rationalising the production of materials in short supply, such as shell steel or chemical glass. But beyond these emergency measures to face the most urgent shortages, the State became involved in the setting up of more permanent organisations for the support of research and cooperation in science and industry, starting with the creation of the Department of Scientific and Industrial Research (DSIR) in 1916. The war-time forced collaboration between firms, and together with the government's new attitude, fostered a climate favourable for the creation, in the immediate post-war period of trade associations, specialised institutes and other such institutions catering to whole industry sectors. One result of this activity was the setting up of such bodies and within individual firms of research laboratories, often with the support of the State.

These new organisations where industrialists and scientists often worked together were propitious for the investigation and the adoption of new techniques such as spectral analysis. The Research Associations for instance, were created by industries themselves at the initiative of the DSIR, who promised to match pound for pound the sums promised by industry sectors for the investigation of projects of potential common benefit. Among the first associations to come into existence were the British Scientific Instrument Research Association (BSIRA) and the Non-Ferrous Metals Research Association.[46]

Twyman, an active member of the BSIRA and of other bodies representing the optical instrument industry, and a supporter of industrial research generally, contributed to enhance the visibility of the instrument industry in the country; and used it as a lever to promote spectrochemical analysis in the metallurgical industry. In particular he succeeded in convincing the Non-Ferrous Metals Research Association to undertake the spectroscopic investigation of metals, a project carried out in part in Hilger's new research department by one of the firm's employees, Donald Smith. The project lasted eight years, and Twyman was satisfied to conclude from it that 'these investigations have had a considerable influence in encouraging the adoption of the spectroscope for routine industrial applications'.[47] Twyman applied similar strategies repeatedly, targeting all relevant organisations, banking on the enthusiasm of industrialists for the scientisation and rationalisation of production to promote spectrochemical analysis.

Success came slowly but surely, gaining momentum in the later 1920s after the giants of the optical industry Carl Zeiss and Bausch & Lomb joined in the movement around 1925. By 1930, Hilger had sold more than 920 quartz spectrographs. From their lists of buyers, it appears that industrial customers were in a large minority, amounting to perhaps a third of total sales, the rest being constituted of research and educational institutions, as well as State, controlling and analytical organisations. The British metallurgical industry was well represented, including The Northern Aluminium Company (West Bromwich), ICI Metals (Birmingham), Improved Metallurgy, Ltd (Avonmouth), Sheffield Smelting Company (Sheffield), Henry Wiggin and Co (Birmingham).[48] Textbooks of metallurgical analysis also increasingly featured spectral analysis, such as William Naish and John Clennell's *Select Methods of Metallurgical Analysis*, which appeared in 1929.[49] In the Second World War, spectral analysis was widely employed, contributing to make the technique, as formulated by Twyman and others, omnipresent. This was shown notably by the fact that the German military indicated the location of the Hilger factory on a London map of crucial places to destroy, in view of their production of spectrographic alloy-testing equipment.

Spectrochemical analysis was thus tested and refined in the 1910s and 1920s, becoming widespread in the 1930s and indispensable in the 1940s. The comparison with the adoption of other forms of chemical instrumentation shows that the rôle played by the Second World War and the subsequent boom of scientific and industrial research for the introduction of instrumentation in organic chemistry was paralleled in the present case by the First World War and the creation of research institutions, which made possible the adoption of spectrochemical analysis in inorganic chemical industries.

Conclusion

This account of the adoption of spectrochemical analysis in the metallurgical industry supports the findings of the historians of the instrumental revolution, in particular concerning the central importance of instrument makers in the process of creating and popularising chemical practices for the use of a family of instrumentation originally associated with the physical sciences. The pro-active *engagement* of Twyman in promoting spectrochemical analysis, when considered together with growing self-awareness of instrument makers in Britain between the two wars, correspond to Terry Shinn's notion of a culture of 'research technology' described in this volume.

In turning the spectroscope into a instrument for routine spectral analysis, Twyman indeed operated a number of translations, of scientific research into routine procedures; of material practices associated with physics into ones acceptable to chemists; and of scientific cultures into industrial ones. In doing so he constantly straddled these different *milieux*, moving incessantly back and forth between them – although the outcome was not always predictable, with physicists enthusiastically taking up his chemical spectroscopes. The instrument maker used his position as a point of contact between the physicists, astrophysicists, chemists and industrialists he supplied to turn himself into a channel for the transmission and translation of practices between these different communities. Twyman and the optical instrument manufacturers certainly illustrate Shinn's idea of 'interstitiality', or the ability to belong to and work with many communities at once.

But Twyman did much more than simply relay techniques from research institutions to industrial firms. He also took an active part in developing methods of spectrochemical analysis, and shaped the field as it emerged. His efforts to build skills into the instruments and the routines he devised were successful in creating a practice of routine spectrochemical analysis requiring little previous training. This made in turn the instrument maker the prime interlocutor of non-scientific users of scientific instruments, raising his status and broadening his sphere of action to an unprecedented extent, notably in his capacity of standard-setter.

But this did not prevent the appropriation of the set-up for a diversity of purposes, making the constant deviation wavelength spectrometer 'generic' in

Shinn's sense. The spectrometer thus displayed in-built possibilities for multiple usage, although in our case, genericity was not so much intended by the manufacturer as imposed by the users.

This tactic however only bore fruit in the context of the inter-war years, during which the introduction of scientific methods in industry and the coordination of state, industry and research institutions accelerated, all developments in which Hilger took an active part. Both the new rôle taken on by instrument makers and the coming into existence of a mass market for scientific instrumentation were made possible by developments resulting from the First World War. The importance of this conflict for the acceptance of spectrochemistry in industry shown here thus serves as a useful complement to the studies which have linked the instrumental revolution in chemistry to the Second World War. The striking parallels between the developments which took place during both wars suggest finally that the emergence of research technologies was not unconnected with the very particular configuration of the network linking the state, science and industry engendered by twentieth century armed conflicts.

Notes and References

1 Y. M. Rabkin, 'Technological Innovation in Science: The Adoption of Infrared Spectroscopy by Chemists', *Isis*, 1987, **78**, 31–54; D. S. R. H. Nuttall, 'Fifty Years of the Hilger Spekker', *Bulletin of the Scientific Instrument Society*, 1987, **15**, 7–9; D. Baird, 'Analytical Chemistry and the "Big" Scientific Instrumentation Revolution', *Annals of Science*, 1993, **50**, 267–290; P. J. T. Morris and A. S. Travis, 'The Rôle of Physical Instrumentation in Structural Organic Chemistry' in *Science in the Twentieth Century*, J. Krige and D. Pestre (eds.), Harwood, Amsterdam, 1997, pp. 715–740.

2 See L. Roberts, 'The Death of the Sensuous Chemist: The "New" Chemistry and the Transformation of Sensuous Technology', *Studies in History and Philosophy of Science*, 1995, **26**, 503–529.

3 G. Urbain, *Einführung in die Spektrochemie*, Steinkopff, Dresden and Leipzig, 1913, p. 8, all translations are mine.

4 H. E. Roscoe, *Spectral Analysis*, Macmillan, London, 1869, p. 110.

5 S. T. Turner, 'Applying a New Science to a New Industry: The Promise of Spectrum Analysis and the Reality of the Bessemer Process of Making Steel', *Rittenhouse*, 1994, **8**, 53–59.

6 F. Twyman, 'Industrial Applications of Spectrography in the Non-Ferrous Metallurgical Industry', *Journal of the Institute of Metals*, 1939, **64**, 379.

7 K. Hentschel, 'The Culture of Visual Representations in Spectroscopic Education and Laboratory Instruction', *Physics in Perspective*, 1999, **1**, 286.

8 For instance the Steinheil catalogues list spectroscopes under the section 'Physikalische Apparate zur Spectral-Analyse, *etc*'. C. A. Steinheil Söhne, *Preisliste über Instrumente für Astronomie und Physik aus der Optisch-Astronomischen Werkstätte von C. A. Steinheil Söhne*, Straub, München, 1894, pp. 36–42; C. A. Steinheil Söhne, *Preisliste über Instrumente für Astronomie und Physik aus der Optisch-Astronomischen Werkstätte von C. A. Steinheil Söhne*, Straub, München, 1900, pp. 12–18.

9 F. Löwe, *Optische Messungen des Chemikers und des Mediziners*, Steinkopff, Dresden and Leipzig, 1925, p. 39.

10 A. C. Menzies, 'Frank Twyman, 1876–1959', in *Biographical Memoirs of the Fellows of the Royal Society*, 1959, 269–279.

11 Anon., 'Obituary Frank Twyman', *Hilger Journal*, *V*, February 1959, 40.

12 The company was incorporated in 1904, with Twyman as its Managing Director until 1946, when it was amalgamated with E. R. Watts and Son. He then became one of the directors of the new Hilger and Watts company until his retirement in 1952. Between 1930 and 1946 he also owned 6/10th of the company's shares. Adam Hilger, Ltd, *Hilger Scientific Instruments 1875–1955*, Hilger and Watts Ltd, London, 1955, pp. 3–9.

13 F. Twyman, 'Reflections on Fifty Years in the Scientific Instrument Industry', *Instrument Practice*, 1950, **4**, 177.

14 See M. E. W. Williams, *The Precision Makers*, Routledge, London and New York, 1994, p. 34.

15 F. Twyman, Mr. Twyman's lecture about Aug. 1944, HILG 3/1, 16. HILG refers throughout to the Hilger archives of the Science Museum Library, followed by the file and folder numbers.

16 F. Twyman, 'Reflections on Fifty Years in the Scientific Instrument Industry', *Instrument Practice*, 1950, **4**, 177.

17 F. Twyman, *On the Development and Present Position of Chemical Analysis by Emission Spectra*, Adam Hilger, Ltd, London, 1927, pp. 13–14.

18 On the physical scientists' culture of precision measurement and its relation with their predilection for the study of instrumentation, see M. Dörries, 'Balances, Spectroscopes and the Reflexive Nature of Experiment', *Studies in History and Philosophy of Science*, 1994, **25**, 1–36.

19 F. Twyman, *On the Development and Present Position of Chemical Analysis by Emission Spectra*, Adam Hilger, Ltd, London, 1927, pp. 13–14.

20 *Ibid.*

21 P. Pellin and A. Broca, 'Spectroscope à Déviation Fixe', *Journal de Physique Théorique et Appliquée*, 1899, S. 3, 8, 19.

22 Wavelength Spectroscope (October 1903), HILG 2/12.

23 Adam Hilger, Ltd, *The Hilger Wavelength Spectrometer (Constant Deviation Type) and Nutting Photometer*, Adam Hilger, Ltd, London, 1919, p. 7.

24 T. H. Blakesley, 'Constant Deviation Spectroscopes', *Proceedings of the Optical Convention*, 1905, p. 54.

25 Frank Twyman, Mr. Twyman's Lecture about Aug 1944, HILG 3/1, 27.

26 G. Guadet, 'Le Prisme Pellin-Broca, Comment l'Agencer? Qui l'a Imaginé?', *Revue d'Optique Théorique et Expérimentale*, 1925, **4**, 493.

27 F. Twyman, 'Reflections on Fifty Years in the Scientific Instrument Industry', *Instrument Practice*, 1950, **4**, 177.

28 Adam Hilger, Ltd, *Catalogue*, Adam Hilger, Ltd, London, 1912, pp. 1–2.

29 F. Twyman, 'Reflections on Fifty Years in the Scientific Instrument Industry', *Instrument Practice*, 1950, **4**, 177.

30 Adam Hilger, Ltd, *Spectrographs for the Ultraviolet*, Adam Hilger, Ltd, London, 1908, p. 2; Brussels Exhibition 1910, (*British Section*) *Catalogue of Mathematical and Scientific Instruments*, London, 1910, pp. 92–93.

31 Main Achievements of Hilger Instruments During the Last Fifty Years (Physical Society lecture 1 July 1944), HILG 3/1, 2.

32 A. F. C. Pollard, 'Notes upon the Mechanical Design of Some Instruments Shown at the Exhibition of the Physical and Optical Societies 1927', *Journal of Scientific Instruments*, 1926–1927, **4**, 187.

33 Adam Hilger, Ltd, *Bulletin of Development Covering the Twelve Months Ending June Thirtieth, 1926*, Adam Hilger, Ltd, London, 1926, p. 3.

34 Von Hippel has pointed to the importance of the user in instrument design. E. von

Hippel, *The Sources of Innovation*, Oxford University Press, Oxford, 1988, especially Chapter 2, pp. 11–27.

35 W. H. Bassett and C. H. Davis, 'Spectrum Analysis in an Industrial Laboratory', *Transactions of the American Institute of Mining and Metallurgical Engineers*, 1923, **68**, 662–669; W. F. Meggers, C. C. Kiess and F. J. Stimson, 'Practical Spectrographic Analysis', *US National Bureau of Standards Scientific Papers no. 444*, 1922–1923, 235–255; A. de Gramont, 'Sur l'Emploi de l'Analyse Spectrographique en Métallurgie: Spectres de Dissociation des Aciers Spéciaux', *Revue de Métallurgie (Mémoires)*, 1922, **19**, 90–100.

36 Adam Hilger, Ltd, *Hilger Instruments in Industry*, Adam Hilger, Ltd, London, 1921, p. 1.

37 *Ibid*, pp. 3–4.

38 Adam Hilger, Ltd, *Spectrographic Outfits for Metallurgical Analyses*, Adam Hilger, Ltd, London, 1930, pp. 8–9.

39 F. Twyman and D. M. Smith, *Wavelength Tables for Spectrum Analysis*, Adam Hilger, Ltd, London, 1923; F. Twyman, *Practice of Spectrum Analysis with Hilger Instruments*, Adam Hilger, Ltd, London, 1931.

40 Adam Hilger, Ltd, *Hilger Spectroscopically Standardised Substances*, Adam Hilger, Ltd, London, 1934.

41 Adam Hilger, Ltd, *The Products and Aims of the Firm of Adam Hilger, Ltd*, Adam Hilger, Ltd, London, 1931, pp. 8–9.

42 T. Shinn and B. Joerges, 'A Fresh Look at Instrumentation: An Introduction' in *Instrumentation Between Science, State and Industry*, T. Shinn and B. Joerges (eds.), Kluwer, Dordrecht, 2001, Ch. 1, pp. 1–13. See also J. K. Taylor, 'The Impact of Chemical Instrumentation on Analytical Chemistry' in *The History and Preservation of Chemical Instrumentation*, J. T. Stock and M. V. Orna (eds.), Reidel, Dordrecht, 1986, pp. 1–13.

43 Cited in J. Donnelly, 'Consultants, Managers, Testing Slaves: Changing Rôles for Chemists in the British Alkali Industry, 1850–1920', *Technology and Culture*, 1994, **35**, 109.

44 Dr H. W. Brownsdon, member of the Institute of Metals, in the discussion following the presentation of F. Twyman, 'Industrial Applications of Spectrography in the Non-Ferrous Metallurgical Industry', *Journal of the Institute of Metals*, 1939, **64**, 394.

45 F. Twyman, 'Main Achievements of Hilger Instruments During the Last Fifty Years', Physical Society lecture 1 July 1944, HILG 3/1, 4.

46 See the *Report of the Committee of the Privy Council for Scientific and Industrial Research*, Eyre and Spottiswoode, London, 1916–1921; R. and K. MacLeod, 'War and Economic Development: Government and the Optical Industry in Britain, 1914–18' in *War and Economic Development*, J. M. Winter (ed.), Cambridge University Press, Cambridge (UK), 1975, pp. 139–203; I. Varcoe, 'Scientists, Government and Organised Research in Great Britain, 1914–16: The Early History of the DSIR', *Minerva*, 1970, **7**, 192–216.

47 F. Twyman, 'Main Achievements of Hilger Instruments During the Last Fifty Years', Physical Society Lecture 1 July 1944, HILG 3/1, 4.

48 Adam Hilger, Ltd, *Hilger Small Quartz Spectrograph*, Adam Hilger, Ltd, London, 1930; Adam Hilger, Ltd, *Some Users of Large Quartz Spectrographs made by Adam Hilger, Ltd*, Adam Hilger, Ltd, London, 1930; Adam Hilger, Ltd, *Some Users of Medium Quartz Spectrographs made by Adam Hilger, Ltd*, Adam Hilger, Ltd, London, 1936.

49 W. A. Naish and J. E. Clennell, *Select Methods of Metallurgical Analysis*, Chapman and Hall, Ltd, London, 1929.

Histories of Baird Associates

DAVIS BAIRD

Son of Baird Associates

In 1936 my father, Walter S. Baird, along with John Sterner and Harry Kelly, founded a scientific instrument company, Baird Associates (hereafter BA). Their first significant product was a transportable three-meter grating spectrograph suitable for the quantitative analysis of metals. Their first sale was delivered to the US Bureau of Mines in 1938, and BA continued to manufacture and sell this instrument well into the 1960s. The company incorporated in 1947. In 1956 it merged with Atomic Instruments, Inc, changing its name to Baird Atomic Inc. In 1978 the name was changed again to Baird Corporation. It was bought by IMO Industries in 1987, and then sold by IMO to Thermo-Jarrell-Ash in 1993, where it now resides.

My father died in 1982, a year after I completed my PhD in philosophy of science. I inherited several boxes full of old documents and objects, which I dutifully stored away to meander through at my leisure. During the course of the 1980s I shifted the focus of my research from the foundations of statistical inference to the philosophy of experiment, particularly the instruments that make experiments possible. Shortly after my mother died in 1987, and I inherited another load of memorabilia, I realised that my father's business was connected to my philosophical research. Baird Associates made the very objects I was trying to understand philosophically! No doubt Freud would have had something to say about my prior blindness to this point. The boxes of memorabilia that I had inherited contained historically valuable documents from the early days at Baird Associates.

These documents are valuable because the history of Baird Associates is a part of the history of a fundamental transformation in science, technology and culture. The 'instrumental revolution', the subject of this volume, has changed forever how science is done, how science and technology interact and the importance of science and technology in culture broadly understood. Issues of capital expenditure and efficiency now are vitally important to science whereas a

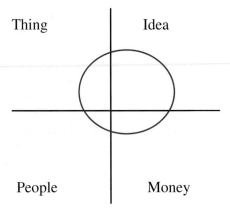

Thing Idea People Money

century ago they were not. Taking a lesson from mass production, automatic, fully 'instrumentalised' methods of analysis have opened up new areas for scientific study. We would not have conceived the human genome project, let alone carried it to completion, without automatic methods of analysis. It is also striking that a private corporation, Celera, competed with publicly funded research to map the genome. The results of the research, at least for Celera, are commodities with which to turn a profit. The instrumental revolution has played an essential rôle in transforming scientific research into a commodity and in eliminating the gap between the pursuit of knowledge and the creation of saleable commodities.

BA did not make the instruments that mapped the genome. But BA, along with other instrument companies as well as academic and governmental laboratories did play a part in the production of our current 'instrumentalised' culture. The company researched, developed, marketed and sold a wide variety of instruments for chemical analysis. Baird and BA have been honoured for their contributions to analytical instrumentation. In 1979 the Scientific Apparatus Maker's Association (SAMA) presented Baird their lifetime achievement award.[1] In 1999, he and BA were honoured as one of the 15 most important corporate and individual contributors to analytical instrumentation at the 50th anniversary Pittsburgh Conference.[2] BA's first grating spectrograph now is in the Smithsonian's Museum of History and Technology.[3]

Baird used to play a game with people. He would draw a pair of Cartesian coordinates and label the four quadrants, 'thing', 'idea', 'people' and 'money'. One was then to draw a simple closed curve on the coordinate plane to describe oneself, presumably thereby to express one's relative interest in these four fundamental categories (figure above).

It is a useful ontology for the study of the instrumental revolution because these are the four fundamental categories involved in this change. New things were made, things that encapsulated ideas and skills previously kept in the exclusive domain of persons. New social networks were developed to create a market for these new things. A new kind of person came to prominence, the research scientist entrepreneur. And all of this demanded a new level of capital expenditure for science, with the result that research has become a commodity.

I am working on a history of BA as a contribution to our understanding of the instrumental revolution of the twentieth century. This is a long-term project, and what I do in this paper is present a collection of stories taken from these boxes of memorabilia, juxtaposing these stories to present the variety and interest of the several histories that encompass Baird Associates. These histories run through Baird's ontology. They include histories of the instruments – the things – BA developed, of BA's corporate culture and of the networks BA cultivated – the people involved – and of BA's financing – money. I close with a discussion of ways BA and the instrumental revolution has changed our ideas of who we are and how we relate to our world.

BA: A History of Devices

Baird Associates went into business '[t]o bridge the gap between the conception of new methods of physical measurement and their practical applications'.[4] They proposed to accomplish this goal by developing 'simple, rugged, accurate, generally usable instruments to make the measurements under consideration',[5] and then to find a market for these instruments making only a 'limited number' of them. Once a market had been established they intended to hand the manufacture of the instruments over to some other firm. BA did not intend to become a manufacturer of instruments.

Financially, this approach was a disaster, for sales of the initial prototype instruments had to carry the entire financial burden of research and development. It took BA a long time to appreciate this economic lesson, as the following case, 15 years into the company's life, makes clear. In 1951 BA contracted with the US Army to make 1519 tank telescopes. BA did not have the finances to underwrite the development of a plant to assemble the telescopes. BA made a deal with another company, American Associates, to provide financing. BA set up American Associates employees in a separate facility, worked all of the bugs out of the design and manufacture of telescopes, and then American Associates underbid BA on the next contract for the telescopes. BA made no net profit on the work, while American Associates made a considerable profit on subsequent contracts.[6] Economic survival forced BA into the manufacturing business. But, it was a lesson learned over a very long period.

While BA is best remembered for instruments used in chemical analysis, they made a diversity of instruments that cannot be characterised under this heading, including a wool dryer for US Customs and metal surface finish standards. The 'associates' characterised themselves as 'industrial physicists'. In 1940 they opened their doors wide for problems or proposals to develop new devices: 'We invite inquiries concerning specialised apparatus made to your specifications. Consult us also on general physical problems'.[7] And people sent inquiries. Here, in a 1940 letter to Walter Baird, is a proposal for a better way to kill and pluck chickens:

When a chicken is struck by lightning or is near a lightning discharge all of the feathers are knocked off clean as a whistle but the meat is not affected. A sudden paralysis of the brain loosens the feather holding muscles... Now, if this blow to the brain can be made sudden and hard enough as in the case of severe electric shock, the feathers are not only loosened, but a subsequent involuntary muscular reaction thrusts them from the body. The problem is to invent a machine which can give this shock under controlled conditions and at not too high a cost.

There might be a series of boxes with spring hinged lids on an endless belt. The chicken is thrust into the box which travels along to a point of contact. Then ... Blitzkrieg! The chicken is dead and de-feathered...

At present I call this apparatus the Blitz Krieg Chicken Assassinator.[8]

BA did not produce this device, and by 1946 they had narrowed their invitation, 'Send us your difficult problems of Analysis, Control or Instrumentation'.[9]

From BA's birth through the 1950s, as BA slowly transformed itself into a manufacturer of (primarily) analytical instrumentation, BA continued to develop a diversity of instruments of which one, two or at most a handful were sold. In 1958, BA Treasurer, Francis Chamberlain, detailed the history of the company's products, including a long list of 'old turkeys'. This is where we find the 'Wool Dryer – made for Customs in Boston – made a few – not commercially acceptable as it took out too much moisture'.[10] Some of the other old turkeys:

1945–1945. Fixed Focus Spectrograph – one for Dr. Saunderson at Dow Chemical...

1946–1947. V-2 Rocket Spectrographic Camera – BA invented and developed – a few were made.

1948–1948. Ultra Violet Monochromator – made two...

1949–1952. Dillon-Murphy X-Ray Projector – invented by Dr Dillon and Dr. Murphy – developed by BA – large expansive lenses made it too expensive – 44 units were sold...

1956–1956. Mixer Amplifier – developed by BA transistor group for broadcasting stations – a dud marketwise.[11]

Old turkeys aside, BA did have its successes, scientifically and to a certain extent financially. BA's best-known line of instruments deployed emission spectroscopy for chemical analysis. Their 1937 spectrograph was an important contribution to quantitative analysis.[12] During the 1930s only eight were sold, but business picked up during the 1940s when another 54 were sold, and the instrument continued in production through the 1960s.[13] In the mid-1940s BA, in co-operation with Dow Chemical, developed a 'direct reading' version of this instrument. Here photo-multiplier tubes and electronics took the place of photographic film and optical densitometers to display virtually immediately the percentage amounts of the various elements of interest in the sample. This was extremely useful in the steel industry where time for analysis was costly.[14] Ten years later BA developed a smaller self-contained version of this instrument that could be used on the foundry floor.[15] The instruments were significant technical

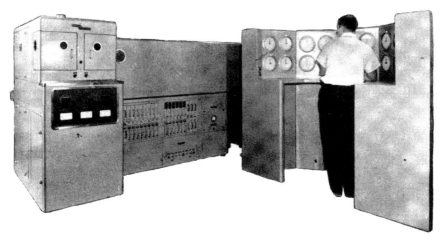

Advertising image of a DR installation, c. 1960. The DR was a 'direct reader', an ultraviolet-visible spectrometer with an automatic concentration readout, hence there was no need to print and interpret spectrograms and line intensities

Walter Baird, John Sterner and Davy Low (first BA employee) inspecting an early infrared instrument, c. 1950

achievements and they contributed to the instrumental transformation of chemical analysis.

BA also developed several instruments that exploited the infrared spectrum. Working with Norman Wright at Dow Chemical and A. H. Pfund of Johns Hopkins University, BA developed a dual beam infrared recording spectrophotometer and a continuous stream gas analyser. The timing was right for these instruments, but BA struggled to compete. At about the same time Perkin-Elmer also developed an infrared spectrophotometer, which, while less precise than BA's instrument, was smaller and cheaper. Over the next ten years BA put its IR instrument through several rounds of improvements in an effort to win market share. In 1958 Chamberlain gave voice to BA's frustrations:

> In 1954 BA stopped making changes long enough to get the cost down and show the only profitable year in the IR line, but thereafter competition induced us to get back in the rat-race of costly changes...
>
> In the meantime, it was clear that Perkin-Elmer, with sales about ten times ours in this field, were not feeling our competition whereas BA was knocking itself out to sell a few.[16]

The continuous stream gas analyser similarly was unprofitable and was discontinued after 1955.

Through the 1950s BA continued to develop products at the cutting edge of current science and technology. They were involved very early in transistor research and sold a line of transistor testers through the 1950s. They also were pioneers in electronic video recording technology, in pattern and character recognition, and in laser technology. Very much in line with BA's initial mission to develop new instruments, and not to focus on manufacture, they established a reputation as research group. In 1962 an internal company report noted that, 'Baird Atomic developed a reputation for being a group of long-hairs who were entirely research oriented'.[17] BA struggled with this. The report continued a page later:

> Accordingly, it was adopted, as an aim of [BA's Research and Engineering Department], that production engineering skills would be developed so that the manufacturing part of a contract would also be given to Baird–Atomic. At that time [mid-1950s] the Company was at the peak of its 'long-hair' reputation and a real effort was required to initiate action toward this goal.[18]

In 1958 Baird bragged in an interview with the Johns Hopkins alumni magazine that BA had 'a laboratory, specialising in physics and electronics, that is better... than ninety percent of the universities in the United States'.[19]

A thorough history of BA has to have the things they made at its core. This is where the company made its biggest mark. This is where the interests of company management were. Treasurer Chamberlain writes:

> Another factor which colours the whole history to date [1958] is the idealism of the founders who believed in quality of things and people above any other economic consideration...
>
> World War II put no fat on the bones of BA but the expansion in the economy that followed brought growth, but not profit. All this time it was hoped that some product would be hit up which would be popular and profitable. Many products were exploited for a short time and a few longer, but in no case was the jackpot opened.[20]

Through its first 20 years BA never hit upon a spectacular success such as Beckman Instruments' pH meter or 'helipot' potentiometer.[21] Many new instru-

The original 'front door' of Baird Associates, and the group of people who were involved from the beginning. Walter Baird is in the middle at the back

ments were made, and much fascinating cutting edge research was pursued. But profits and capital were elusive. It was making things, not money that appealed first to the founders and subsequent management. This is where the heart of the company lay.

BA: A History of Company Culture and Conflict

I write of 'the heart of the company' and 'company culture'. This clearly was not static, but evolved over time, although for many years the founders' socialisation in academic science and engineering played a formative rôle. The transformation from a couple of academic entrepreneurs, when this was virtually unheard of, into a moderately large company with subsidiaries in several other countries, is an important dimension to the history of BA.

At the start in July 1936, BA consisted of its co-founders, Baird, Sterner and Kelly. They rented a building in Harvard Square (20 Palmer Street, where part of the Harvard Coop stands today) and they lived on the top floor. The company's office was on the second floor and the company's shop was on the first floor. Only Baird devoted all his time to BA. Sterner and Kelly kept 'day jobs' to provide capital for the company. For lack of funds, Kelly opted out of this arrangement and out of the partnership in April 1937.

BA consisted of three guys who thought they might be able to have some fun and 'independence of operation'[22] making instruments. In October 1936, Baird was on a trip to Delaware to sell Du Pont an instrument. Kelly was at his day job in Norwalk Connecticut. Sterner, the only associate left in Cambridge, wrote his fellow associates:

Mugs:

This is a joint letter for both of the absent members of the firm of Baird Associates Unlimited Unincorporated Insolvent...

Hope Meester Bear made out all right at Duponts. Are they willing to sell out to us?... Was informed by Sears and Roebuck that we are not listed in Bradstreet and Dun, which is an obvious oversight on the part of either Mr. Bradstreet or Mr. Dun or both. I said neither is Mae West and she's solvent but the credit dep't was unimpressed and promised only to "try" and get our acct through. I doubt if they will, so I looked into the compressor situation. Found an outfit in town named Gerard Electrical which will sell us the equiv. of Sarah Roebuck's compressor for about $60... I have been flirting with Poisseuille's Law [*sic*] and am wondering if 2 1/2 cu. ft. per min. is enough. Harry get out the slipstick and figure: air; nozzle cross-section 3 or 4 mm. (effective x-section); length of nozzle constriction order of 10 cm.; pressure at pump around 100#/in^2 [pounds per square inch]; cap of tank 32 gal.; pumping speed 2 1/2 cu.ft./min; put in a factor of two for safety. Please note that the units are mixed up to mislead you and dont forget the difference between a dyne a pound and a bottle of sloe gin. Please give me answer by mail immediately and get the decimal point right (10 or 0 quiz)...

Nowforchristssakefaithnow [*sic*] will youse guys please get on the ball and don't forget to bring me a tuxedo to get sick all over on Dec. 12 annodominiinthenameofthefatherthesonandthebairdassociatestherats [*sic*]...

[Signed in type] Disillusioned.[23]

Baird's trip to Du Pont sealed their first sale. They delivered their first product, an X-ray tube, one month later for which they were paid $250.[24]

BA continued scraping along for the rest of 1937, 1938 and 1939. The associates were just Baird and Sterner and whatever part-time help they could afford. On 15 December 1939, BA hired its first full time employees, master machinist, Harry Wolff, and a high school graduate, Davy Low, who continued to work for BA until his death in 1983.[25]

With the war in Europe business picked up and this prompted BA's first major change. The company moved. The partners left to residences in the suburbs, and the company moved to larger quarters on University Road (still in Harvard Square). Contemplating the move, Baird wrote in his diary:

The BAs are just about ready to move.... The whole philosophy of the business is changing from an intimate personalised contact with customers and employees we are reaching a more formal attitude. We hope to keep the relationship between the customers and ourselves; we cannot be too sure about the employer relationships. For the first time there will be a wall between the office and the shop. Naturally we'll do our best to retain as much of the old relationships as possible but the old slap happy days are gone.[26]

Two and a half weeks later the company, with all nine employees had moved.

An aerial view of the Baird Associates factory, Cambridge, Massachusetts, c. 1950

As BA grew Baird and Sterner had to focus on developing new business and this meant both developing new instruments and selling them. It also meant managing. In April 1941, BA developed regulations governing vacations and sick leave.[27] In May 1942 BA formalised shop and office procedures covering the receipt of orders, scheduling jobs, ordering materials, receiving materials and invoices and handling time cards.[28]

At the close of 1942 BA had 13 employees. By March 1943 BA had more than 60 employees. There was stress. In August 1943, Baird responded to a written set of concerns from BA staff:

> The question of policy six months ago and policy now is indefinite to a certain extent. There is no doubt that each of us has grown up considerably in the last six months, and because of that, outlooks have changed in people as well as in the group. It is a fundamental policy of the BAs that this be true, not only now but for the indefinite future. The BAs are, and I hope will always be, a group which will tackle any problem fundamentally from an experimental point of view. The more things we try, the better are we able to determine wheat from chaff. It is expected of every member of the business that he make an attempt to grow with the business, and that means not only in his particular activity, but in all related activities.[29]

Two aspects to the company emerge from this memo. First, clearly the staff had concerns about how things were run. Second, the company sought to maintain the atmosphere of a research lab. Problems should be tackled 'fundamentally from an experimental point of view'.

The company was incorporated in 1947, and this brought further changes. Still, there remained a certain level of kidding around; the agenda for the first formal directors meeting of the newly incorporated Baird Associates required '[s]ide arms and flasks to be on the table'.[30] Ten years later BA had more than

450 employees at the University Road plant and two affiliates. Without elaborating, it is clear that the company that had begun as a kind of academic lab had transformed into a moderately large organisation requiring careful attention to structure and management. While still involved in research on instrumentation, their stated goals in 1955 were '1. To achieve a good profit. 2. To grow'.[31] Company culture had to change as the company itself changed. Both to understand BA and to understand the culture BA integrated itself into, we need to understand the development of the company's culture.

As a History of Social Networks

One aspect of Baird Associates that makes it an interesting company to study is the manner in which BA navigated competing demands from the various constituencies that BA both depended on and served. BA sat at the intersection of several communities with divergent cultures. BA's founders were trained to do scientific and engineering research. BA drew from this culture for its intellectual and commercial livelihood. BA's market consisted of industrial, governmental and academic laboratories employing people trained in wet chemical analysis, not instrumental analysis. BA and the other new instrument making firms had to create demand for their new approach to analysis. BA's commercial livelihood also required BA to participate in the capitalist world, struggling for market share and profits. By the late 1940s roughly a quarter of BA's gross income came from contract work done for various military branches of government;[32] BA was a part of the young but voracious 'military–industrial complex'. These different domains put significant strains on the way business was conducted.

The most significant fact about the market for the instruments BA sold is that it did not exist when the company was founded. There was some use of prism spectrographs for chemical analysis, but this was not routine. The metals industries by and large did not use spectrographs to analyse their product or 'the melt' during production. Thus Baird Associates, along with the other firms that appeared about this time, had to create this market. In part this was a process of establishing new social networks that would connect instrument makers with people in industry who were in a position to see the value to the instruments being developed.

Elsewhere I have written about both the Massachusetts Institute of Technology [MIT] summer spectroscopy symposia organised by George Harrison and the Pittsburgh Conference.[33] MIT's summer spectroscopy symposia commenced in July 1932 and ran every summer until 1941 when the war intervened. These symposia created networks necessary for establishing a market in spectrographic instrumentation. BA's first spectrograph sale resulted from their participation in the 1937 conference.

After the war a variety of other venues for the exchange of information on chemical instrumentation appeared. Undoubtedly the most successful of these was the Pittsburgh Conference ['Pittcon']. From modest beginnings (250 registrants) the conference has grown to enormous proportions: the 1998 conference had 28 118 attendees and 1217 exhibiting companies.[34] Pittcon's growth attests to its success in connecting the various constituencies involved in the instrumental transformation of analysis. Analysts, whether from the private or public sector, could exchange information on their needs and desires with instrument makers and researchers. Baird Associates was one of 14 companies to exhibit at the first, 1950 Pittcon, and has continued to exhibit at this conference, at least up to and including 1999.[35]

Individuals who both understood analytical needs and instrumental capabilities were well placed to serve as an intellectual conduit for the exchange of information between analytical labs and instrument makers. Mary Warga, a 1937 PhD from the University of Pittsburgh with a specialisation in emission spectrochemical analysis, was a key figure in connecting industrial laboratories with new instrumental techniques. During World War II she trained nearly all of the laboratory workers in the Pittsburgh steel mills in spectrographic analysis.[36] With the formation of the Pittsburgh Conference, she took an active rôle in connecting instrument makers with industrial laboratory personnel.

Pittcon and the MIT spectroscopy conferences and people like Mary Warga helped establish markets for spectrographic instrumentation. BA's connection with these people and institutions was essential for finding, indeed creating, buyers for their wares.

I close this section with two stories that illustrate the conflicting cross-currents that BA navigated. The Naval Research Lab contracted BA to develop instruments to go in the nose cone of the V2 rockets that had been captured from Germany at the end of World War II. The idea was to get spectrographic information about the upper reaches of the atmosphere and the constraints in doing so were demanding. The instrument had to fit inside the limited space of the rocket and operate automatically. It couldn't be heavy, but it had to be rugged enough to withstand impact from 60 miles of free fall. Baird described the work in his 1979 SAMA Award speech:

> White Sands [Missile Range, New Mexico]. The war is now over. No longer do we have WPA, priorities, ration books, and the blackout...
>
> One of the spoils of the war was a collection of 25 to 50 V2 rockets. These, for some reason, came under the control of the Navy. We worked for the Navy...
>
> Learning to launch the V2s was not easy. Several blew up on the pad. Some wouldn't go at all. One took off and landed near Alamogordo. One ended in Mexico.
>
> The dream then was to use these rockets as a platform to carry instruments to measure what went on above our most dense atmosphere. The tension was exciting. Everyone involved knew we were opening up a new world...
>
> ...Gather in the control center at the launch time. Wait with stop watches in hand. We listened for the boom. Sound travels 1100 feet per second at 2500 feet. BOOM!
>
> And we were off in any vehicle which could go over the desert. We find the debris.
>
> Later we find that we have the spectrum of the ozone layer fifty miles above the earth.[37]

In his unpublished memoirs Baird describes several of the difficulties in more detail. Among those who had worked at White Sands for a long time one rocket was known as 'Sir Launchalot'. Five or six times Sir Launchalot had been brought to the launching pad but failed to launch, once in front of 600 of Washington's 'big brass'. Sir Launchalot was a hybrid. We were learning about rocketry by examining and duplicating the parts in the German rockets. As soon as one of our parts was replaced by the original German part, eight or nine years old, Sir Launchalot finally took off: 'It was just absolutely gorgeous, and that was humiliating'.[38] We still had a lot to learn. Our parts were not good enough yet.

The second story, 10 years later just post-Sputnik, concerns importing apparatus from the Soviet Union for teaching physics. At the time BA was affiliated with the Ealing Corporation, an importer of scientific apparatus:

> One day Ealing's president, Mr. Grindle, saw on the cover of *Physics Today* a picture of a Russian student working on a strange piece of equipment. Within days he was in Moscow… He was shown through a number of warehouses full of teaching apparatus, mostly for physics.
>
> He placed an order for ten of each of some fifty different items. The shipment arrived, was priced out, freight, customs. And when marked up 100 percent was still so inexpensive, even our 'build it yourself' couldn't compete.
>
> A show was set up. All the concerned people in Cambridge [Massachusetts] were invited. Selling prices were on each apparatus. Some of the professors working on the new teaching program PSSC practically cried. The equipment was the center of interest at the annual Physical Society show.
>
> Then came the backfire. Remember!! The cold war was still on. Sales agreements with a number of USA companies for Baird equipment were cancelled. Others threatened to. Not long after the whole project just died.[39]

Several of the devices ended up among the toys with which my father hoped to excite my interest in science.

Among other things, these two stories illustrate the tension between the international or 'universal' character of science[40] and the rôle science had come to play in issues of national security. Should it have mattered that BA, through Ealing Corporation, was importing inexpensive, engaging and effective teaching apparatus from the Soviet Union? Was the work done for the Office of Naval Research done simply to gather new, previously inaccessible information about the ozone layer? In both cases the universal character of science, where our search for understanding nature transcends national and social boundaries, comes into conflict with science's rôle as tool of national sovereignty. BA tried to play both sides of this fence, with limited success.

BA: A History of Financing

BA needed money and this always was in short supply. When Baird got the first order from Du Pont he took it to the local bank seeking a loan to buy the necessary materials and machinery:

I was turned down cold. Next Stop – The Household Finance Company. There I obtained a personal loan (36 per cent a year).[41]

'The discovery that he needed collateral ... was one of the great shocks of his life'.[42] BA continued to depend on loans from friends and friends of friends and finance companies. Many times a cheque would arrive and be spent the same day to pay off creditors and employees.[43] When they were producing just a few instruments each delay in finishing, delivering and being paid for the instrument hurt. Sterner wrote a friend, Stan Walters (who subsequently came to work for BA), in December 1937:

This delay bitches up our finances but we have things fairly well under control, except for Barbour Stockwell who is getting pretty impatient (the amount hanging over from last July and we have to depend on them for much of the work on the spectrograph). So if you still run across someone who wants to lend $200 for two months grab him and offer him plenty of interest.[44]

Walters himself lent the firm money. But even he had to wait for payment:

Dear Stan: You probably are calling me all kinds of pretty names for not sending you the money for your next installment. Answer is 'we ain't got it'. We are again in the midst of another of our bottle-necks.[45]

It took BA six years to repay Harry Kelly the $500 he had contributed to the partnership between July 1936 and April 1937.[46]

The first important financial backer of BA was Herbert White, who had owned and had run the University Press of Cambridge, Massachusetts. By 1940 White had retired and the University Press Building was vacant. BA sought to rent the space, but they could not afford to pay for it. White took a liking to Baird and Sterner. He fixed the rent to 4% of BA's gross sales. This arrangement continued until White's death in the mid-1960s. White also agreed to lend BA their first significant capital, $6000, for which he received 6% interest.[47]

During the late 1930s, through mutual friends, Baird and Sterner met Georges Doriot, a professor at the Harvard Business School. Doriot had an unusual interest (for the time) in small businesses. After the war, in 1946, Doriot joined forces with Karl Compton, then President of MIT, and Merrill Griswold, an investment banker. Together they created the American Research and Development Corporation (AR&D). The purpose of AR&D was to provide capital for small high technology firms in the Cambridge/Boston area. AR&D was one of the first attempts to promote high technology companies through investment funds, an early experiment in venture capitalism. BA was AR&D's third (February 1947) investment, $225 000 in exchange for BA stock. Coinciding with BA's incorporation, this was the first major infusion of capital into the company.

Government contracts also were essential to BA's survival. As of March 1947, roughly a quarter of BA's gross income ($130 000) came from research and development work done on contract with the government. In addition to BA's V2 rocket work, BA developed tourmaline crystals for high-speed electronic switching and highly sensitive bolometers for the Office of Naval Research. A full history of BA financing must include a discussion of the contract work the firm did for the Government.

BA worked to change the character of the company from a research lab focused on exciting new science and technology to a manufacturer with a mission to make a profit. This was a struggle as the previous 1962 reference to BA as a 'bunch of long hairs' suggests. In 1958 Baird told the Hopkins alumni magazine this,

> We decided to emphasize the sales end of the business.... I don't know how other managers feel about this kind of thing, but from my point of view the only way to change the spots is to change the animal, too. When I became chairman of the board we brought in a new young fellow, Dr Davis R. Dewey, II, ... to be the president. His attitude very definitely is oriented toward the marketing side of things, though he is trained in sciences, too'.[48]

This, unfortunately, did not turn out to be a happy change of management. By the early 1960s Baird and Dewey were not speaking to each other. Davy Low, BA's first employee had to hand carry written messages back and forth between the two leaders of the firm. In 1963, after a three-month suspension of both executives and an investigation by an outside consulting firm, Dewey was fired and Baird reinstated as president.[49]

The history of financing is central to the history of BA, and to the history of chemical instrumentation. One of the lessons that this history teaches is that, perhaps ironically, much of this history has little directly to do with chemical instrumentation and a lot to do with chance, intellectual curiosity and business manoeuvres. BA ended up pursuing a great diversity of business opportunities at some distance from chemical instrumentation. We see the same result in the recent biography of Arnold Beckman and the instrument firm he founded.[50] Tidy academic analytical categories, such as chemical instrumentation, do not well map onto the messy history they attempt to denote.

BA: Ideas of Science, Technology and Culture

The evolution of BA's name itself tells a tale. BA began as Baird Associates and became Baird Associates, Inc when the partnership incorporated in 1947. The idea behind the name, 'Baird Associates' was literal. Baird Associates was a group of people associated with Walter Baird. Entries in Baird's diary through the early 1940s refer to 'the BAs': the 'Baird Associates'. The idea was to form a group of people with scientific and engineering skills who were interested in making new instruments. In 1956 the name changed to Baird Atomic Inc when the company merged with Atomic Instruments, Inc. Two factors drove this choice. BA had come to be known by its initials, 'BA', and for some time the firm had promoted a marketing connection between its nickname and the phrase 'better analysis'. Thus, 'Baird Instruments, Inc' or 'Analytical Atomic, Inc' or some other variant would not have worked.

In 1978 the 'Atomic' was dropped. In the 1950s 'atomic' had positive connotations. It reminded us of how technology, and especially nuclear technology, had won the war against fascism. It made us think of a future with unlimited and cheap power, of a modern future where science and technology provided rational solutions to our problems. By the 1970s the valence for 'atomic' had changed. Now the word reminded us of nuclear war. Nuclear winter was on people's minds, and the counter-culture movement of the 1960s had tarnished the idea of rational solutions provided by science and technology.

These changes were difficult for my father. I know this personally for my own quiet teenage rebellion put me at odds with him over just these issues. He concluded his 1979 SAMA acceptance speech reasserting his confidence in science and technology:

> I have great faith in technology – past, present, and especially, future. Too many people and too vocal don't know how to count their blessings.
>
> I have yet to run into a scientist or engineer who would spend his effort, knowingly, in a way to harm himself, wife, children, or another. The Doomsday Sayers, called by any name, usually have never had a course in physics, much less economics. Technical problems can be solved but, like everything else, at some cost.[51]

He went on to suggest that we use the space shuttle to send our nuclear waste 'back to the Sun from whence it came'. 'Based on our safety record in space travel, it should be very good – much better than for building a bridge, a dam, or mining coal'.[52] We should build, he said, 10 or 20 nuclear power plants on the glaciers of eastern Alaska or western Canada: 'Power lines are inexpensive compared to wars, pipelines, and delays. I don't think many weekend "protesters" will make a trip to the construction site. They couldn't find the place, even if they flew over it'.[53] Again, leaning on Alaska, he proposed to put any two-time felon in an unfenced colony near Fort Yukon: 'I think I'd like to hire any person who could find his way to "civilization"'.[54] Finally, he complained about Government regulation:

> I am reminded of Atwater Kent in the 1920s. Faced with continued strikes and demands, he said, 'I've had enough', closed his plants, leaving thousands out of work.
>
> Is it impossible, improper, to consider a major shutdown of business to show the 'public' sector from whence comes their bread and butter?[55]

Losing the 'atomic', losing the public's faith in science and engineering was hard for Walter Baird and many of the engineers and scientists who felt 'they' had won the Second World War and provided post-war prosperity. My generation didn't know how to count our blessings. Perhaps this is true, but only partially so as Three-Mile Island, Chernobyl, and the Challenger space shuttle demonstrate. Nonetheless Baird Atomic became Baird Corporation and instrument making companies adapted to new cultural norms.

At a deeper level we can see profound changes brought about by the ascend-

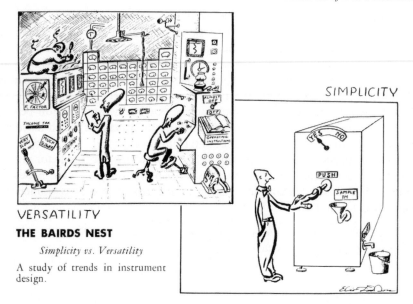

THE BAIRDS NEST

Simplicity vs. Versatility

A study of trends in instrument design.

Simplicity and versatility

ancy of instrumental measurement and control. The cartoon above appeared on the last page of a 1951 Baird Associates magazine, *Better Analysis.*[56]

The cartoon contrasts two ideals for instrumental analysis. Versatile analysis requires several operators with huge brains fiercely concentrating. One of the operators is deeply aggravated because something in the complex machine is not working. By contrast, simple analysis requires one calm and contented operator. There is one button to push and a simple 'yes' or 'no' output. The other message is that instruments should be 'push-button simple' to operate. They should provide unambiguous answers to questions posed (notice there is no output in the versatile instrument and one can either turn it 'off' or 'almost off'). They should be quick and financially efficient (notice that the versatile instrument either operates 'very slow' or 'much slower' and has an 'income tax' input). Human judgement should not be required. The operator of the simple instrument has a much smaller brain than his versatile counterparts.

Here is an ideal for how we should solve our problems. Roughly speaking, we should be able to submit the problem to an instrument, push a button, and have the instrument tell us 'yes' or 'no'. Costs should be reduced, for human judgement is expensive. On its face this seems an obvious ideal for instrumental analysis. But what do we sacrifice? Versatility according to the cartoon. What does this mean? In a sense we sacrifice a fine level of discrimination with subtle options that require careful and difficult human judgement. In exchange we get simplified cost-efficient analysis. Instruments, not humans, make the discriminations and judgements. Cost considerations might require us to build less versatile or precise instruments than instruments operated by 'big-brained humans'. But the sacrifice in precision is more than compensated by the gain in reduced cost and efficiency.

Observations based on a single cartoon are not sufficient to sustain a general claim about the rise of an ideal for instrumentation. Elsewhere I have argued this point at greater length.[57] Here I only aim to suggest that the instrumental revolution of the twentieth century prompted quite deep changes in our understanding of how we should relate to our world. We came to new ways of trading off costs against precision. The things made by these instrument makers changed our ideas and our way of solving our problems. If we want a deep understanding of these changes we have to consider the ideals of the people that made these things and the forces they worked with in their corporate environment.

The last observation I offer about BA's history concerns two competing notions of research. An idealistic notion of research with a long pedigree sees research as a joint and shared effort to understand the world and our place in it. Its aim is not primarily utilitarian and its products are not commodities with which to profit, but gifts shared among the community of inquirers. Robert Merton labels this the 'communistic' character of science.[58] Universities evolved from monasteries, communities of people held together in faith. The university has been a community of inquirers held together by faith in the goodness of finding out about the world and in the adequacy of their methods to accomplish this end. Academic researchers typically have not been paid for their contributions and where they have been paid the 'academic value' of the contribution is diminished; textbooks are worth less than research monographs. This notion of research runs counter to the emerging notion of our 'information-based economy' that the results of research are commodities with which profits can and should be made. Witness Celera Corporation. Witness the legal strains currently engulfing intellectual copyright and information-based or conceptual-based patent law.[59] During the nineteenth century the ethos of science held that no 'true man of science' would seek a patent on his scientific discoveries. Not so now. Edison, entrepreneur capitalist, was not a true man of science and his contemporary H. A. Rowland quipped of Edison's electric light that the 'spark of Faraday blazes at every street corner'.[60]

BA's difficulties finding capital and making profits arose out of the struggle between these two notions of research. At the outset anyway, one can understand the nature of BA well by seeing it on analogy with academic research. In the academy, by and large, the published paper is the unit of research progress. At BA the working prototype instrument was the unit of progress. BA sought to extend the community of academic scientific research to the private sector. This ran afoul of the practical need to support this work by manufacturing, marketing and selling. Baird and the other instrument making pioneers of the mid-twentieth century were unusual because they were research scientist entrepreneurs. The title of the alumni magazine piece about Baird is 'Businessman-Scientist'.[61] Now the direction of cultural imperialism runs the other way. A cutting edge conceptual category for university professors is 'academic entrepreneur'. BA tried to move the academic model to business with mixed success. Now the business model is moving into the academy.

The instrumentation revolution must shoulder part of the responsibility for the incursion of the business model on the academy. Instrumentation research aimed at making new things rather than asserting new propositions developed in the private sector where the academic model didn't work. Marketing, profit margins, competition and capital were powerful forces for turning instrumentation research into commodities. Another step turns academic research into commodities. Reflecting in 1975, Baird wrote:

> [C]ertain things could have been done better. They could have been done with more of an idea with respect to money. But here was a period that was absolutely exciting in terms of ... producing new and interesting stuff. ... I guess the difference, looking backwards, is that I was much more interested in science and the improvement of science and what science could do than I was interested in making money. Now that may sound kind of peculiar but, nevertheless, I think if you go back over all these years and look at all of our annual reports, you will find that somehow each year we ended with a little bit of plus and a hell of a lot of excitement.[62]

Concerns about market-share and money are exciting too. But they offer a different kind of excitement than the improvement of science and finding out what science could do. This difference marks the struggle between the 'communistic' and the 'capitalistic' notions of science that we have inherited from the instrumentation revolution. We also see here a shift in my father's 'ontology curve' from a curve with a predominance in 'thing' and 'idea' to one with a predominance in 'money' and 'people'.

Notes and References

1 Scientific Apparatus Maker's Association, 1979 Award, Scientific Apparatus Maker's Association, Phoenix, Arizona, 1979.
2 J. Poudrier and J. Moynihan, 'Instrumentation Hall of Fame', *Made to Measure*, supplement to Today's Chemist at Work and Analytical Chemistry, 1999, 10–38.
3 D. F. Walsh, 'The History of Baird Corporation: A Broad Perspective on the Progress of Industrial Spectroscopy', *Applied Spectroscopy*, 1998, **42**, 1336–1350.
4 Baird Associates, Statement of Organization and Aims, Baird Associates, Cambridge, Massachusetts, 1938.
5 *Ibid.*
6 F. Chamberlain, *Baird–Atomic, Inc: Principal Products, Past and Present (1936–1957)*, Baird–Atomic, Cambridge, Massachusetts, 1958.
7 Baird Associates, *Grating Spectrographs, 1940: Bulletin #5*, Baird Associates, Cambridge, Massachusetts, 1940, 1–4.
8 F. Brinser, correspondence to Walter S. Baird, unpublished, 1940.
9 Baird Associates, *Advertising Bulletin #22: Spectrographic Source Power Unit*, Baird Associates, Cambridge, Massachusetts, 1946.
10 F. Chamberlain, *Baird–Atomic, Inc: Principal Products, Past and Present (1936–1957)*, Baird–Atomic, Cambridge, Massachusetts, 1958.
11 *Ibid.*
12 D. Baird, 'Baird Associates Commercial Three-Meter Grating Spectrograph and the Transformation of Analytic Chemistry', *Rittenhouse*, 1991, **5**(3), 65–80.

13 Baird Corporation, *Installations of Optical Emission Instruments*, Baird Corporation, Bedford, Massachusetts, n.d.

14 D. Baird, 'Encapsulating Knowledge: The Direct Reading Spectrometer', *Foundations of Chemistry*, 2000, **2**(1), 5–46.

15 D. Baird, *Thing Knowledge: A Philosophy of Scientific Instruments*, University of California Press, forthcoming.

16 *Ibid.*

17 Baird Atomic, Inc, *Review of Research and Engineering Department Operations for the Board of Directors of Baird–Atomic Inc*, Baird–Atomic, Inc, Cambridge, 1962.

18 *Ibid.*

19 L., R., *Businessman-Scientist*, *The Johns Hopkins Magazine*, 1958, **9**, 11–22.

20 F. Chamberlain, *Baird–Atomic, Inc: Principal Products, Past and Present (1936–1957)*, Baird–Atomic, Cambridge, Massachusetts, 1958.

21 A. Thackray and M. J. Myers, *Arnold O. Beckman: One Hundred Years of Excellence*, Chemical Heritage Foundation, Philadelphia, 2000.

22 L., R., *Businessman-Scientist*, *The Johns Hopkins Magazine*, 1958, **9**, 11–22.

23 J. Sterner, correspondence to Walter S. Baird and Harry Kelly, unpublished, 1936.

24 Baird Associates, financial statement by Willard Helburn, Baird Associates, Cambridge, Massachusetts, 1936.

25 Baird Corporation, 'David A. Low 1915–1983', *Spectrum*, 1983, **21**(3), 1; Baird Atomic, Inc, *1939–1969: The Years at University Road*, Baird Atomic, Inc, Bedford, Massachusetts, n.d.

26 W. S. Baird, diary, 1941.

27 Baird Associates, 1941, #496.

28 Baird Associates, Tentative Shop and Office Procedures, Baird Associates, Cambridge, Massachusetts, 1942.

29 W. S. Baird, Staff Memorandum, Baird Associates, Cambridge, Massachusetts, 1943.

30 W. S. Baird, Report to Directors of Baird Associates, Inc, for Board Meeting April 3, 1947, Baird Associates, Cambridge, Massachusetts, 1947.

31 Baird Associates, Inc, The Operation of Baird Associates, Baird Associates, Cambridge, Massachusetts, 1955.

32 Baird Associates, Inc, *History, 1947*, Baird Associates, Inc, Cambridge, Massachusetts, 1947b, 1–3.

33 D. Baird, 'Analytical Chemistry and the "Big" Scientific Instrumentation Revolution', *Annals of Science*, 1993, **50**, 267–290; D. Baird, *Thing Knowledge: A Philosophy of Scientific Instruments*, University of California Press, forthcoming.

34 J. Wright, *Vision Venture and Volunteers: 50 Years of History of the Pittsburgh Conference on Analytical Chemistry and Applied Spectroscopy*, Chemical Heritage Foundation, Philadelphia, 1999.

35 *Ibid.*

36 F. A. Miller, 'In Memoriam: Mary Elizabeth Warga 1904–1991', *Optics and Photonics News*, March 1992, 6.

37 W. S. Baird, acceptance speech: 1979 Scientific Apparatus Makers Association Award, not published; in the possession of Davis Baird, 1979.

38 W. S. Baird, memoirs, unpublished, 1975.

39 W. S. Baird, acceptance speech: 1979 Scientific Apparatus Makers Association Award, not published; in the possession of Davis Baird, 1979.

40 R. K. Merton, 'The Normative Structure of Science', *The Sociology of Science: Theoretical and Empirical Investigations*, N. Storer (ed.), University of Chicago Press, Chicago, 1973, 267–278.

41 W. S. Baird, acceptance speech: 1979 Scientific Apparatus Makers Association Award, not published; in the possession of Davis Baird, 1979.

42 L., R., *Businessman-Scientist, The Johns Hopkins Magazine*, 1958, **9**, 11–22.

43 Baird Corporation, 'How Baird Began: The Men, The Idea, The Struggle to Succeed', 1986, *Spectrum*, **26**(1), 6–7.

44 J. Sterner, correspondence to Stan Walters, unpublished, 1937.

45 J. Sterner, correspondence to Stan Walters, unpublished, 1938.

46 H. Kelly, correspondence to Walter S. Baird and John Sterner, unpublished, 1937; Baird Associates, correspondence to "Whom it May Concern" with handwritten notes detailng repayment of loan from Harry Kelly, unpublished, Cambridge, Massachusetts, 1939–1943.

47 H. H. White, correspondence to Baird Associates, unpublished, 1941; J. Sterner, personal interview, 1990.

48 L., R., *Businessman-Scientist, The Johns Hopkins Magazine*, 1958, **9**, 11–22.

49 J. Sterner, personal interview, 2000.

50 A. Thackray, and M. J. Myers, *Arnold O. Beckman: One Hundred Years of Excellence*, Chemical Heritage Foundation, Philadelphia, 2000.

51 W. S. Baird, acceptance speech: 1979 Scientific Apparatus Makers Association Award, not published; in the possession of Davis Baird, 1979.

52 *Ibid.*

53 *Ibid.*

54 *Ibid.*

55 *Ibid.*

56 Baird Associates, Inc, *Better Analysis*, Baird Associates, Inc, Cambridge, Massachusetts, 1951, **2**, 1–12.

57 D. Baird, 'Analytical Instrumentation and Instrumental Objectivity', in *Of Minds and Molecules: New Philosophical Perspectives on Chemistry*, N. Bhushan and S. Rosenfeld (eds.), Oxford University Press, Oxford, 2000, 90–113.

58 R. K. Merton, 'The Normative Structure of Science', in *The Sociology of Science: Theoretical and Empirical Investigations*, N. Storer (ed), University of Chicago Press, Chicago, 1973, 267–278.

59 S. Shulman, *Owning the Future*, Houghton Mifflin Company, Boston, 1999.

60 A. D. Moore, 'Henry A. Rowland', *Scientific American*, 1982, **246**(2), 150–161.

61 L., R., *Businessman-Scientist, The Johns Hopkins Magazine*, 1958, **9**, 11–22.

62 W. S. Baird, memoirs, unpublished, 1975.

Production Control Instruments in the Chemical and Process Industries

STUART BENNETT

Department of Automatic Control & Systems Engineering, University of Sheffield, Mappin Street, Sheffield S1 3JD, England, Email: S.Bennett@sheffield.ac.uk

Introduction

In a major petrochemical complex, hundreds of production control devices are used and it is arguable that without the precise control of operations such controllers provide, many chemical production processes would not be feasible. Many types of production control devices are used in the chemical industry, but of the ones which utilise feedback control over 95% are the three-term (Proportional, Integral and Derivative) control devices, or more briefly, PID controllers. They are controllers which produce as a control output a summation of values proportional to the error, the integral of the error, and derivative of the error.

The concept of PID control emerged, in several disparate fields, in the 1920s, but practical industrial controllers were largely the result of work in two major industrial instrument companies, the Foxboro Company and the Taylor Instrument Companies. The first industrial PID controllers were implemented during the late 1930s using pneumatic technology; during the 1950s and 1960s electronic versions of the controller were introduced and since 1980 the majority have been implemented using digital computers. The story of the PID controller can be told as the history of a concept and how instrument companies and users adapted to different technological implementation of the concept. But there is also another story: developing the technology necessary to implement effective automatic control was not a simple task, so why was it done? The advertisements of the manufacturers of industrial instruments, particularly in the USA and in the first forty years of the twentieth century, point to the influence of Taylorism and the

Scientific Management movement. Managers determined the 'one best way' and automatic controllers implemented it without question.

And what of the companies who made the provision of instruments to industry their major business? Did they occupy the space categorised by Terry Shinn as 'research-technology'? They clearly saw themselves as providing what now would be termed 'enabling technology', that is they supported and enabled the use of other technologies. They also saw themselves as selling solutions: 'dictate a letter telling us your problem' said a 1925 advertisement of the Taylor Instrument Companies, 'and we will find you a solution'. By the second half of the century these solutions were integrated systems of measurement, display, recording and control. Inevitably these solutions were technology driven since the companies wanted to sell products as part of the solution.

Background

During the latter part of the nineteenth and early part of the twentieth century complex changes were taking place in the ownership and control of industry and commerce, changes which impacted on the organisation of work and working practices. Work had to be managed as part of a 'system' and the 'system' determined not only what work should be done but how it should be done: Frederick Winslow Taylor's 'the one best way'. Advertisements for instruments which could monitor what workers did – 'its presence causes careful firing and steady steam' claimed one – had appeared in journals such as *Engineering Management* from the late 1890s but numbers and frequency of such advertisements increased following the publication of Taylor's views and the growth of the Scientific Management movement.[1] In 1914 the Bristol company ran a series of advertisements containing a picture of a bowler hatted supervisor and with captions such as 'keep track of work the Bristol way', 'now he sees why that last heat cost more'; and in 1924 the Taylor Instrument Companies promoted their temperature controllers as providing the 'sixth sense of industry' which could remove the uncertainty 'that always exists when human senses are gambled on'; using instruments 'guesswork is eliminated'.[2]

In the USA, the industrial market for indicating, recording and controlling instruments grew rapidly from about 1919 to 1929 and, even during the Depression years, although the absolute value of instrument sales declined, sales relative to general manufacturing investment increased.[3] The sales figures covered a wide range of instruments and it is not possible to make a precise assessment of relative sales of the various types, but it is clear that the predominant requirement in industry was for instruments for indicating, recording and controlling temperature. The high profile of temperature measurement was a reflection of the size of the metal processing industry in the United States in the early decades of the twentieth century. In the period 1923–1929, sales to the chemical industry increased by ninefold and to the petroleum industry by fivefold, but although it is not possible to identify the variables being measured it is clear that in these industries temperature, pressure and concentrations are important quantities which influence the yield and speed of operation. This rise in the use of instruments coincides with the introduction of many continuous processes which required, for efficient operation, precise control of process variables.

Independent of the type of variable being measured, the majority of the sensors in use produced either a small change in electromotive force (emf) or a small change in pressure; the technology used to build indicating, recording or controlling instruments was thus constrained by the need to amplify these small changes and to make the changes visible. In both cases the first stage of the process was to convert the emf or the pressure change into mechanical movement – by means of a galvanometer for the emf and by a sensitive bellows or modifications of the Bourdon tube techniques for pressure changes – and this in itself was sufficient to provide an indicator instrument. Recording, particularly for movement obtained from a galvanometer, was more difficult since the drag of any marking device on paper would distort the measurement (photograph techniques were not suitable for routine production measurements). The Callendar recorder, manufactured by the Cambridge Scientific Instrument Company, which used a servo-mechanism to link the pen to the movement of the galvanometer needle, was an effective and reliable laboratory instrument. However, it was insufficiently robust for industrial use and the most widely used recorder of this type was the one devised by Morris E. Leeds and produced by the Leeds & Northrup Company from 1911 until the 1950s. William H. Bristol with his pressure recorder of 1889 had in essence solved the recorder problem for instruments based on small pressure changes, although extensive detailed engineering work was necessary on choice of materials, methods of winding helical tubes and of precisely centering the tube, before accurate and reliable recorders could be manufactured on a large scale.[4]

The Bristol and Foxboro companies are examples of instrument companies which, from their inception, saw their instruments as being used for industrial processes rather than for laboratory use. By way of contrast, the Taylor Instrument Companies, formed by the merger of companies which had manufactured instruments such as liquid-in-glass thermometers for laboratory and general use, is an example of an established company which saw the potential of the industrial market.

Understanding the Control Problem

In the chemical industry, as in other industries, the purpose of control is normally to hold some measurable quantity of a process (temperature, pressure, flow rate, pH value) constant when the process is subject to unpredictable (in value) disturbances. Disturbances can be of three types: random, for example, due to variations in the compositions of the raw materials or to minor ambient temperature changes; a result of a change in demand from an upstream or downstream process – such changes are referred to as load changes and typically require a change in the operating point of the process; and slow, small changes arising from drift in the process behaviour. A controller with proportional action can deal effectively with the random disturbances but load changes and slow drift require integral action (in the chemical and process industries integral action was called 'reset').

A controller combining proportional and integral action was patented by Morris Leeds in 1920 but the Leeds and Northrup Company did not succeed in building a practical controller of this type until 1929. Controllers incorporating the two forms of action were built by the Dow Chemical Company from about 1923 onwards for internal use. Both these companies used electric controllers and the control action was obtained using two electric motors and mechanical linkages, and application was limited to processes for which slow control action was effective.

The practical controllers being sold between 1905 and 1930 were all controllers with on–off action because the only practical form of power amplification available was the relay or switch. On–off control action is simple and cheap to implement but it gives rise to 'hunting', that is the quantity being controlled oscillates between limits set by the dead-band of the relay.

Most of the early on–off controllers were electric because the pneumatic relay tended to stick and thereby load the measuring system. Through Edgar H. Bristol's invention of the flapper-nozzle amplifier in 1914, the Foxboro Company was able to build fast acting pneumatic controllers and also combined recorder-controllers. Pneumatic devices were attractive to many industries since their supporting infrastructure required mechanical skills such as pipe fitting and metal working which were readily available; they also had the advantage of intrinsic safety in explosive atmospheres, an important advantage in the rapidly growing petrochemical industry. However, by the late 1920s, if they were not to lose their market to companies producing electric controllers, they needed to develop a controller with at least proportional action, but preferably proportional plus integral (reset) action.

In a publicity brochure issued in 1929 the Foxboro Company argued, defensively, that the adoption of 'throttling action', that is proportional action, would reduce the accuracy of the controller, but behind the scenes they were working to produce such a controller. First public signs of the activity were two patent applications – one in the name of Clesson E. Mason and the other of William W. Frymoyer. Mason was an application engineer working in the Tulsa, Oklahoma, office of the company and Frymoyer was Works Superintendent at the Foxboro plant. In 1929, Mason moved to the main offices in Foxboro and became head of control research in 1930, and in September of that year filed a patent for a pneumatic controller which incorporated a negative feedback link from the output back to the input of the flapper-nozzle amplifier. The effect of the feedback link was to convert the very high gain non-linear flapper-nozzle amplifier into a stable, linear amplifier. The device also had additional pneumatic elements which provided reset actions and thus it provided PI (Proportional, Integral) control action. It is interesting to note that this negative feedback amplifier was developed simultaneously but independently from the now much more widely known electronic negative feedback amplifier of Harold Stephen Black.[5]

The Foxboro Company launched the new controller in September 1931 under the name *Stabilog*. Sales were poor at first: 1932 was the worst year of the

A Foxboro PID controller, c. 1946

Depression for sales of industrial instruments and furthermore, the customers did not understand the new controller. Re-launched in 1934, it was accompanied by a brochure explaining in simple, non-mathematical terms how the controller worked and why it gave much better control. Rival companies were quick to see the benefits of the new control method: the Taylor Instrument Companies brought out its so-called *Dubl-Response* unit in 1933 and the Tagliabue Company responded in 1934 with its *Damplifier* controller. Taylor's challenge was the most significant as the *Dubl-Response* unit incorporated a feedback link from the position of the main control valve, hence including all of the controller components within the feedback loop.

PI control is excellent at compensating for the actual difference between the set point and measured value (proportional action) and for, over a period of time, correcting a persistent error, but it could not compensate for a potential error. Human operators controlling a process were known to anticipate changes in the measured variable and to adjust the process to compensate for these anticipated changes. The need for anticipation arises, for example, when there is a delay in an instrument registering the change in temperature because of thermal insulation between the sensor and the source of the temperature change. Ralph Clarridge of the Taylor Instrument Companies found a way of introducing anticipatory action into a pneumatic PI controller during 1935 and the company added it as a special order when their engineers thought it necessary for a particular application. It appeared as a standard option, called 'pre-act', in the company's 1939 version of the *Fulscope* controller. The Foxboro Company developed their version of anticipation, which they called 'hyper-reset', during 1937–38 and it became part of the standard *Stabilog* controller around 1940.

Development of Theoretical Understanding

Prior to the 1930s, academics and the major professions paid little attention to the development of process controllers: it was, perhaps, this lack of attention which led A. Ivanoff to write that 'the science of the automatic regulation of temperature is at present in the anomalous position of having erected a vast practical edifice on negligible theoretical foundations'.[6] Ivanoff's statement is correct if we interpret 'theoretical' as implying the modelling and analysis using differential equations, but we should not be misled into thinking that the 'practical edifice' had no foundations: it was constructed on what we now might call 'intelligent control', that is heuristic control based on observation of the human operator.[7] Inventors such as Morris Leeds and Elmer Sperry (and many others) had an intuitive understanding that on–off and proportional control actions were not adequate. Some of the manufacturers were aware of the need for a theoretical basis for design, '[N]otwithstanding our activity and success', reported I. Melville Stein of Leeds & Northrup, in 1929, 'we have never made a comprehensive study of the general problem of automatic control'.[8] The language of the report is non-mathematical, just as is that used by John Grebe and his colleagues in 1933 when explaining control of chemical processes,[9] but both demonstrate a clear and thorough understanding of many of the problems of automatic control. Mason in his 1930 patent used differential equations to explicate the operation of his invention and the explanation of the operation of the *Stabilog* given in the Foxboro brochure of 1934, although not in mathematical form, demonstrated an understanding of the principles.

What Ivanoff identified was not a lack of understanding but a lack of an agreed set of concepts and language with which to categorise and discuss the behaviour of automatic control devices: what was the connection between engine governors, servo-mechanisms, automatic steering of ships, automatic stabilisers on aeroplanes, telephone repeater amplifiers, pneumatic and electric process control devices? They are now all classed as 'feedback control devices' but the word 'feedback' was then a neologism, appearing in 1920, in *Wireless Age*, to describe undesirable parasitic connections in a wireless amplifier, that is what is now called 'positive' feedback.[10] A detailed theoretical justification and a formal mathematical description of a PID controller had been given by Nicholas Minorsky in the *Journal of the American Society of Naval Engineers* in 1922.[11] But why would an engineer working in the instrument companies or the process industries have looked at a paper on the automatic steering of ships in this journal? Why, also, would any instrument engineer (or patent lawyer working for them) have thought that the filing, in 1928, of a patent application with the title 'Wave translation system' was relevant? Manoël de Mayo ('Major') Behar writing in *Instruments* in 1930 observed that not only was 'the body of knowledge … widely scattered' but also that what was known 'was jealously guarded' and that which was 'made public has been not only fragmentary but highly particularized and narrowly applicable'.[12]

The first drawing together of important ideas from several sources came in

Major Behar. Reproduced from Instruments and Automation, (*August 1958*), *p.1351*

1934 with the publication, in the *Journal of the Franklin Institute*, of Harold Locke Hazen's paper on the design of fast acting servo-mechanisms for the differential analyser and other analogue computing devices.[13] Hazen had worked with Vannevar Bush and, in trying to develop a theory to support the design work, had carried out a thorough survey of the available literature on control systems which he tried to codify.[14] Douglas Hartree, then at the University of Manchester was a regular visitor to MIT where Hazen's work was done and he arranged for a differential analyser to be built at the University of Manchester. One of the first uses of this machine was to simulate the effects of time-lags in process control systems: work done in collaboration with A. Callender, of ICI. Details of this work were given in a paper at a 1936 conference on automatic control organised by the Chemical Engineering Group of the Society of Chemical Industry.[15]

The major work, however, began in the same year with the push, led by Ed S. Smith, to form an Industrial Instruments and Regulators Committee of the American Society of Mechanical Engineers (ASME).[16] Prior to this initiative most information relating to industrial instruments and their use appeared in the journal *Instruments*, whose editor, Major Behar, was an enthusiastic and tireless proponent of the use of automatic control. Smith actively sought contributions for publication in the *Transactions of ASME*.[17] Full recognition of the importance of instruments in science and industry came in 1942 when the American Association for the Advancement of Science chose the subject of instrumentation for one of its Gibson Island conferences; attendance at these conferences was by invitation only and no proceedings were published. A list of attendees and speakers reveals that this meeting should have facilitated an interchange of ideas and methods between the different groups interested in automatic control.[18]

Consolidation 1940–1955

By 1940 the value of the PI and PID controller had been demonstrated but much still needed to be done before it was widely adopted by industry. There were three main issues: how to find the appropriate settings for the controller (having a simple means for adjustment in the field was useless if there was no easy way of finding the best settings); persuading designers to produce plants which were controllable; and re-design of the controller to make it less dependent on complex and fragile mechanical linkages.

The first issue was quickly addressed. In 1942, J. G. Ziegler of the Sales Engineering Department and N. B. Nichols of the Engineering Research Department of the Taylor Instrument Companies set out two procedures for finding the appropriate controller parameters.[19] They assumed that the dynamic behaviour of typical processes could be characterised by three parameters, the gain (K), the so-called time constant of the process (T) and the time lag of the process (L). They gave procedures for finding the values of the above parameters (system identification) and some rules, the so-called Ziegler–Nichols rules, for finding, from the parameters, the proportional gain and integral and derivative action times which will produce a particular behaviour (performance criterion) from the process. These rules were simple in concept, but in practice they were not easy to apply because the settings obtained depended not only on the plant characteristics but also on the interaction between the derivative and integral actions of the controller. Controllers manufactured by different companies had different forms of interaction (and some none at all). For example, in 1951, in a detailed analysis of some commonly used controllers, A. R. Aikman and C. I. Rutherford of ICI identified five principal types of interaction.[20]

The growing awareness of the power of the frequency response approach led in the early 1950s to investigations of its use in process control applications.[21] For example, ICI built a frequency response analyser to obtain plant data and to examine how frequency response ideas could be used to find controller parameters.[22] The academic work was summarised and explicated in two papers by Geraldine A. Coon published in *Control Engineer* in 1956.[23] An interesting feature of Coon's work was that it was based on simulations run on an analogue computer and that she investigated controller behaviour for a variety of process characteristics. Most control text books still give the Ziegler–Nichols rules for tuning PID controllers, however, there are over one hundred other rules: the additional rules arise from setting different performance requirements and from making different assumptions about the plant model.

A second, less well known, paper by Ziegler and Nichols appeared a year later in which they comment that too often process plants do not work as expected. The engineers realise that some factor has been neglected but cannot identify what is missing. 'This missing characteristic' they argue 'can be called 'controllability' – the ability of the process to achieve and maintain the desired equilibrium value'. Their argument was that instrument and process lags can make a plant difficult to control and close attention must be given to minimising such lags.[24] Controllability did not emerge in theoretical work until the 1960s but Ziegler

Advertisement for the Bristol Company, 1953. Reproduced from Review of Scientific Instruments, *July 1953, vol 6, p. xi*

and Nichols were not alone among instrument engineers in recognising its practical importance and in realising that the location of instruments to reduce measurement lags was an important factor in determining the controllability of a plant.

The third problem was more difficult to deal with, particularly as engineering efforts and resources were diverted to the war effort. After the end of the war the leading companies, Foxboro and Taylor, made minor changes to the existing designs, improving the mechanics and the methods for adjusting the controller parameters, but major changes did not appear until the 1950s. The Foxboro Model 58 *Consotrol* range which appeared in the early 1950s was the result of a major re-design in that it incorporated a clever force balance arrangement.[25] This was not the first force balance-type controller – for example, the Leeds & Northrup Company's pneumatic *Micromax* of 1944 used a force-balance arrangement – but this was the first such instrument from the leading pneumatic controller company.

During the same period the Taylor Instrument Companies introduced their replacement for the Fulscope range, *Transet Tri-act* controller: the principle of which had been described by Ralph E. Clarridge in 1950.[26] The *Tri-act* contained two flapper-nozzle amplifiers: the first amplifier had a fixed proportional gain and a variable pre-act setting, the output of which was then fed to a second stage with variable proportional gain and variable reset action. Clarridge argued that 'conventional' controllers gave a large overrun during start up; by introducing the two stage process, it became possible to adjust the reset (integral) gain without affecting the derivative action. This enables the controller to be tuned to give good performance in respect of both load disturbances and set-point disturbances, whereas with the majority of previous controller implementations the controller had to be tuned for either load disturbances or set-point disturbances. The Clarridge structure is still found in many digital controllers today.

During the 1950s another feature began to appear in pneumatic controllers, namely anti-wind-up. Under certain process conditions, the integral term can become very large; in process terminology, it 'winds up'. The phenomenon is often caused by some part of the system becoming saturated and when the cause of the saturation is removed the wind-up effect results in a very slow return to normal operation. Industrial instrument manufacturers were aware of the phenomenon and incorporated devices to reduce its effect, but the exact nature of these appears to have been treated as a trade secret.[27]

Electronics

A Department of Scientific and Industrial Research report, in 1956, observed that the 'pneumatic type' of automatic controller 'is technically the most advanced and many reliable designs are available' with over 90% of the market, but that there was a growing interest in electrical and electronic controllers.[28] One explanation for the interest was the expectation that the techniques developed during the 1930s in the telephone industry which had applied successfully for gun and radar control during 1939–1945 could be applied in industry. But there are also deeper and more subtle reasons. Cross-fertilisation which occurred during and immediately after the war produced a realisation that changes in pneumatic pressure, voltage, current and the movement of mechanical linkages were all conveying information: they were, in the language used by communication engineers, 'signals'. Obtaining plant measurements, using the measurement to find the control action and transmitting this action to the plant is 'signal processing'. By using electrical analogies for mechanical and pneumatic operations, design techniques developed by electrical engineers could be used, and a common language for control began to take shape. This resulted in a greater insight into process dynamics and the dynamics of measuring systems, controllers and actuators, and the realisation that, by converting the pneumatic signal from the measuring system to an electrical quantity, the signal could then be processed using electrical devices. Manipulations in the electrical domain were much less constrained by the physical restrictions inherent in using pneumatic and mechanical components.

Instrument companies had been experimenting with and producing controllers incorporating electronic amplifiers since the late 1930s, and A. J. Young, writing in 1955, described six electronic PID controllers.[29] By 1957, G. P. L. Williams of the George Kent Company was able to report that the new electronic instruments were capable of performing all the functions previously available only with pneumatic instruments: these included, in addition to PID, the ability to carry out 'addition, multiplication, squaring and other mathematical operations'.[30] He also noted that the instrument manufacturers were fully aware of the possibilities of transistors, and new products using transistors were being developed. There had been deep suspicion among process engineers about

Modern control room, 1956. Reproduced from Instruments and Automation, (*October 1956*), *p. 2007*

the reliability of electron tubes and it was the appearance of transistorised instruments such as the Foxboro *Consotrol* range, in 1959, that made electronic PID controllers acceptable to the industry.

Prior to the arrival of the transistorised controller, the main driver for the adoption of electronics had been the manager's desire for information: 'know at your desk what's happening ... throughout the plant'. 'The plant' was the way in which the Bristol Company expressed this in 1936,[31] but by the 1950s a large petroleum refinery could have '500 controllers, ... 1500 indicators and 800 recorders', central control rooms with display panels '100 to 200 feet long', and with hundreds of instruments overwhelmed operators with information.[32] Electronics seemed to offer a cheaper means of bringing information from the plant to a central location but it also offered a means of selecting critical information, for example, variables which were out of limits; summarising information; and easy connection to newly developing digital read-out and logging systems.[33] In 1955 the journal *Instruments* introduced a regular section on digital automation and process logging systems based either on a digital computer or on technologies associated with digital computing and during 1959 *Instruments & Control* carried descriptions of 67 digital data logging systems. Between 1955 and 1959 discussions on how digital computers might be used for industrial process control and then descriptions of their use there began to appear in the literature.[34]

Automatic control was firmly established in a wide variety of industries with the chemical and petroleum industries being two of the leading users (in the USA between 1947 and 1957 these industries accounted for almost half of the total sales of instruments to manufacturing industries).[35] However, these were largely independently controlling separate parts of the plant and overall performance of the plant still depended on human operators adjusting operating settings for the plant. In complex plants even the best operators could not achieve consistently good performance.

Digital Computers

The digital computer offered the prospect of being able to manipulate large quantities of data in short periods of time and the Cold War provided the leverage to extract, particularly in the USA, public funds for the development of digital computers designed specifically for on-line control. Such computers, intended for use in airborne control systems, began to appear by 1953. These computers differed from the computers designed for data processing in that they provided interrupt mechanisms and a much wider variety of interface units. In addition the Cold War focused attention on mathematical and computational methods for solving certain types of optimisation problems. By the latter part of the 1950s both computing technology and computer-based optimisation methods were sufficiently developed for some industrial companies to consider using them. Chemical and petrochemical companies were in the vanguard of this change, although the major drive came from the Ramo-Wooldridge Corporation who, in 1957, approached the Monsanto Company with a proposal to apply an RW-300 computer to control a chemical plant. The company, which had formed a systems engineering section in 1955 to investigate control techniques, was receptive to the idea and went ahead with a supervisory control system for an ammonia plant. It chose an ammonia plant because the process was well known so that disclosing details of operation to Ramo-Wooldridge personnel would not reveal major process secrets; a reasonable improvement in performance would provide a return which might justify the cost of the installation; and the plant was modern and thus well instrumented.

This project took over three years to complete and no figures were released with regard to economic performance. There was a report that the additional instrument costs needed for computer control were above the initial estimate and that the computer 'makes new demands on the accuracy of such measurements. In manual operation, absolute accuracy of measurement has never been deemed a major virtue since it is usually more important to hold consistent values of variables than to know the true values. Computers, however, may contain relationships requiring true and accurate information...'.[36] The 'relationships' referred to are those in the mathematical model of the plant which was used as a basis for calculating optimum settings and both the accuracy of measurements and the accuracy of the mathematical model are limiting factors in all production control systems which seek to compute optimal settings.

The first industrial plant on which closed loop control by digital computer was achieved was the catalytic polymerisation unit at Texaco's Port Arthur, Texas, plant on 15 March 1959. This system used supervisory control, that is the computer replaced the operator functions, not the existing automatic controllers which controlled the individual control loops on the plant. It was designed to minimise the reactor pressure and to find the optimal distribution between the feeds to the five reactors, to use the measurement of catalyst activity to control the hot-water input flow and also to find the optimal re-circulation. In doing the above the computer system supervised 72 temperatures, 26 flows, 3 pressures and 3 compositions.

Other early installations, as in the above, simply added a digital computer onto existing analogue controllers and the computer either directly adjusted the various set-points for the controllers or advised the operators on the settings. ICI at its soda ash plant at Fleetwood took a radical new step in that the complete analogue control system for the plant was replaced by a single Ferranti Argus 200 digital computer. The digital computer directly read the measurements from 224 sensors and controlled 129 valves. The name Direct Digital Control (DDC) was coined to emphasise the fact that the computer controlled the process directly. The system went live in November 1962 and ran for three years.

This was not the first DDC system since the public mention of the project in 1961 prompted engineers at the Monsanto Chemical Company to quickly investigate direct digital control and, in collaboration with TRW [Thompson–Ramo–Wooldridge] Computers, they installed a trial DDC system on an ethylene unit at Monsanto's Texas City plant which went live in March 1962 but which was run on a trial basis for only three months.[37]

Interest in computer control grew rapidly and by 1971 there were forty-one manufacturers of process control computers. The established industrial instrument companies responded rapidly to the threat to their markets. As early as 1960, Bailey, Foxboro (with RCA), Leeds & Northrup (with Philco) and Minneapolis–Honeywell were offering computer-based systems. The existing instrument companies had the advantage not only of knowing the market and having a good understanding of process control, but also of having developed interface technologies such as analogue-to-digital signal conversion for use with the data logging systems which they began to introduce in the mid-1950s. However, they all lagged in the market behind TRW in terms of the number of installations until the mid-1960s.

What proportion of these installations included DDC is difficult to ascertain: however, judging by the number of papers which began to appear in the late 1960s, the inclusion of DDC in computer control schemes must have been increasing. The advantages of DDC, it was argued, were that the cost-per-loop did not increase linearly with the number of loops as with analogue control, and also that making modifications only involved changing the code and not rewiring as was required with analogue systems. There were, however, concerns about the reliability of digital computers and many of the early DDC schemes included back-up analogue controllers for critical loops, and electronic PID controllers were quickly modified to provide automatic change-over to analogue back-up should the digital computer fail to update the controller output within a specified time interval. This strategy was strongly promoted by the established instrument manufacturers since it maintained their market for analogue devices. The established companies also developed special process control languages for programming the digital computer: these languages enabled the user to set up the system as if it were an analogue system. The cost of the process computers in this period (1960–1970) was high and could only be justified for installation on large plants.

The development at the end of the 1960s of the large-scale integrated circuit (LSI) opened up new possibilities: 'by 1975' predicted Anthony Turner of Motorola, in February 1970, 'LSI circuits will probably be the basis of digital computers', continuing 'analogue controllers should gradually evolve into digital devices, providing accuracy at low cost' and they 'will be relatively simple to combine into multipoint configurations, which can be applied to optimise unit processes on a local basis'.[38] He proved to be right for, in 1975, Honeywell's Process Control Division announced its Total Distributed Control Architecture (TDC). This provided a communication network to which instruments and small computers could be connected. No longer was it necessary to bring many cables back to a central control room: dual co-axial cables provided the ring along which all the data could be transmitted and computer screens replaced the walls of instrument displays.

Direct digital control raised questions about algorithm implementation and tuning and many practical features built into pneumatic and electronic controllers had to be re-discovered. The flexibility of software implementation meant that the non-interacting form of the PID algorithm could be used but issues of bumpless transfer, set point changes, and integral action wind-up had to be faced, as did tuning to take into account the effects of sampling. There were also additional complications arising from limited precision arithmetic. These problems were solved by companies such as Foxboro, Leeds & Northrup and others. They also, during the 1980s, attacked the problem of tuning the controller by developing controllers which tuned themselves and by developing 'intelligent' tuning aids and self-tuning controllers.

The early work on self-tuning PID controllers was done at the Foxboro Company by Edgar Bristol (a grandson of William H. Bristol) and led to the EXACT system introduced in 1984. The EXACT controller measures certain features of the transient response of the process when a large disturbance occurs and adjusts the controller parameters on the basis of heuristic rules applied to these measurements. Other self-tuning controllers were introduced during the latter part of the 1980s which used model-based re-tuning methods, that is they use system identification techniques to continually update the parameters of a model of the process and the tuning parameters are calculated from the model characteristics. These types of self-tuners can be used on processes with unpredictable variations; a simpler approach (gain scheduling) can be used when variations in performance can be predicted, for example, when the operating point of a process is changed, the dynamics may also change and require different controller settings. In one hundred years we have moved from a simple on–off control device to one which can observe the changes in the dynamic behaviour of the process it is controlling and adjust its control action in the light of these observations.

Modern computer control room. Courtesy of R. Maisonneuve, Publiphoto Diffusion/Science Photo Library

Conclusion

The PID controller is just one of the many forms of controller which have been, and continue to be, used in the chemical and process industries. Its importance lies in its universality. Correctly tuned it gives excellent control for a very wide class of processes; badly tuned it still provides some level of control. It can be implemented in a range of technologies: the dominant technology for the first sixty years of the century was pneumatics and now microprocessor-based implementations are almost universal. The concept of this particular form of control lies in the behaviour of a human operator and this form of control was observed in applications as widely varying as steering a ship and control of the temperature of a steam-generating boiler. The difficulty for the first forty years of the twentieth century was how to implement this form of control action given the technologies available.

The question arises that if the human operator was effective why seek a technological replacement? The answer is twofold. Firstly, the change in the organisation of the firm with the growth of a professional management class, which was strongly influenced by Taylor and 'Scientific Management', led both to a demand for instruments (indicators and, preferably, recorders) which told the managers what was happening and also for automatic controllers which could be set to provide the 'one best way' to do the job. In this sense the automatic controller is 'socially constructed'. But there were other factors: efficiency of operation, particularly for new processes, and changes in processes used to manufacture chemicals and other products resulted in operations which could not feasibly be controlled by the unaided human being. Thus the second reason for developing automatic controllers was to facilitate new technological systems.

The leading manufacturers of industrial instruments were influenced by both factors; they were not simply instrument manufacturers but were systems engineering companies. They made and sold instruments but they also sold solutions to problems. 'Put your problems up to us' was the invitation in a Taylor Instrument Companies catalogue of 1926, 'take us into your confidence ... since many applications call for special treatment of existing conditions and we must know these conditions to handle your requirements intelligently'. Field engineers worked closely with customers and became aware of, and expert in, a wide range of measurement and control problems. They communicated information about problems, often with suggestions for solutions or details of improvisations which they had made, to the head office of the company. They also carried out field trials of ideas produced by engineers working in the research departments. They were, to use a modern term, 'system integrators' and as such were keenly aware that all elements in the system, process, measuring system, controller and actuator had to match. Their understanding of chemical and other processes and their ability to sell complete solutions enabled them to dominate the market even if their products did not contain the latest technology.

The instrument companies also knew who made the investment decisions. Much of their advertising was aimed at persuading production managers that spending money on instruments would given them more control. 'Not only is a plant process itself enhanced by computer control' argued R. Blake, the sales manager for Kent Automation Systems, 'but the overall management also becomes more efficient ... by having to hand up-to-the-minute information on plant status'.[39] This process continues: several companies now offer software which uses the internet to provide real-time access to instruments located on plants anywhere in the world, managers can find out, from their desks, what is happening and can even change plant settings.[40] But were products designed and developed to meet managers needs? Did managers want centralised control rooms? Did they want large expensive computer systems? Did managers make use of the information generated? Did they use the controllers to ensure that plants were operated efficiently? Or were managers and the instrument companies simply feeding off each other in an endless cycle of novelty? It was clearly in the interests of both groups to portray the new systems as being successful and better than the old ways. It is only in the last decade that it has begun to be widely accepted and recognised that information is not the same as knowledge and that what managers need (and want) is knowledge: as yet the digital computer cannot generate this knowledge without human assistance.[41]

The companies, through their technical brochures and through the contributions of their staff to trade and professional journals, did much to educate users about the use of PID control. However, they also retained much knowledge within the companies. In the academic community the PID controller has received little attention until recently. There was the assumption, in the 1960s and 1970s, that all the problems had been solved and it was only when a leading control engineer, Karl Åström, took an interest in the topic, and thus made it respectable, that interest grew. The humble PID controller which had served the process industries so well for 60 years had, for the first time in its history, an international conference devoted solely to it in 2000.[42]

Do the several thousands of process instrument companies fall into Terry Shinn's category of 'research-technology'? The key features of research-technology, he argues, are genericity, interstitiality and metrology.[13] Process instrument companies certainly qualify on the grounds of genericity in that they developed key components and methods that could be applied in many industries and for different measurements. Trade catalogues often, on the surface, seem to offer narrow niche products, but detailed comparison of the instruments reveals that much is identical apart from the sensor element. They also qualify on the test of interstitiality. Practitioners moved between manufacturers and users and between the professional worlds of the physicist, chemist and engineer. The ASME offered the subject a home with the formation of an Industrial Instruments and Regulators Committee in 1936 and supported work on instrumentation for many years. In the UK, after much discussion and negotiation between professional bodies representing physicists and engineers, a separate professional body, the Society of Instrument Technology, was formed in 1944.[44] And with regard to metrology, one of the first acts of the ASME Instruments and Regulators Committee was to try to establish standards and an agreed nomenclature. On the basis of Shinn's criteria, the products of the leading companies and the activities of their professional staff (and some of the professional staff of the companies using industrial instruments) would seem based to fall within the scope of research-technology.

Notes and References

1 F. W. Taylor, *Principles of Scientific Management*. Harper & Brothers, New York, 1913. Taylor's views had first appeared in 1903 in a paper, 'Shop management', *Transactions of the American Society of Mechanical Engineers*, 1903, **24**, 1337–1456.

2 *Engineering Management*, 1924. See also S. Bennett, 'The Industrial Instrument – Master of Industry, Servant of Management': Automatic Control in the Process Industries, 1900–1940', *Technology and Culture*, 1991, **32**, 69–81.

3 S. Bennett, 'The Use of Measuring and Controlling Instruments in the Chemical Industry in Great Britain and the USA During the Period 1900–1939', in *Determinants in the Evolution of the European Chemical Industry, 1900–1939: New Technologies, Political Frameworks, Markets and Companies*, A. S. Travis, H. G. Schröter, E. Homburg and P. J. T. Morris (eds.), Kluwer Dordrecht, 1988, pp. 215–237.

4 S. Bennett, 'The Development of Process Control Instruments 1900–1940', *Transactions of the Newcomen Society*, 1992, **63**, 133–164; P. H. Sydenham, *Measuring Instruments: Tools of Knowledge and Control*, Peter Peregrinus, Stevenage, 1979; A. J. Williams, 'Bits of Recorder History', *Transactions ASME, Journal of Dynamic Systems, Measurement and Control*, 1973, **1**, 6–16.

5 S. Bennett, *A History of Control Engineering 1930–1955*, Peter Peregrinus, Stevenage, 1993; D. A. Mindell, 'Rethinking Feedback's Myth of Origin', *Technology and Culture*, 2000, **41**, 405–434.

6 A. Ivanoff, 'Theoretical Foundations of the Automatic Regulation of Temperature',

Journal of the Institute of Fuel, 1934, **7**, 117–130, disc. 130–138.

7 K. M. Passino, 'Bridging the Gap Between Conventional and Intelligent Control', *IEEE Control Systems Magazine*, 1993, **13**, 12–18.

8 I. M. Stein, Notes on General Theory of Control – 385, Leeds & Northrup Company, 1929, in Hagley Museum & Library, Leeds & Northrup Papers, Accession No. 1110, Reel #7, Volume 11.

9 J. J. Grebe, R. H. Boundy and R. W. Cermak, 'The Control of Chemical Processes', *Transactions of American Institute of Chemical Engineers*, 1933, **29**, 211–255.

10 *Oxford English Dictionary*, 2nd edition, 1989. It is ironic that the words 'feedback' and 'feedback control', unless qualified, now imply negative feedback.

11 S. Bennett, 'Nicolas Minorsky and the Automatic Steering of Ships', *IEEE Control Systems Magazine*, 1984, **4**, 10–15; N. Minorsky, 'Directional Stability of Automatically Steered Bodies', *Journal of the American Society of Naval Engineers*, 1922, **34**, 280–309.

12 M. F. Behar, 'The Engineer's Handbook of Industrial Instruments', *Instruments*, 1930, **3**, 425–462.

13 H. L. Hazen, 'Theory of Servomechanisms', *Journal of the Franklin Institute*, 1934, **218**, 283–331.

14 L. Owens, 'Vannevar Bush and the Differential Analyzer: The Text and Context of an Early Computer', *Technology and Culture*, 1986, **27**, 63.

15 A. Callender and A. B. Stevenson, 'The Application of Automatic Control to a Typical Problem in Chemical Industry', *Society of Chemical Industry, Proceedings of Chemical Engineering Group*, 1936, **18**, 108–116. See also A. Callender, D. R. Hartree and A. Porter, 'Time-lag in a Control System', *Philosophical Transactions of the Royal Society of London*, 1936, **235**, 415–444; and D. R. Hartree, A. Porter, A. Callender and A. B. Stevenson, 'Time-lag in a Control System – II', *Proceedings of the Royal Society of London*, 1937, **161**, Series A, 460–476.

16 S. Bennett, 'The Emergence of a Discipline: Automatic Control 1940–1960', *Automatica*, 1976, **12**, 113–121.

17 E. S. Bristol and J. C. Peters, 'Some Fundamental Considerations in the Application of Automatic Control to Continuous Processes', *Transactions of the American Society of Mechanical Engineers*, 1938, **60**, 641–650; E. D. Haigler, 'Application of Temperature Controllers', *Transactions of the American Society of Mechanical Engineers*, 1938, **60**, 633–640; C. E. Mason, 'Quantitative Analysis of Process Lags', *Transactions of the American Society of Mechanical Engineers*, 1938, **60**, 327; C. E. Mason and G. A. Philbrick, 'Automatic Control in the Presence of Process Lags', *Transactions of the American Society of Mechanical Engineers*, 1940, **62**, 295–308; E. S. Smith, 'Automatic Regulators, Their Theory and Application', *Transactions of the American Society of Mechanical Engineers*, 1936, **58**, 291–303, disc. 59; E. S. Smith and C. O. Fairchild, 'Industrial Instruments, Their Theory and Application', *Transactions of the American Society of Mechanical Engineers*, 1937, **59**, 595–607; A. F. Spitzglass, 'Quantitative Analysis of Single-capacity Processes', *Transactions of the American Society of Mechanical Engineers*, 1938, **60**, 665–674.

18 A photograph taken by J. G. Ziegler together with a list of attendees and speakers appeared in *Instruments*, 1943, **11**, 337.

19 J. G. Ziegler and N. B. Nichols, 'Optimum Settings for Automatic Controllers', *Transactions of the American Society of Mechanical Engineers*, 1942, **64**, 759–768.

20 A. R. Aikman and C. I. Rutherford, 'The Characteristics of Air-operated Controllers' in *Automatic Manual Control: Proceedings of the Cranfield Conference*, A. Tustin (ed.), Butterworths, London, 1952, pp. 175–187.

21 G. H. Cohen and G. A. Coon, 'Theoretical Considerations of Retarded Control', *Trans. ASME*, 1953, **75**, 827–834.

22 A. R. Aikman, 'The Frequency Response Approach to Automatic Control Problems', *Transactions of the Society of Instrument Technology*, 1951, **3**, 2–16.

23 G. A. Coon, 'How to Find Controller Settings from Process Characteristics', *Control Engineering*, May 1956, 66–76; G. A. Coon, 'How to Set Three-term Controllers', *Control Engineering*, June 1956, 71–76.

24 J. G. Ziegler and N. B. Nichols, 'Process Lags in Automatic Control Circuits', *Transactions of the American Society of Mechanical Engineers*, 1943, **65**, 433–444.

25 A. J. Young, *Process Control*, Instruments Publishing Company, Pittsburgh, 1954.

26 R. E. Clarridge, 'A New Concept of Automatic Control', *Instruments*, 1950, **23**, 1248–1292.

27 K. J. Åström and T. Hägglund, *PID Controllers: Theory, Design and Tuning*, Instrument Society of America, Research Triangle Park, NC, 1995, p. 82. Two of the few descriptions can be found in papers published in the house journal of the George Kent Company. See S. J. Clifton, 'Instrumentation and Automatic Control at Uskmouth Power Station', *Instrument Engineer*, 1954, **1**, 103–109; and F. J. Cunningham, 'Instrumentation and Automatic Control of Open-hearth Furnaces', *Instrument Engineer*, 1956, **2**, 11–19.

28 DSIR, *Automation*, Her Majesty's Stationery Office, London, 1956.

29 A. J. Young, *Process Control*, Instruments Publishing Company, Pittsburgh, 1954. The controllers were made by Evershed & Vignolles (UK), Hartman & Braun and Schoppe & Faeser (German), and in the USA Leeds & Northrup, Manning, Maxwell & Moore and The Swartwout Company.

30 G. P. L. Williams, 'Trends of Development in the British Instruments Industry', *Instrument Engineer*, 1957, **2**, 75.

31 Bristol-Company, 'Know at Your Desk What's Happening... Throughout the Plant', *Instruments*, 1936, **9**, A10.

32 E. Ayres, 'An Automatic Chemical Plant', in *Automatic Control, Scientific American* (ed.), G. Bell & Sons, London, 1955, p. 45.

33 C. E. Mathewson, 'Advantage of Electronic Control', *Instruments and Automation*, 1955, **28**, 258–265.

34 An annotated bibliography listed over 80 such publications, see E. M. Grabbe, 'Computer Control Systems – an Annotated Bibliography' in *Proceedings of 1st IFAC World Congress, 1960, Moscow*, Butterworths, London, 1961, pp. 1074–1087.

35 The market for scientific and process control instruments was large, an estimated $4500 million in the USA in 1960, see *Instruments and Control Systems*, 1960, **33**, 54.

36 Comments by Grant E. Russell, Manager, Systems Engineering Section, Research and Engineering Division, Monsanto Chemical Company reported in Editorial, 'Monsanto Unveils Integrated Computer-controlled Process', *Instruments and Control Systems*, 1960, **33**, 1888–1893.

37 T. M. Stout and T. J. Williams, 'Pioneering Work in the Field of Computer Process Control', *IEEE Annals of the History of Computing*, 1995, **17**, 6–18.

38 A. Turner, 'Computers in Process Control: A Panel Discussion', *Instruments and Control Systems*, 1970, **43**, 81–85.

39 R. E. Blake, 'Advantages to be Gained from Process Control by Computer', *Electronics and Power*, March 1977, 219–221.

40 Advantech, advertisement, *Plant and Control Engineering*, 19 April 2001, p. 12, see also www.advantech.com.

41 The work on intelligent agents which is emerging from the artificial intelligence

community is seeking to provide software entities which can generate knowledge and which can use this knowledge so as to be pro-active rather than simply responsive. For a discussion of agency see A. G. J. MacFarlane, 'Information, Knowledge and the Future of Machines', *Measurement and Control*, 2001, **34**, 9–13, 52–55, 81–85.

42 PID'00, IFAC Workshop on Digital Control: Past, Present and Future of PID Control, Terassa, Spain, 5–7 April, 2000.

43 T. Shinn, 'Research-technology Instrumentation: The Place of Chemistry', in this volume.

44 S. Bennett, 'The Society of Instrument Technology Ltd: The Early Years', *Measurement and Control*, 1994, **27**, 135–140.

Impact of Instrumentation on Chemistry

Tools, Instruments and Concepts: The Influence of the Second Chemical Revolution

PIERRE LASZLO

Cloud's Rest, Prades, F-12320 Sénergues, France, Email: clouds-rest@wanadoo.fr

Introduction

My standpoint is that of an experimental scientist and of a spectroscopist. As an organic chemist, I have contributed to the methodology of organic synthesis.[1] As a specialist in nuclear magnetic resonance,[2] I did work also in developing new methodologies,[3] for instance in devising uses of the sodium-23 nucleus for probing molecular and biomolecular interactions.

Chemical instrumentation is such a nice, such an important topic to take up because its history offers so many avenues of investigation, so many questions begging for an answer. Some examples are: Why did dielectric measurements vanish from the laboratory? Why has the laser had such a relatively small impact on chemistry? How should we explain the recent rebirth of calorimetry? Why was the refractometer such a long-lived tool and, conversely, why did it recently disappear from laboratories?

I have just referred to a 'tool'. This brings up what I hold to be a key distinction, that between the tool and the instrument. Etymologically, 'tool' derives from an old Germanic root *tou- with the meaning 'to do, to make'. It is an action word. 'Instrument' derives from the Latin *instruere*, also an action verb, meaning 'to build, to equip, to furnish'. This verb is akin to *struere*, 'to build in piles', itself a cognate of the Latin word that gave us 'structure'. The Latin name *instrumentum* thus had the meaning of something that serves to equip or furnish.

I offer the following demarcation between the two categories, of a laboratory tool and an instrument.[4] None of these distinctions is absolute, but their accumulation is convincing. First, as a rule, the tool is less costly than the instrument.

Second, a tool is available all the time, whereas an instrument requires a sign-up sheet. Third, the form of their output also differentiates them: use of the tool is transparent in meaning; it will often provide the research worker with data on a piece of paper to be pasted into the laboratory notebook without further elaboration. Conversely, the output from an instrument shows a need for interpretation. Fourth, tools are located in the lab proper, while instruments are looked after in special rooms, or even in dedicated centres.

Examples of tools are: the balance;[5] the thermometer; the hair dryer; the rotary evaporator; the chromatographs;[6] and the still.[7] Examples of instruments are: the mass spectrometer;[8] the nuclear magnetic resonance spectrometer;[9] the synchrotron. Another interesting historical question, begging for a narrative, is a current change: X-ray diffractometers, in quite a few chemistry departments and institutes, abandon the category of 'instrument' for that of 'tool', a shift which R. B. Woodward had predicted several decades ago.

To define a chemical instrument, there are two opposed viewpoints, the loose and the precise. In a loose manner, of little use to the serious historian, chemical instrumentation can be defined as the set of devices and procedures related to chemicals in their analysis, synthesis, production, properties, *etc.* In a more precise manner (in what one might term a Wittgensteinian definition), chemical instrumentation is the set of devices found in a representative chemistry laboratory. Unfortunately, this second definition has been rendered useless. There is no longer a standard chemical laboratory, if such a thing ever existed (a point to which I shall return).

I see at least a dozen goals for a history of chemical instrumentation and I shall list them, in no particular order:

1. to tie instruments (their use, their design and construction) to the elaboration of concepts;
2. to relate instruments to the monitoring of phenomena, as embodied in the suffix *-scope* (as in electron transmission microscope for instance), and thus to discoveries;
3. to infer boundaries between disciplines and sub-disciplines from the presence (or the absence) of some given apparatus in scientific work;[10]
4. to relate specific instruments to the histories of disciplines and sub-disciplines. An example would be the use of say phosphorus-31 NMR, or that of multinuclear magnetic resonance, as both are related to the rise of organometallic chemistry;
5. to develop a history, based on scientific instruments, of the interface between chemical science and the instrument makers of both types, craftsmen and manufacturers;
6. glassblowing and glassblowers assume a central place in such a history, and deserve their own history, with a remarkable continuity over the centuries. For instance, it would be interesting to focus on a device such as the Dean–Stark water trap and to relate it to chemical discoveries and creations;
7. to chronicle instrument acquisition by both individuals and institutions as

it relates to social networks of patronage and power; to scientific fashions; to symbolic values, status first and foremost;

8. to delineate lines of forces in the complex web of entangled relationships between chemistry and physics. The term 'physical methods' is revealing, in such a context;[11]

9. the converse is also true. Chemistry can be and is a service science. This is even more so for a sub-discipline such as analytical chemistry, with respect to industrial chemistry or medicinal chemistry. Since quite a few contributions to this conference bear on this very point, I do not have to belabour it further except to comment that this is only a part, a restricted and often a rather minute part in the history of chemical instrumentation;

10. just like furniture in a castle, instruments in a laboratory have a function of internal decoration. As such, their history is related to other historical topics, laboratory architecture on one hand, laboratory visits by both colleagues and dignitaries on the other hand;

11. a textual history of instruments cannot be neglected. One of its dimensions is documenting the rôle of instrumental references in published work, in the two communication channels of the iconography and of the Experimental Part, regarding the citing of instruments resorted to, or the absence thereof;

12. the history of *instrumentalification* is an integral component of any history of chemical instrumentation. Two instances that come to mind are addition and elimination reactions for establishing stereochemical relationships; and recourse to chemical shift data from internal samples of liquids such as methanol or ethylene glycol for temperature measurement inside an NMR sample;

13. a key part of the history of chemical instrumentation deals with errors induced by the instrument. A history of the artefact is highly desirable. Obviously, the *in vivo/in vitro* dualism within clinical studies relates to such a history. A related point is the blind spot of any instrument. For instance, mass spectrometry of organic molecules is, to a good approximation, blind to stereochemistry (in the 1980s, when Fred McLafferty and I were jointly teaching a graduate course in organic spectroscopy at Cornell, I could seldom resist rubbing it in!); and X-ray diffraction is somewhat impervious to atomic motions.

Technology and the Advancement of Science

In assessing the rôle of instrumentation in the laboratory (in the chemical laboratory especially) one should first ascertain if the scientific research under study was more technique-determined or problem-determined. There is a closely related question, since chemistry is both an industry and a science, of industry-determined *versus* science-determined methodologies.

Which brings up the status of analytical chemistry, a sub-discipline of chemistry whose spectacular rise in the nineteenth century was due to a large extent to

Liebig's university laboratory in Giessen. During the twentieth century, while the training of analytical chemists has continued to be a prerogative of academia, their employment has been predominantly the province of industrial and government laboratories – a mundane consideration, but one highly relevant here. Indeed, while industrial laboratories as a rule give sole responsibility of the instrumentation to analytical chemists,[12] precisely the opposite holds true in most academic laboratories in chemistry: the specialists (whether organic chemists, inorganic chemists, atmospheric chemists, nuclear chemists, *etc.*) enjoy hands-on direct use of instruments, and the analytical chemists are kept out, whether from an interface layer or from an interpretative capacity. There are issues here, worth investigating in more depth, of division of labour, of social pecking orders[13] and prestige and, most important, of networks of power.

To return to the industry–academia dichotomy, let me bring up a contemporary issue. It will document the field of conflicting forces just mentioned. This is molecular modelling. It thrives currently. The pharmaceutical industry, after investing heavily in other methodologies which did not give it the broad access to the devising of new drugs it sought (infrared vibrational frequencies, Hansch partition coefficients, calculated electronic distributions in drug molecules, mass spectral fragmentation patterns and so on), is once again taking up Emil Fischer's lock-and-key metaphor. It pours money into the detailed study of the docking of a drug molecule at a receptor site embedded in a protein host. Considerable financial investments by pharmaceutical companies have allowed a technology of molecular modelling to flourish. Since everyone, academic chemists included, loves a pretty picture, we are inundated with such representations.[14] I am not sure that, at least so far, there have been any scientific or even industrial returns from this industry-determined methodology.

The Affect from the Second Chemical Revolution

Joachim Schummer's central thesis is that the 'ontological attitude' of chemists has switched from material substances to molecular structures; that this switch occurred during the second half of the twentieth century; and that recourse to spectroscopic instrumentation has had this profound impact. This is an interesting proposition and there is much to commend it.

As a first comment, logically, one ought to query whether chemists, taken as a professional social group, have an 'ontological attitude'. Do they need one? Do they value it? To the contrary, chemists pride themselves as artisans, as practitioners of a traditional handicraft where observation and experimental skill take precedence over theory and metaphysics. The protracted battle over the introduction of atomic theory, during most of the nineteenth century, has endowed some of the most conservative chemists with a durable allergy to non-positivistic thinking. Not so long ago, the 'classical *versus* non-classical ion' controversy was such an episode, in which Herbert C. Brown lambasted Saul Winstein and his supporters for their conceptualising of organic structure in any but traditional terms.[15]

While I agree with Schummer that a conceptual switch has indeed been brought about by the increased reliance upon instrumental analysis, I would urge that the substance–structure dichotomy should not be construed as clear cut to chemists, who tend to blur the two concepts into a continuum, nor therefore as fundamental to historians of modern chemistry. The practice of the mixed melting point is emblematic of the successful synthesis of a target molecule. It has endured even in the age of organic spectroscopy. It illustrates the absence of a marked contrast between the notions of substance and of structure in the mindset of twentieth century chemists.

The second point is the periodisation. Schummer claims, at least implicitly, that the substance-to-structure conceptual switch occurred between 1950 and 1970. Actually, the seeds for it were sown much earlier, at the turn of the twentieth century. Those primitive physical methods of structural elucidation, molecular refraction (Ludwig Lorenz in Copenhagen and Hendrik Lorentz in Leiden introduced their joint formula in 1869), and half-a-century later the parachor (developed by Samuel Sugden in London in 1924) – incidentally, they go much of the way in explaining the institutionalised inclusion of the index of refraction among the canonical physical properties of a substance – were the physical means available and in full use, in the time of Wilhelm Ostwald and until the young Robert B. Woodward devised his rules for relating UV–visible spectra to the structure of conjugated organic molecules. The historical conceptual continuity is obvious there: analysing data from physical measurements on a molecule as the sum of increments, each associated with a given group of atoms.

Furthermore, the identification of a substance from a spectrum in empirical, 'fingerprint' manner, also started in the nineteenth century. Let me give just one concrete example: *L'Agenda du chimiste*, a French predecessor to the *Handbook of Chemistry and Physics*, shows in its 1887 edition no fewer than five pages of absorption spectra (about thirty per page) for chemical species as diverse as alizarin blue, Bengal rose, cochineal, methemoglobin, perchromic acid or potassium permanganate. The absorption bands are displayed as broad and generally featureless humps, in six different spectral ranges. Surely, this qualifies as spectroscopic identification of organic compounds, as early as 1887.[16]

Schummer's contribution raises a host of fascinating questions, such as the nature of chemical innovation (does churning-out new compounds for burial in the *Chemical Abstracts* registry truly qualify as such?); the conservatism of some groups within the chemical community as exemplified, for instance, by the half-century which elapsed before organic chemists (Woodward was also a leader there, starting in the 1960s) started using X-ray diffraction; and the constant drive toward miniaturising the amounts of reaction products: only a few milligrams are deemed sufficient nowadays in organic synthesis, in harmony with the performance of chromatographs and spectrometers both, following the lead of virtuosos of practical laboratory work, such as that by W. Clark Still at Columbia University.

Carsten Reinhardt tells convincingly the story of the advent of mass spectrometry as a potent physical tool for elucidating the structures of organic molecules. His main thesis, to put it in a nutshell, is that chemical industry, the

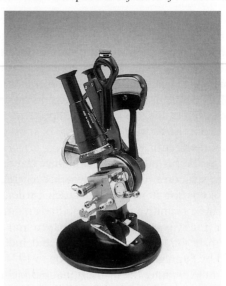

Bellingham and Stanley Abbe refractometer, 1956. Science Museum collections, inventory number 1998–22. Science Museum/Science and Society Picture Library

petrochemical industry in particular, provided the need and the money for building the instruments, while physical organic chemistry provided the concepts. Both are indisputable.

Nevertheless, Reinhardt's account needs some complements. The first concerns the second chemical revolution, brought about by the introduction of physical instrumentation for determining the structure of molecules. This revolution was ushered in by the research university, the novel institution which appeared in the aftermath to World War II and as a follow-up to *Science, the Endless Frontier*, Vannevar Bush's famous report to President Truman in 1945. Only within American research universities were resources made available in the 1950s by Federal granting agencies for the purchase, or the construction, of such large instruments. Until then, chemists could and did use only benchtop tools of relatively little cost, such as microscopes, polarimeters or refractometers. Thus, it should come as no surprise if applications of mass spectrometry to organic molecules first appeared in the United States, at the interface of industry and academia.

Carsten Reinhardt does not adjudicate between two sociological factors, the appearance of a new instrumental methodology as fallout from World War II; and the dominance of the chemical community by physical organic chemists and the attendant semi-empirical approach, during a very brief period, in the 1960s. The tales which remain to be told are those, on one hand, of the technology which made it feasible to produce and to maintain the ultrahigh vacua required by mass spectrometry; and, on the other hand, those of the institutional histories of chemistry departments in American universities, such as Harvard or Princeton, which contained in the late 1940s and during the 1950s strong personalities such as James B. Conant or Hugh S. Taylor who trained talented co-workers and who pushed physical organic chemistry to the forefront of the discipline (for a short while).

Yet another void in Reinhardt's otherwise admirable history concerns mention of a key factor, books, in the acceptance and in the spread of a new methodology within chemistry during the second half of the twentieth century. While word of mouth was sufficient for the leaders of the profession to equip their laboratories with new types of expensive instruments, published papers did not provide enough know-how to the beginner. To a very large extent, the three protagonists in Reinhardt's story, *viz.* Fred W. McLafferty, Klaus Biemann and Carl Djerassi owed their enormous influence on budding mass spectrometry to the publication, very early on, of remarkably didactic, truly seminal books whose appearance was extremely timely.[17] One should not underestimate the importance of the printed page. I have argued elsewhere[18] that physical organic chemistry itself owed quite a share of its success in the 1960s to the remarkable textbook published by Edwin S. Gould.[19]

There are at least three 'social' (or 'societal') issues implicitly raised by Reinhardt's paper. The first has to do with the list of publications which, as is all too well known ('publish or perish'), provides in a single number an index to the scientific 'worth' of an author which academia uses henceforth for admission to its ranks and promotion. Authors of papers presenting mass spectrometric elucidation of the structure of natural products, such as Biemann and especially Djerassi, were extremely prolific; they could boast of long publication lists. This aspect of the introduction of mass spectrometry as a tool for organic chemistry undoubtedly was one of the reasons for its success, by the rapid transformation of an initial bridgehead gained into a field conquered.

The second fascinating issue is that, all too briefly alluded to by Reinhardt, of the tactical alliance between natural products chemists and physical organic chemists. One would like to know, in the depth and detail which only history can provide, the material interests which made them collaborate rather than compete; the features in recruitment and training of graduate students and post-doctoral fellows which turned out to be complementary; and whether these two groups did unite to stem the rising tide of synthetic organic chemistry, a rise that was inevitable, given the powerful support of this third sub-discipline by the pharmaceutical industry.

The third issue is the notion of the 'technical expert', apparent in the quote by Hinshelwood in Reinhardt's paper. The rise of the specialist, in mass spectrometry or in nuclear magnetic resonance, was perceived as a threat by some of the more traditional organic chemists, whether they were synthetic organic chemists or members of the natural products structural school. Thus, there was a pressure to constrain these new talents within the ghetto of their speciality, lest they would wrest control of the discipline from the elite which then prevailed.

This book focuses on instrumental techniques as they impact on science, chemistry specifically. This topic and its students both run the risk of self-absorption within the technique and to the detriment of the science. There are few things more intoxicating to a technical expert than wanting to improve the technique he or she has mastered, aiming at goals such as further increases in resolution or sensitivity. It is very easy to get trapped in this way.

But how to delineate a demarcation between internal and external aspects of a technique, the latter identifying with the scientific applications? It is desirable to do so, lest the very purpose of this book becomes muddled. The criteria characteristic of technical narcissism are borrowed from the market economy. Whenever an instrumental technique claims an improved return with respect to investment (whether of time or money), whenever it advertises an ability at being standardised, systematised or automated, one can be assured that science is the loser.

Let me quote just one recent illustrative example. Bradford W. Gibson, Irwin D. Kuntz, Gavin Dollinger *et al.*, *PNAS*, 2000, **97**, 5802, have proposed a new procedure, based on cross-linking followed by mass spectrometry for identifying folding patterns of proteins. To quote these authors, their method is 'many times faster than the current standard techniques, it can be performed on much less material, automation is eminently possible, and the means already exist to apply it more generally in the field of protein structure analysis'.

Use of Instruments and Construction of Concepts

Before continuing to outline in broad strokes the virgin terrain remaining to be explored by historians of chemistry, two caveats are in order. I have already hinted at one. The erroneous perception of a sudden change, instead of an underlying slow evolution, can be a temptation. It ought to be resisted. The history of organic spectroscopy, originating in the 1880s rather than in the 1950s, illustrates the merits of Braudelian *longue durée*.

Conversely, historians occasionally carry over concepts from one period into another, in which they have become anachronistic. Structure theory is a case at hand. It had operational meaning in the time of Kekulé. But its meaning had changed radically by the time of Woodward.

Not only concepts are dated. Words are also rooted in history. Far from being innocent, they are weighed with semantic baggage. A case at hand is 'reification': its use is inseparable from its introduction by Hegel, followed by Karl Marx's indictment of capitalistic dehumanisation of the workers and by the ensuing reiteration throughout the Marxist canon. Such a loaded philosophical term should not be applied, I would argue, to the posited switch in the perception by professional chemists, from substance to structure.

I also have to oppose Leo Slater's statement that: 'The Woodward (UV) Rules foreshadowed Woodward's later work with Roald Hoffmann, the Woodward–Hoffmann Rules for which Hoffmann shared the 1981 Nobel Prize with Kenichi Fukui'. I beg to differ. The UV Rules were steeped in the empirical tradition of molecular refraction and the parachor. Expressed with additive increments, they were quantitative and belonged with physical organic chemistry. The Woodward–Hoffmann Rules are selection rules. They are qualitative, they are yes/no logical statements, in the spirit of quantum chemistry and of molecular spectroscopy *à la* Gerhard Herzberg.

Woodward was a genius and I consider him the epitome of twentieth century synthetic chemistry and, arguably, the best chemist in the century. He had his shortcomings though. For instance, his claimed 1944 total synthesis of quinine, jointly with William von Eggers Doering, is questionable; as was made clear recently by Gilbert Stork's first stereoselective total synthesis of the natural anti-malarial.[20]

Such shortcomings should not detract from Slater's excellent anthology of quotations from the Master, R. B. Woodward. I single out for consideration the self-portrait:

> Second, the tendency to allocate to specialists the responsibility for the making, and in particular the interpretation, of physical measurements, is to be deplored. The capacity of the physical specialist to place his results properly in the context of an organic chemical investigation is often narrow and unrealistic, and the organic chemist will find himself magnificently rewarded, who takes pains to be himself in a position to understand and interpret the physical aids he wishes to use. In any event, physical methods, and the principle that they should be used whenever possible, are now part of our armamentarium, and we expect no surcease of further developments in this direction.

This 1956 statement by Woodward can be paraphrased as follows:

  (i) do not endow experts with the responsibility for obtaining and interpreting physical data on organic molecules;
 (ii) organic chemists should make themselves knowledgeable in the new techniques;
(iii) they alone can make full and realistic use;
(iv) the future of the new physical methods is wide open.

This prophetic announcement by Woodward unfortunately went unheeded. I shall return at the end of my paper to the division of labour he denounced and to some of its lasting consequences.

Other contributors to this volume equate physical organic chemistry with the study, by largely empirical means, of the large and rather complex molecular architectures of organic chemistry. In the 1960s these molecules were too large to be approached otherwise; they were unwieldy and unyielding to application of quantum chemistry. While such conventional wisdom has a great deal of truth, it remains an oversimplification. The two episodes I am about to recall both offer lessons to the historian. They bear on the issue at hand, which can be rephrased as physical organic chemistry (or likewise biophysics) resisting a simple, positive definition.

My first story deals with use of nuclear magnetic resonance for the sodium-23 isotope to monitor the state of the sodium cation Na^+ in biological tissues. The early investigators, a couple of dozen in number, all reported superposition of two resonances, one sharp and one broad. Their knee-jerk reaction, from their training as spectroscopists, was to resolve these overlapping absorptions into their two components, responsible, they discovered, for approximately 60% and

40% of the total intensity. Thus, all these experts rushed to the published conclusion of the presence in or around cells of two distinct populations of sodium ions, 40% in the free, hydrated state; and 60% in a more slowly reorienting state, because of attachment to biopolymers. Such an inductive, 'physical organic' or 'biophysical' reading of the evidence turned out to be totally wrong! The appearance of the sodium absorption was simply, from quantum-mechanical first principles, a consequence of slow modulation for the two allowed separate transitions $-\frac{1}{2} \rightarrow +\frac{1}{2}$ and $\pm\frac{1}{2} \rightarrow \pm\frac{3}{2}$. In this particular case, and in spite of the complexity of the system, only the deductive approach from quantum mechanical first principles was tenable. The inductive approach, characteristic of physical organic chemistry, turned out in this case to be flawed and misleading.

The lessons to the historian are: (i) recourse to instruments provides scientists with 'revelations' about nature; (ii) understanding such revelations is always a matter of interpretation, is theory-laden; (iii) it is exceedingly naïve to equate complexity of the system under study and the need to apply an inductive and empirical or semi-empirical approach; (iv) scientific errors include the falling into traps from commonly shared myopias or collective blinders.

And indeed those leading scientists, of lasting stature and achievement, those who had had the vision to seek from the start the prowess of new instrumental methodologies such as mass spectrometry and NMR, were smart enough not to rely on the experts in one instrumental method or the other, and to give themselves insider knowledge. For instance, within the province of synthetic organic chemistry, *a priori* at the furthest remove from quantum mechanics and molecular spectroscopy, R. B. Woodward quietly and closely followed progresses in molecular orbital theory – a familiarity he would draw upon in elucidating the selection rules for electrocyclic reactions, after enlisting the help of Roald Hoffmann. Likewise, E. J. Corey knew enough quantum mechanics to conceptualise the phenomenon of virtual coupling in nuclear magnetic resonance, after enlisting the help of Jeremy I. Musher. To both Woodward and Corey, the construction of concepts (the Woodward–Hoffmann rules and virtual coupling) combined keen observation, a broad-based training in *both* elementary quantum mechanics and physical organic chemistry, irrespective of the apparent complexity of the molecules concerned.

My second story is that of the construction, within organic chemistry, of the concept of enantiotopic and diastereotopic atoms (and groups of atoms). It resulted directly from instrumental observation: ^{1}H nuclear magnetic resonance showed non-equivalence of protons, for instance the two hydrogens in a CH_2 methylene group would give rise to distinct resonances. At first, such observations surprised organic chemists. They went against conventional wisdom, which as is often the case was only ingrained error: so-called 'free rotation' making such atoms or groups thereof equivalent on a time average (which is utter nonsense). Organic chemists had convinced themselves that a methylene group, or the two methyl groups within an isopropyl group, were undifferentiated units. However, Kurt Mislow and Morton Raban quickly made coherent sense of all such observations and introduced an adequate terminology.[21] They had to do so,

Kurt Mislow. Photograph courtesy of Kurt Mislow and Princeton University

because the community of organic chemists was confused and/or did not pay attention, although the notion had been commonplace among biochemists and enzymologists for many years.

The lessons to the historians are here: (i) the presence of barriers to knowledge transfer at times between sub-disciplines, in this case biochemistry and organic chemistry; (ii) the conceptual clarification introduced by Mislow led to practical consequences in the laboratory for both wet chemistry (enantioselective reactions) and spectroscopy, such as the devising of chiral shift agents and the NMR *instrument* replacing the polarimeter *tool* for measuring enantiomeric excesses and so-called optical yields; (iii) yet again, scientific errors include falling into traps from commonly shared myopias or collective blinkers; (iv) enrichment of synthetic organic chemistry from the input by physical organic chemistry, with the introduction of a whole new technology of enantioselective reactions based upon their monitoring by nuclear magnetic resonance.

The Showcase and Approval by the Peers

During a visit to a laboratory by an outsider, whether a colleague or a dignitary, instruments, indeed often found to be idle, serve as status symbols. Why are expensive instruments kept idle and even sometimes mostly serve just for show? What are the reasons for this paradoxical behaviour? The Braudelian *longue durée* is especially pertinent here: the tradition goes back to displays and demonstrations in aristocratic courts and salons during the eighteenth century, and earlier.

In the adversarial process which scientific communication has now become, the demonstrative rhetoric from instrumental data has acquired pride of place;

since it represents and even sometimes quantifies the quality of the evidence. One might document the switch in a journal such as *Journal of the American Chemical Society*. It occurred at the end of the nineteenth century. Prior to that, the instrumental iconography of the chemistry article had the documentary function of depicting some apparatus, the way in which to put it together and to use it. Approximately at the turn of the twentieth century and contemporary with the beginnings of organic spectroscopy, illustrations switch from representing the instrumentation to depicting data, displayed in graphs of various sorts, after it has been collected by means of such instruments.

It has remained so to this day. In a chemistry paper, instrumental data is the foremost piece of iconography, presenting illustrative evidence for evident rhetorical purposes.[22] Someone ought to study it in-depth, and to take a good look at the statistics.[23,24] Joachim Schummer, in this volume, has carried out such a useful quantitative and diachronic study of *Liebigs Annalen*, documenting the exponential rise[25] of so-called physical methods, such as NMR[26] and mass spectrometry,[27] during the second half of the twentieth century.

Lacking the complementary synchronic study, I have dipped at random and sampled issue number 12 of volume 39 of *Angewandte Chemie* for the year 2000, in the English language version. The 38 papers (reviews, articles and communications, and excluding the book reviews) run from page 2043 to page 2178. First, the table of contents has its own iconography, an innovation which *Angewandte Chemie* was among the first journals to pioneer: chemical formulae (68%) molecular models (24%), graphs (5%) and one scheme (2.6%). Looking at the papers themselves, the results entirely conform to what any chemist might have used their intuition to predict, *viz*; an overwhelming predominance of formulas, both connectivities and conformations: 129 figures. Calculated molecular models take up 15 figures in that issue. Regarding other instruments besides the computer,[28] pride of place goes to X-ray solid-state structures, in 17 figures. The rest of the iconography is very diverse: two X-ray powder diffraction patterns, one EXAFS (X-ray absorption fine structure) and two XANES (X-ray absorption near-edge structure), three IR spectra,[29] four 1D NMR and two 2D NMR spectra,[30] one EPR spectrum, one UV–visible spectrum, five fluorescence emission spectra, one cyclic voltammogram, one mass spectrum, two chromatograms, a couple of molecular orbital diagrams, a potential energy hypersurface, a few miscellaneous graphs such as binding curves and Michaelis–Menten kinetics… The finding is a broad diversity of tools. In other words, there is, at least at present, no such thing as a standard chemistry laboratory with a standard chemical instrumentation. Each figure in a published paper gives us insight, is a window into the 'ideal laboratory' as conceived by its main author.

Which brings up another point: what kind of instrumental results did those scientists choose to feature in their papers? My strong suspicion, based on a combination of insider knowledge and of experience, is that you do not use quite the same instruments to impress a visitor to your laboratory and a journal referee. In the former case, you will strive for maximum performance (resolution, sensitivity, *etc.*), you will advertise your 'grantsmanship', you will proudly show off the in-house contraptions. In the latter case, you will opt for conformity, normalcy and standard operating conditions. You will want to intimate easy reproducibility: 'any idiot should be able to repeat the work' is the implicit message.

A concept both administrative and architectural bridges academic science and industrial science. This is the instruments centre. The intent is to save money by having specialists run specialised equipment for the community. Sometimes, they serve a department or an institute; and sometimes they give service to a whole geographic region. They represent an administrative solution to the problem of access to the high(est) performance instruments, to that of their acquisition and maintenance.

The Location of Instruments in the Practice of Chemistry

Such instrument centres (an NMR laboratory, a mass spectrometry laboratory or an X-ray laboratory) are indeed differentiated architecturally, administratively and even socially from the chemistry laboratories nearby. They are 'factory-like', to borrow Peter Galison's analogy, who has chronicled for high-energy physics the constitution of similar areas (what he calls 'trading zones').[31] Indeed, such instrument centres look very much alike, whether in an academic, a governmental or an industrial setting.

Instrument centres in chemistry buildings serve as an interface, not so much between experimenters and theorists (as may be the case in physics), as between experimenters and instrument makers, directly or indirectly (whenever one deals with a commercial instrument). I refrain from extending to them the Galisonian qualification of a 'trading zone': chemistry is considerably more secretive than physics, probably because of proprietary knowledge stemming from funding or from consultancy with the pharmaceutical industry. Hence, there is surprisingly little exchange of information between the users, say, of an NMR centre.

Indeed, the rôle of such centres, even more than providing access by chemists to sophisticated machines, is their insulation from chemistry. The interface doubles up as a fire wall. There is a notion that mainstream chemistry remains a material science, with its proper language (that of chemical formulas) and that instruments are both precious and a necessary evil which, this unformulated but powerful doctrine holds, should not contaminate chemistry.

This is reflected in university departments and institutes by the coexistence of two communities, operating under a system of 'apartheid': on one hand, chemists wearing as distinctive signs lab coats and now safety glasses, handling glassware, having served an apprenticeship, steeped in a tradition of craftsmanship; on the other hand, hybrids between engineers and technicians, with areas of expertise such as electronics or the writing of software, prevented from competing for academic promotion since they are kept from teaching on the minus side, but who, conversely on the plus side, are regularly poached by industry.

Mainstream 'wet' chemistry thus displays an ambivalent relationship towards its own instrumental practice. The interaction is at the same time deeply symbiotic while it is often perceived and resisted as parasitic. This is even more of a paradox given the huge financial investment.

Varian's A-60, introduced in 1961, was the first NMR spectrometer designed for routine use. A stable, reproducible, user-friendly and affordable system, the A-60 could be operated in a chemistry laboratory by an organic chemist with little knowledge of the physics of NMR. Image taken from A-60 The New NMR!, *a sales booklet published by Varian. Photograph courtesy of Varian Inc.*

How can such a schizoid perception be explained? I have had the opportunity, as a spectroscopist, of raising this very question numerous times with colleagues, in synthetic organic chemistry especially. The answer many of them give is all the more interesting for being historical, at least at its face value. When pushed, they are likely to tell you that chemical structures are likely to endure, whereas instrumental methodologies are ephemeral, one fashion displaces the previous one.

Such answers are revealing. Chemical structures are like Platonic archetypes. Of course this evokes the age-old dichotomy, going back to Plato, between pure intellectuals in their ivory tower, and the manual workers. This very distinction was represented in the architecture and in the topography of chemical institutes of the nineteenth century: glassblowers were confined to a separate workshop. It has endured throughout the twentieth century and the glassblowing workshop has spawned the instrument centres.

Exceptions: there are none. There may be apparent exceptions and they turn out to be, or to have been tools, rather than instruments (according to the distinction I made at the outset). In the 1960s, Varian Associates put on the market a wonderful machine aimed at the organic chemists, the A-60 spectrometer. Numerous breakthroughs resulted from the close interaction between graduate students and this particular tool, an interesting story in itself. Currently, graduate students in a number of chemistry departments entertain a similar intimate relationship with the likewise fruitful X-ray diffractometer.

Yes, 'articulated knowledge and machinic performances are reciprocally tuned to one another'.[32] Which brings up a few final points.

Conclusion

I started this paper with an attempt at defining an instrument of science. Thus, I feel duty-bound to close it with another attempt at a definition, that of science. If we look to science as a text, instruments are inscribing devices. They provide traces, in the form of spectra, plots, graphs and grams, *etc*. Indeed, a key word and concept here is that of 'interpretation'. The chemist's task is to interpret instrumental readings; and to translate them into the language of chemistry.

Chemical science is thus defined by what it does, *i.e.* transforming data about matter into chemical formulas. Formulas are the morphemes in the language of chemistry.[33] In the same textual metaphor, any new instrument will enable a partial rewriting of the existing knowledge: the text of science in truth is a palimpsest,[34] constantly being erased in part for the purpose of being rewritten. Instruments, if taken as a group, constantly redissolve entire blocks of the existing knowledge. They undermine the cumulative nature of scientific knowledge.

What then is the rôle of the scientist in using an instrument? If he or she is creative, this is on account of non-standard use of such an instrument. If he or she is passive, a slave to the instrument merely recording data as a routine and 'stamp collecting', then this is a contribution to the passing of this particular instrument. Its future extinction, in favour of an altogether different type, will bring about a restructuring of knowledge, a redefinition of disciplinary boundaries, a redistribution in the division of labour and a revised curriculum for the education and training of young scientists. Replacement of the polarimeter with the NMR spectrometer, along the convoluted path documented above (from instrument to concept, and back to instrument), is such an example.

Thus, my first metaphor, science as text, dissolves into a second metaphor, that of science as a Darwinian evolution outdating each instrument type in turn and bringing about its extinction. The definition of science, from its use of a subset of the general class of instruments, makes other metaphors yet spring up. The emphasis on the visual, on *scope*-instruments, suggests the admittedly sexist (and Baconian) metaphor of the unveiling of nature. 'Acteon Viewing Artemis Bathing' was a classic allegorical scene for a painter. The 'scientist-as-voyeur' monitors a phenomenon, *i.e.* he/she is both witness and controller. If creative, the voyeur scientist is intent upon discovering a significant detail, in like manner as the art historian Giovanni Morelli made use for his attributions of the rendering by the artist of say an ear lobe.

And what then is specific to use of instruments by chemists? The dematerialisation;[35] and the detemporalisation: often the naive realism associated with the static and rigid molecular object leads to the time dimension being erased. Each picture of a molecule is in a time frame oblivious of both antecedent and subsequent instants. Often, chemists will import from physics their tool-box and they will pay negative duty by black-boxing the physics in the process! Often, chemists will satisfy their urge to bring the invisible into visibility and visualisation with their tools of trade, chemical reactions: revealing a latent image and then fixing it, photography-like, as in many an instance of chromatography.

Acknowledgment

I wish to thank the organising committee for having invited me to a very enjoyable conference. Peter Morris, in particular, besides having been a most genial host, was a most attentive and professional editor, to whom I feel most grateful.

References

1 P. Laszlo (ed.), *Preparative Chemistry Using Supported Reagents*, Academic Press, San Diego, CA, 1987.

2 P. Laszlo and P. J. Stang, *Organic Spectroscopy. Principles and Applications*, Harper's Chemistry Series, Stuart Alan Rice (ed.), Harper and Row, New York, 1971.

3 P. Laszlo (ed.), *NMR of Newly Accessible Nuclei*, Academic Press, New York, 1983.

4 R. Bud and D. J. Warner (eds.), *Instruments of Science. An Historical Encyclopedia*, The Science Museum and the Smithsonian Institution, New York and London, 1998.

5 R. G. W. Anderson, 'Balance, Chemical', in *Instruments of Science. An Historical Encyclopedia*, R. Bud and D. J. Warner (eds.), The Science Museum and the Smithsonian Institution, New York and London, 1998, pp. 45–47.

6 P. Morris, 'Chromatograph', in *Instruments of Science. An Historical Encyclopedia*, R. Bud and D. J. Warner (eds.), The Science Museum and the Smithsonian Institution, New York and London, 1998.

7 R. G. W. Anderson, 'Distillation', in *Instruments of Science. An Historical Encyclopedia*, R. Bud and D. J. Warner (eds.), The Science Museum and the Smithsonian Institution, New York and London, 1998, pp. 182–184.

8 J. H. Beynon and R. P. Morgan, 'The Development of Mass Spectrometry. An Historical Account', *International Journal of Mass Spectrometry and Ion Processes*, 1978, **27**, 1–30.

9 T. Lenoir and C. Lécuyer, 'Instrument Makers and Discipline Builders: The Case of Nuclear Magnetic Resonance', *Perspectives on Science*, 1995, **3**, 276–345.

10 P. J. T. Morris, A. S. Travis and C. Reinhardt, 'Research Fields and Boundaries in Twentieth-century Organic Chemistry', in *Chemical Sciences in the 20th Century. Bridging Boundaries*, Carsten Reinhardt (ed.), Wiley-VCH, Weinheim, 2001, pp. 14–42.

11 E. A. Braude and F. C. Nachod, *Determination of Organic Structure by Physical Methods*, Academic Press, New York, 1955 and 1962.

12 D. Baird, 'Analytical Chemistry and the "Big" Scientific Instrumentation Revolution', *Annals of Science*, 1993, **50**, 267–290.

13 N. C. Russell, E. M. Tansey and P. V. Lear. 'Missing Links in the History and Practice of Science: Teams, Technicians and Technical Work', *Hist. Sci.*, 2000, **38**, 237–241.

14 D. Dowling, 'Experimenting on Theories', *Science in Context*, 1999, **12**(2), 261–271.

15 S. Weininger, 'What's in a Name? From Designation to Denunciation – the Nonclassical Cation Controversy', *Bulletin for the History of Chemistry*, 2000, **25**, 123–131.

16 *Agenda du Chimiste à l'Usage des Ingénieurs, Physiciens, Chimistes, Fabricants de Produits Chimiques, Pharmaciens, Essayeurs du Commerce, Distillateurs, Agriculteurs, Fabricants de Sucre, Teinturiers, Photographes, etc.*, Hachette, Paris, 1887.

17 F. W. McLafferty, *Interpretation of Mass Spectra*, W. A. Benjamin, New York, 1966; K. Biemann, *Mass Spectrometry, Organic Chemical Applications*, McGraw-Hill Book Co, New York, 1962; H. Budzikiewicz, C. Djerassi and D. H. Williams, *Interpretation*

of Mass Spectra of Organic Compounds, Holden-Day, San Francisco, 1964.

18 P. Laszlo, *Miroir de la Chimie*, Collection Science Ouverte, Le Seuil, Paris, 2000.

19 E. S. Gould, *Mechanism and Structure in Organic Chemistry*, Holt, Rinehart and Winston, New York, 1959.

20 G. Stork *et al.*, 'The First Stereoselective Total Synthesis of Quinine', *J. Am. Chem. Soc.*, 2001, **123**, 3239–3242.

21 K. Mislow and M. Raban, 'Stereoisomeric Relationships of Groups in Molecules', in *Topics in Stereochemistry*, N. L. Allinger and E. Eliel ((eds.)), Interscience–Wiley, New York, 1967, Vol. 1.

22 A. Cambrosio and P. Keating. 'Of Lymphocytes and Pixels: The Techno-visual Production of Cell Populations', *Studies in History and Philosophy of Biological and Biomedical Sciences*, 2000, **31C**(2), 233–270.

23 J. Schummer, 'Scientometric Studies on Chemistry, I. The Exponential Growth of Chemical Substances', *Scientometrics*, 1997, **39**, 107–123.

24 J. Schummer, 'Scientometric Studies in Chemistry, II. Aims and Methods of Producing New Chemical Substances', *Scientometrics*, 1997, **39**, 125–140.

25 J. D. de Solla Price, *Little Science, Big Science*, Columbia University Press, New York, 1963; J. D. de Solla Price, *Science Since Babylon*, Yale University, Press, New Haven, CT, 1975.

26 R. G. Lawler, 'Spectrometer, NMR', in *Instruments of Science. An Historical Encyclopedia*, R. Bud and D. J. Warner (eds.), The Science Museum and the Smithsonian Institution, New York and London, 1998, pp. 556–558.

27 K. A. Nier, 'Spectrometer, Mass', in *Instruments of Science. An Historical Encyclopedia*, R. Bud and D. J. Warner (eds.), The Science Museum and the Smithsonian Institution, New York and London, 1998, pp. 552–556.

28 P. J. T. Morris and A. S. Travis. 'The Rôle of Physical Instrumentation in Structural Organic Chemistry', in *Science in the Twentieth Century*, J. Krige and D. Pestre (eds.), Harwood, Amsterdam, 1997, pp. 715–740.

29 Y. M. Rabkin, 'Technological Innovation in Science: The Adoption of Infrared Spectroscopy by Chemistry', *Isis*, 1987, **78**(1), 1–54.

30 R. Freeman, *Spin Choreography. Basic Steps in High-resolution NMR*, Spektrum, Oxford, 1997.

31 P. Galison, *Image and Logic. A Material Culture of Microphysics*, University of Chicago Press, Chicago, 1997.

32 A. Pickering, *The Mangle of Practice: Time, Agency and Science*, University of Chicago Press, Chicago, 1995.

33 P. Laszlo, *La Parole des Choses, ou le Langage de la Chimie*, Collection Savoir, Hermann, Paris, 1993.

34 G. Genette, *Palimpsestes. la Littérature au Second Degré*, Poétique, Le Seuil, Paris, 1982.

35 P. Laszlo, 'Chemical Analysis as Dematerialization', *Hyle-International Journal for the Philosophy of Chemistry*, 1998, **4**, 29–38.

The Impact of Instrumentation on Chemical Species Identity From Chemical Substances to Molecular Species*

JOACHIM SCHUMMER

Institute of Philosophy, University of Karlsruhe, D-76128 Karlsruhe,
Germany, Email: Joachim.Schummer@geist-soz.uni-karlsruhe.de

Introduction: The Ontological Issue of Identity

Every science is about certain objects or entities. Talking about objects in an unambiguous and consistent way requires a notion of identity. Usually, scientists take that for granted or implicitly use a specific notion without further reflection. The notion of identity thus becomes part of their habitual ontological commitment. Sometimes, however, ontological issues emerge in the forefront of scientific research, such as the well-known ontological problems of quantum mechanics. While philosophers of science have much debated the quantum mechanical issues, every classificatory science is more involved in ontological issues than physics. That is particularly true for chemistry, since chemists are deeply involved in classificatory problems. Just imagine the painstaking efforts chemists have made to distinguish and characterise the millions of substances that exist today, most of them being white powders indistinguishable to the naked eye.

Many crucial episodes in the history of chemistry are related to identity issues. Take for instance the alchemical quest for making precious metals such as gold. Many sceptics argued that, for ontological reasons, artificial gold cannot be the same as natural gold; even if all empirical properties are the same, artificial and natural substances are strictly distinct. They referred to a non-empirical concept of substance identity, for which they found support in Galen's and Avicenna's odd interpretation of Aristotle's distinction between natural things and arte-

*This paper is a revised version of a chapter of *Chemical Relations* (book in preparation).

facts.[1] A second, and related, example is the debate between vitalists and anti-vitalists in the nineteenth century. Marcellin Berthelot's enormous efforts in synthetic organic chemistry provided rich material evidence for his anti-vitalist claim that there is no ontological difference between natural and artificial organic substances.[2] A third example, which eventually paved the way for the modern concept of elements, comes from eighteenth century Swedish mineral chemistry. Against the prevailing assumption that nickel, cobalt and manganese were modifications or mixtures of the few accepted metals, Axel Cronsted and Torbern Bergman argued that they had to be considered as distinct metals, because they possessed individual and constant properties and could not be made from the other metals.[3] As we will see later, Bergman first formulated and consistently applied a rather modern concept of substance identity. The most important example, however, comes from the change of the concept of elements. A change of substance identity, in Aristotelian terms, could become a simple change of the state of aggregation, such as ice to liquid water.[4]

Let us turn now to philosophical formulations of the concept of identity. According to the famous law of Leibniz, two entities (x and y) are identical, if and only if they possess exactly the same properties (ϕ):

$$x = y \equiv \forall \phi \{\phi (x) \equiv \phi (y)\}$$

There has been much debate in philosophy about which (kind of) properties shall be considered in this definition. A deep-seated tradition prefers a version of the Principle of Identity, according to which all properties, including space-time coordinates, must be the same to hold identity, *i.e.* numerical identity. Others distinguish between so-called 'intrinsic' and 'extrinsic', 'pure' and 'impure', and so on properties to derive various weaker concepts of identity. In the present context, we need not bother much about these sophisticated distinctions, mainly because they are irrelevant for the question of identity in chemistry and tailored to some other specific areas (as philosophers are accustomed to ignoring chemical issues). Instead, we can confine ourselves to the more liberal concept of relative, qualitative or species identity. If we take material samples as the primary entities of chemistry, chemical species identity can be defined in terms of properties that are regarded as essential for the sameness of chemical species. Two material samples are chemically identical if and only if they possess all the same essential properties. If they differ in only a single essential property then they belong to different species.

Which properties, then, count as chemically essential? The problem is far from being trivial. As we will see later, the scope of essential properties even has an impact on what we count as a chemical entity. Furthermore, as the identity issue is an ontological problem, about the abstract structure of the world, it should be distinguished clearly from the epistemological problems of whether and in what manner chemical identification can actually be achieved. Chemical identification presupposes a concept of species identity. Such a concept is also presupposed when a sample of a new species is found or produced because its novelty must be substantiated by establishing the identity of that species in terms of its essential properties.

The concept of species identity, while being prior to epistemological questions, need not be fixed once and for ever. As we will see, the concept has changed several times in the history of chemistry. Since the present paper focuses on these changes, it is about the ontological attitude of chemists, rather than about ontology in the received view.

The main topic of the paper is the tremendous impact of twentieth century spectroscopic instrumentation on chemical species identity. In order to understand this, we need to consider how species identity was defined and determined in classical modern chemistry. Moreover, since there has not been an explicit discourse about the concept of species identity in chemistry, the ontological attitude of chemists must be derived from their actual practice as reported in the literature. I will mainly concentrate on how chemists have established the identity of new species in synthetic organic chemistry.[5]

Historical Steps Towards Canonical Substance Characterisation

I will use the term 'classical modern chemistry' for a scientific approach which gradually superseded both metaphysical and mere observational attitudes in favour of experimental and operational approaches, before the rise of spectroscopic methods. As we have already seen, chemistry was largely bound to a metaphysical concept of substance identity until the late eighteenth century, if not later in some regards. Whatever observational properties two material samples had in common, they could be considered different because of different underlying metaphysical principles. Two material samples may, indeed, look the same, but turn out to behave quite differently in certain contexts. On the other hand, two samples may have a different appearance (*e.g.* because of different modification forms) but behave similarly in chemical contexts.

The classical modern approach to substance identity was first formulated and consistently applied by Torbern Bergman. He defined substance identity in terms of composition which was operationally related not only to experimental analysis but also to experimental synthesis: 'In investigating the principles of a body, we must not judge them from a slight agreement with other known bodies, but they must be separated directly by analysis, and that analysis must be confirmed by synthesis'.[6] As James Llana has pointed out, Bergman clearly disregarded the philosophical elements for the purpose of substance identification. Instead, composition had to be formulated in terms of simple substances, *i.e.* the outcome of final experimental analysis which then must be the starting point of experimental synthesis. As a central part of the so-called chemical revolution, Bergman's approach, which was basically restricted to minerals, was later enlarged by Lavoisier and others and put on the sound basis of chemical elements as the final products of chemical analysis.[7] It was the task of the succeeding generation to develop experimental standard procedures for elemental analysis particularly of organic substances,[8] and it took even more generations to establish the required synthetic capacities for organic substances.

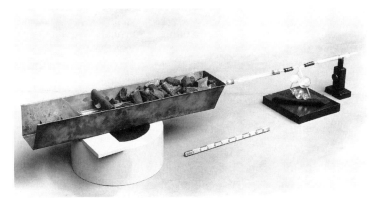

Reproduction of Justus Liebig's celebrated combustion train for organic analysis. Science Museum collections, inventory number 1914-308. Liebig claimed his apparatus allowed 400 combustions to be made in a year, a great saving of time on earlier methods. Science Museum/Science and Society Picture Library

In Bergman's, as well as many earlier chemical writings, we find a more or less implicit notion of *purity* of substances. If chemists spoke of substance identity, be it in the metaphysical or in the chemical sense, they usually related this concept only to 'pure substances'. Surprisingly, there is very little known about the history of that most fundamental chemical concept.[9] However, we have good reasons to believe that the standard methods for purifying liquids and solids remained much the same from the mid-eighteenth century onwards, mainly distillation, sublimation and recrystallisation. Carl Scheele, for instance, frequently used solubility and crystal form or boiling point as qualifying properties of his reagents,[10] which suggests that he routinely recrystallised or sublimed his solids and distilled his liquids. And since there is, for basic reasons, no way to define the concept of substance purity other than by referring to operational methods of purification,[11] we may assume that our concept of purity was already (implicitly) well established in the eighteenth century.

Hence, purified material samples were the objects for which species identity was determined, and this was done by providing the elemental composition based on both experimental analysis and synthesis as well as by some auxiliary properties. That approach remained basically the same until about 1950. Let us see now how chemists usually characterised, and thereby fixed the identity of new chemical substances in the first half of the twentieth century.

Canonical Characterisation of New Substances

If one examines the chemistry journals published between 1900 and 1950, one finds a schematic method of characterising new substances. This 'canonical form of substance characterisation', during the period of the classical chemistry, consisted of six categories:[12]

1. detailed description of preparation from starting materials, including puri-

fication method (distillation, recrystallisation) and yield;
2. results of elemental analysis including empirical formula (occasionally molecular weight);
3. melting point or boiling point (including pressure if vacuum distillation is applied);
4. visual characteristics (crystal form, colour);
5. solubility in various solvents;
6. some exemplary chemical reactivities.

Surprisingly, from a systematic point of view, neither one nor all of these properties suffice to fix chemical species identity in an unambiguous way. Imagine you have a million data sets of that kind and want to decide if each constitutes a distinct chemical species. Given the imprecision of qualitative properties 4 to 6 and the limited precision of quantitative data 2 and 3 as well as the possibility of isomers, we would expect to find many doublets and triplets in a database of distinct chemical substances. Moreover, since there can be more than one way to produce a certain substance, property 1 would make us distinguish substances that we otherwise consider to be the same. Pragmatically, these problems might have been negligible in the nineteenth century, but by the mid-twentieth century nearly a million chemical substances had been found.

From a systematic viewpoint, the problem is even more serious, because the number of physical and chemical properties is unlimited.[13] Thus, if we extend the canonical set and consider all physical and chemical properties as essential properties, there are infinitely many properties to be considered. Two samples belong to the same chemical substance, if all of their essential properties are the same. Even if they differ by only one essential property, they belong to different chemical species. Hence, since it is practically impossible to determine infinitely many properties, we can never conclude, on a logically sound basis, that two samples belong to the same chemical substance. All identity claims in chemistry, based on an open set of essential properties, are necessarily only provisional.

Wilhelm Ostwald was the only chemist who fully appreciated the challenge this problem presented to the fundamentals of chemistry. The way he tackled the issue on the first page of his *Grundriß der Allgemeinen Chemie* is telling. He simply invented a first law of nature (*Naturgesetz*): 'If two substances correspond in some properties, then they correspond also in all other properties'.[14] Ostwald should have known that this is too good to be true. Given a finite level of measurement precision and a certain range for 'some' quantitative values of material properties, simple mathematics tells us how many possible substances can be distinguished by these properties; and how many doublets we should expect among a given number of substances. Even if we refer to the 'qualitative' chemical properties, the case of different isomers that correspond in many but not all properties disproves Ostwald's 'law of nature'. The 'law' was simply an anachronistic reference to the canonical form of substance characterisation. By the time Ostwald formulated his law, the mere quantity of known chemical substances had already practically undermined such simplistic solutions.

In fact, chemists did not trust their canonical characterisations, let alone

Ostwald's 'law'. This is well documented in every chemical paper where *practical substance identification* was required on a safe basis. If chemists wanted to prove that the outcome of a new reaction was chemically identical with an already known substance, they did not rely on comparing characteristic properties. Until the 1960s, instead of comparing data, the usual way to prove identity with an 'authentic' substance was by mixing the two samples physically and examining mixing effects, like melting point depression. For that purpose, either the 'authentic' substance was reproduced according to the literature or even a sample was physically handed over from a colleague, which was usually acknowledged in the paper. That is to say, chemists actually relied on an *operational concept of substance identity* based on mixing samples instead of comparing essential properties.

We may conclude that canonical characterisation served to fix chemical species identity only within a certain context. It allowed chemists to identify a chemical substance among the other reaction products of the described preparation procedure. Chemists familiar with the corresponding substance classes and types of reaction may have used more or less implicit knowledge that helped them grasp substance identity by tacitly excluding implausible candidates. The specific form in which this knowledge had become explicit was chemical structure theory, an ingenious approach to the chemical identity issue.

Support from Chemical Structure Theory

Nowadays we are inclined to consider chemical structure theory as the central step of entering the microcosm of molecules and atomic structure. However, such a view tends to overlook the specific chemical problems that structure theory helped to solve. As regards the problem of chemical species identity, structure theory was an invaluable means, particularly in the realm of organic chemistry where the rapid proliferation of substances had caused enormous problems.

Let us first recall the identity problem. Since there are indefinitely many characteristic properties in which chemical substances can differ, one has to determine and compare indefinitely many properties of two samples in order to prove their substance identity, which is impossible. Hence, all identity claims in chemistry based on an open set of characteristic properties are necessarily only provisional.

How did chemical structure theory help to solve the identity problem? The fundamental (either metaphysical or methodological[15]) assumption of structure theory was that there is exactly one characteristic chemical structure for every chemical substance. By that assumption, chemical species identity was transferred to a theoretical level, *i.e.* substance identity should correspond to structure identity and *vice versa*. As a consequence, experimental structure determination of a substance was at the same time the fixing of chemical substance identity.

Of course, chemical structure determination was also made with the help of empirical properties, above all, the chemical properties. The ingenious way to determine substance identity *via* structure determination consists in the *careful selection of a few chemical properties among the infinitely many characteristic properties.* There are neither definite selection rules nor a fixed canon of properties, such as the canonical characterisation. Instead, for each chemical substance it is up to the chemists to select and find those chemical properties that allowed an unambiguous assignment of a certain structure. Sometimes only a few reaction properties are necessary to infer the structure of a new substance from other already well defined compounds. By contrast, large research groups worked for years on the structure determination of a single substance, by devising sophisticated steps of controlled decomposition, partial synthesis, or modification. That made structure elucidation intellectually satisfying for many chemists.

The number of properties for fixing substance identity thereby shrank from infinity to a carefully selected set. Moreover, substance identification no longer required the physical presence of a reference sample, nor the comparison of physical data, but only the comparison of two structures obtained by independent procedures of structure determination. When the first complete records of known structures were set up and ordered in a systematic way, chemists could easily check whether their new products were really new or identical with already known substances.

One might wonder why the canonical characterisation of new substances remained an obligatory part of every chemistry paper. Chemical structure determination only gradually became a constitutive part of chemical papers, and only in organic chemistry. One can even find papers in the 1950s in *Liebigs Annalen* with careful listings of canonical characterisations, but without any constitutional formula. I can only assume that chemical structure elucidation remained too 'hypothetical' to comprise the exclusive basis for substance identity, at least for many chemists. Furthermore, parts of the canonical characterisation had their own values. They presented the 'uninterpreted' experimental properties which formed the basis of the structure determination. They met the laboratory requirements of relative substance identification within experimental contexts. In addition, other chemists needed to know how to carry out the synthesis in order to reproduce the results as well; and as such, the preparation was a central part of a scientific paper.

The Social Side of Substance Identity Claims

As I have argued elsewhere, the production of new substances has been the central chemical activity during the past two centuries. In this context, the issue of substance identity also has an important social dimension. For ordinary chemists, the production of new substances is their main contribution to the progress of chemistry. Unlike other scientists, chemists are not only authors of ideas and papers, they are also creators of new kinds of material entities. However, they have to determine the identity of a certain entity before they can claim certain priority rights. Thus, fixing the identity of a new substance means at the same time making an unambiguous claim to an original contribution.

Everybody familiar with the history and sociology of priority claims in science[16] knows how important clear rules are for scientific progress. This is particularly true if one has hundreds of thousands or even millions of such claims, instead of some hundred spectacular 'discoveries'. In addition, priority claims to new substances also play a central rôle for patent systems, and thus need a sound juridical basis. In order to avoid priority struggles and parallel research one needs a powerful classification system, based on established criteria of species identity and able to incorporate indefinitely many new species. Moreover, it should make it easy to check whether a certain species is already known.

The situation was particularly difficult in the mid-twentieth century. The two world wars as well as nationalistic trends split the scientific publication and documentation systems. Moreover, the number of chemical substances grew from about 100 000 in 1900 to nearly one million in 1950, about 95% being organic compounds. *Beilstein*, the main reference handbook of organic chemistry, did not meet the needs of chemists. The literature prior to 1929 was not covered until 1956, and even then, no attempt was made to distinguish stereoisomers. Furthermore, the section covering the literature of the period 1950–1959 was not even completed until 1987. Of the one million organic substances produced before 1959 and covered by *Beilstein's Centennial Index* of 1992, we find hardly any spectroscopic characterisation. Instead, the identity is determined roughly by the insufficient canonical characterisation, including constitutional or configurational information if available. Checking if a certain substance has already been covered by *Beilstein* is a very ponderous procedure and requires sophisticated or even implicit knowledge in many cases. Moreover, such a check was obviously not considered to be reliable by chemists, since they continued to 'prove' identity with authentic substances by mixing the samples until the 1960s.

In the early 1960s, the *Chemical Abstracts* Service (CAS) realised that chemists needed a better system that provided quick and reliable 'identity checks'. CAS 'developed an algorithm for generating a unique, unambiguous machine-language representation of the two-dimensional structure of a chemical compound, together with a method for recording additional data, such as stereochemistry. This algorithm became the foundation of the CAS Chemical Registry, a computer-based system that automatically identifies structural diagrams and assigns to each a unique CAS Registry Number.[17] Established in 1965, the registry system soon became the definitive reference source for checking and claiming substance identity.

Beilstein and the CAS system followed completely different strategies of database management. Following the natural history tradition, *Beilstein* collected as many properties as possible for each substance. That was meant to help future generations of chemists, but it completely ignored the needs of contemporary chemists concerning substance identification. When the *Beilstein Centennial Index* with one million substances appeared in 1992, the CAS system had already registered another 10 million substances, mainly synthetic organic compounds not isolated from natural sources. Since the CAS system was kept up-to-date, chemists could easily use the system to lay claim to the novelty of their substances. I suspect that, unlike CAS, the *Beilstein* team failed to recognise the issue of substance identification and its social significance for making novelty claims.

The Impact of Instrumentation: Introductory Remarks

Classical chemical structure theory was largely based on chemical properties. From a logical point of view, these differ from physical properties in that they establish relations between chemical substances, between reaction partners and reaction products connected by chemical transformations. Systematically arranging all chemical knowledge yields a network structure in which every substance is related to every other substance by direct or indirect links. In such a network, the identity of each substance corresponds to its network location defined by its characteristic relations to other substances. As I have argued elsewhere, classical chemical structure theory reproduces that network on a theoretical and sophisticated level, such that the relations between substances correspond to relations between chemical structures.[18] If the identities of chemical substances are defined by relations with each other, as in the classical approach, then chemical substances are relational entities.

Spectroscopic methods, on the other hand, provide physical properties, mainly electromagnetic properties, on various sophisticated levels. These physical properties do not describe relations between different chemical substances, but the response of an isolated material sample to electromagnetic fields. However small the logical difference between chemical and physical properties may appear to non-philosophers, it became important in the second half of the twentieth century when spectroscopic methods gradually replaced the classical chemical approach of substance identification. The success story of instrumentation in chemistry, to which we now turn, is also a success story of physical properties, in the course of which the concept of species identity was modified and adapted to physical properties. At the preliminary end of that story, namely the present, not only the concept of species identity but also the kind of species have changed, from chemical substances to quasi-molecular species. In order to analyse this process on a sound empirical basis another historiographic method is required.

A Preliminary Note on Historiographic Method

Historians of chemistry, as well as chemists, are strongly challenged by the enormous amount of scientific work produced during the twentieth century. Nearly everything is now a hundred times larger than in 1900. We have now a hundred times more chemists, chemical papers, and chemical substances than at the beginning of the twentieth century.[19] Nearly any development plucked from this explosive growth can be presented as a 'scientific revolution' in one or the other inflationary meaning of that term, despite the fact that most developments show a steady annual growth rate of about 5–6%. Moreover, historians of science need to refer to primary sources that are expected to be in some sense representative of the subject under study. Against the current background of some three million chemists and more than 700 000 chemical publications a year, every selection runs the risk of being arbitrary. Nowadays, we have no Boyle, Lavoisier, Berzelius or Liebig whose opinion about chemistry might count as representative. For basic reasons, there is definitely no chemist with an overview of about more than 10% of his field, the average overview being less than 1%.[20] Thus, historians of twentieth century chemistry have to focus on particular events, small circles, or extremely narrow scientific topics, if they wish to apply traditional historiographic methods in a serious manner.

For the topic of the present paper, standard historiographic methods are unsuitable. To understand how the development of instrumentation has affected the ontological attitudes of ordinary chemists, it does not suffice to refer to some selected writings. Even if there were an explicit discourse among a handful of chemists about such things, which is actually not the case, these opinions could hardly count as representative. In order to understand the attitudes of ordinary chemists, the millions of working chemists, we have to make a *random* selection in the statistical sense. The advantage of statistical methods is that we can clearly indicate in what sense the results are representative, including an error estimate. All we need to do is:

1. to develop reasonable categories
2. make a random selection of chemical papers of a given type; and
3. perform a document analysis of the papers with the help of the categories.

I have used this method in an earlier paper to successfully analyse such difficult issues as the aims that chemists are following in their research.[21]

For the purpose of the present paper, I analysed hundreds of randomly selected papers published in *Liebigs Annalen der Chemie* during the past 150 years in the above-mentioned method. The only pre-selection criteria were that the papers should be from different authors/research groups and should belong to the field of synthetic organic chemistry; *i.e.* preparation and characterisation of *new* substances should be a central part of the paper. This excludes papers from natural product chemistry, in which the natural sources serve to identify the compounds. Among chemistry journals, *Liebigs Annalen* has traditionally a high standing in synthetic organic chemistry, and a reputation for keeping strict and conservative standards in the presentation of results. Until recently, authors were mainly from Germany and some other European countries, usually belonging to an 'elite' with instrumentally well-equipped laboratories. As for the categories, I mainly noted what properties and what instrumental methods were used to characterise the new substances.

The Rise of Spectroscopic Characterisation Since 1950

It is now well-known that spectroscopic methods caused fundamental changes in the sub-discipline called 'analytical chemistry' around 1950.[22] There was a lag before there was any corresponding impact on synthetic chemistry, as one might expect. Before 1950, chemists followed the canonical characterisation in such a uniform manner that I can omit the quantitative results. Only after 1950 did a major change begin, which is still taking place today. The top line in the graph on page 199 provides an approximate idea of what has happened during the past 50 years. The average number of spectroscopic methods (including X-ray diffraction) used for the characterisation of new organic compounds rose from nearly

*The Varian CFT-20, developed in 1969, was the first routine Fourier-transform (FT)
spectrometer for C-13 NMR. With a dedicated minicomputer, it was designed for the
typical laboratory budget, and during the early 1970s, led to the widespread use of C-13
NMR in organic chemical laboratories. Photograph courtesy of Varian Inc.*

zero in 1950 to 4.5 in 2000, with the most rapid growth taking place in the 1960s.

In 1950, the only spectroscopic method used to any extent (about 35%) was
visible and ultraviolet absorption spectroscopy, which been readily available
since about 1940.[23] Interestingly, this method has never become a standard
procedure, but was used in the following decades only for some 30% of substances whose colour (dyes) or electronic structure attracted some interest among
chemists. Not so for all the other methods presented in the graph opposite.[24]
Infrared spectroscopy (readily available since about 1943) and even more H-
NMR spectroscopy (readily available since about 1962) rapidly became standard
methods and remain so today. Already in 1970, some 80% of all new organic
substances were characterised by both infrared and H-NMR data. Since then,
the much more expensive methods of mass spectroscopy (available for qualitative organic analysis since the late 1950s[25]) and C-NMR (since the late 1960s, see
Figure) have been steadily growing to around 80% today. The most recent trend
in organic chemistry is the use of X-ray diffraction for crystallographic analysis
of new compounds, which is now applied to around 50% of all compounds. We
will come back to this strikingly late adoption of crystal analysis by organic
chemists as compared to inorganic chemistry (single crystal diffraction goes back
to 1912, powder diffraction to 1918).

Of course, the graph presents only rough trends in terms of the application of
general methods. During the decades, every method has been considerably
improved, which is reflected in the presentation of spectroscopic data. For
instance, NMR data, while being confined to a few chemical shift data in 1970,
are now presented as extensive listings of shift data and spin–spin coupling
constants, occasionally combined with the double-resonance technique. The
main refinements of IR spectroscopy, Fourier transform technique and the

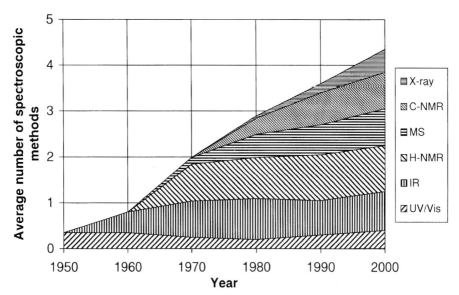

Average number of spectroscopic methods (including X-ray diffraction) used for the characterisation of new solid organic compounds in (full) papers of Liebigs Annalen (2000: European Journal of Organic Chemistry). *Random selection of 20 papers at the beginning of each decade (for each method ε ⩽ 11–18%, α = 0.1). Note that the 'thickness' of each layer represents the relative frequency of each method; a thickness of 1.0 means that 100% of the papers apply the corresponding method*

application of laser (pulses), improved both precision and short-term applicability. The establishment of high-resolution mass spectroscopy caused an extraordinary proliferation of data, from a single peak, the molecular weight, in the 1970s to long series of decomposition peaks today. All spectroscopic data are presented now as listings or tables of values, usually in the 'Experimental' section that has grown correspondingly, while the more informative spectrographic plots had only a short career in the 1970s.

Nowadays, the number of spectroscopic data determined for each new compound has grown to such an extent that chemistry journals have been forced to put firm restrictions in place, in particular, for crystallographic data. Thus, the 1995 'Instructions for Authors' of *Liebigs Annalen* states that 'X-ray structural analyses will only be accepted if they contribute to the solution of a chemical problem and if the crystallographic features are unique'. Even then, 'The structural description should be restricted to selected significant parameters'. Additional data are to be deposited at the Fachinformationszentrum (FIZ), Karlsruhe, for possible retrieval by individuals at a later date. More recently, some authors refer to private World-wide Web pages where interested readers may find more data. Since many chemistry journals are now publishing a parallel internet version, they allow authors to deposit additional data on the journal's website. However, it is far from clear whether this policy meets the needs of readers rather than the needs of authors. In any event, there is a tremendous

A modern infrared spectrometer. Courtesy of Geoff Lane/CSIRO/Science Photo Library

pressure from the authors' side to publish as much data about their new substances as they can.

Parallel Changes

In parallel with the rise of spectroscopic methods, other changes in the characterisation of new substances have occurred. The short career of the refraction index of liquids is quite interesting, because that was the only conventional optical property (in a rudimentary form, the refractive index of several liquids was measured by Ptolemy in the second century). At its peak in 1960, more than 60% of organic chemists reported the refraction index of their liquids. Since then, the reporting of the refractive index has declined steadily to less than 10% today. In addition, most of the properties listed in the classical canonical characterisation have been increasingly disregarded. As one might expect, the visual characterisation of crystal form does not play much of a rôle nowadays (about 30% in 2000 as compared to 90% in 1950), whereas colour (or to be more accurate, the absence of colour) is still reported by some 60%. Nowadays, chemists typically use phrases such as 'a white solid' (not 'white crystals') or 'a colourless oil', which means nothing more than a successful purification. Reports about exemplary reactivities and solubilities have become rare or reported only implicitly, for instance, noting the solvent used for recrystallisation/chromatography or by further synthetic steps.

Melting points are still used for the characterisation of solids (80%), whereas boiling points for liquids have declined to less than 20%. It is sometimes even impossible to recognise whether the new substance is a liquid or a solid! The reason behind that is a fundamental change in purification methods in organic chemistry. Since about 1970, chromatography has become an important alternative for both liquids and solids. Today, about 80% of the liquids and more than 60% of the solids receive their final purification by a chromatographic technique. Nonetheless, only a few chemists report chromatographic retention values to characterise their liquids (currently about 20%) or solids (< 10%), probably because of low reproducibility and precision.

Of the canonical characterisation methods, only the detailed (but now much more condensed) description of the preparation procedure and the results of the elemental analysis remained practically unchanged. However, elemental analysis appears to have lost its rôle of providing a characteristic chemical property. As the 1995 'Instruction for Authors' of *Liebigs Annalen* suggests, it is only used to establish the purity of the sample, and thereby validating the spectroscopic data: 'The *purity* of all new compounds must be verified by elemental analysis'. (Emphasis in the original.)

Towards Spectroscopic Substance Identity

Let us turn now to the central question – how the instrumental development has affected the concept of chemical species identity. The way chemists have implicitly dealt with this issue suggests three more or less successive approaches. The first two are adaptive strategies; the third one, in some sense a consequence of the second one, has led to a fundamental ontological change.

The Fingerprint Approach

At the earliest stage, until the early 1970s, the canonical characterisation of new compounds remained relatively robust. Spectroscopic data, in particular infrared data, were considered as additional characteristic features of new compounds.[26] First of all, spectroscopic data are simply a series of optical material properties, *i.e.* absorption coefficients at different wavelengths. Given a certain precision, spectrographic plots provide sufficiently rich information to distinguish between millions of pure substances. More than any property of the canonical characterisation, including the refraction index that was temporarily established for that purpose, such a plot meets the requirement for determining substance identity in an unambiguous way, it provides a 'fingerprint', as chemists began to call it.

Such fingerprints, or the reduced form of a set of characteristic peak data, were soon applied to substance identification. Instead of determining mixed melting points with 'authentic samples', some chemists began to prove identity by comparing the spectra. The first obstacle was of course the lack of spectra of known substances. When a considerable number of spectra were recorded, the next problem was how to match a new spectrum with thousands of spectra of the library. As early as the 1950s, before the rise of computers, a mechanical device was already constructed for that purpose.[27] For each substance, the characteristic infrared data were coded by punched holes on a paper card. The complete set of cards were then machine-sorted, such that two cards with same holes could

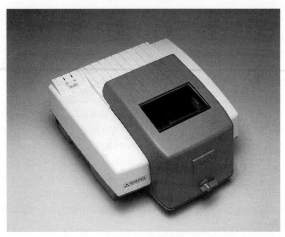

*Nicolet 205 FTIR spectrometer, c. 1992. Science Museum collections, inventory number
E1999.6785. The introduction of FT (Fourier-transform) techniques transformed infrared
spectroscopy. In contrast to the early FTIR spectrometers, the Nicolet 205 is easy to use.
The spectra obtained from it can be checked immediately with infrared spectra libraries
stored on disk. Science Museum/Science and Society Picture Library*

automatically be identified. However, the mechanical device never found its way
to the standard equipment of chemical laboratories.

It is well known that the fingerprint approach has experienced an extraordinary revival since the digitisation of spectrometers. Modern standard spectrometers of all sorts are now routinely equipped with computer databases
(so-called libraries) that allow 'fingerprint identification' while recording the
spectra.

The Structure Determination Approach

We have seen that chemical structure determination was the classical method of
overcoming the insufficiency of the canonical characterisation. The identity of a
chemical substance was fixed by collecting as many chemical properties as
needed for an unambiguous structural interpretation. This was largely accomplished by assigning 'functional groups' to a 'carbon skeleton' according to
certain chemical properties. Very soon, chemists realised that both infrared and
NMR spectroscopic data provided useful clues for that purpose. The rapid
establishment of these two methods, in contrast, for instance, to ultraviolet–visible spectroscopy,[28] occurred because they provided characteristic signals
for many 'functional groups'.

However, this process did not constitute a sudden replacement of the original
chemical method. On the contrary, the method of structure determination
remained basically the same from a logical point of view, chemical and spectroscopic properties were both used in the same way. If the carbon skeleton and the
various functional groups are determined, very often there are still many different constitution formulas possible. A typical structure determination procedure
proceeds by step-wise exclusion of the various possibilities, until only a single

possibility remains. Chemists use all sorts of arguments, be it from chemical or spectroscopic properties,[29] to exclude structural possibilities, as long as the arguments are, in some sense, sound. Thus, spectroscopic properties could be easily incorporated into the chemical approach of structure determination and thus substance identification.

Examining the way chemists argued for their structures, the rôle of spectroscopic data in structure determination has gradually changed together with the type of structure to be determined. Surprisingly, as late as 1960 all new substances were characterised by *constitution* formulas. In contrast to natural product chemistry, *configuration* formulas only became the standard form in synthetic organic chemistry in the late 1980s, with the rise of enantioselective synthesis. Indeed, it is only in the past couple of years, that *conformation* structures have played a significant rôle (more than 10%), mainly in combination with crystallographic analysis by X-ray diffraction.

Up to the 1980s, synthetic organic chemists typically used spectroscopic data in constitution analysis only to confirm structures. After the constitution was determined by means of traditional chemical methods, they argued this result was confirmed by analogy with spectra of known (structurally determined) compounds. At least this is how they explicitly presented their arguments in the main part of their papers. In addition, it very soon became a standard presentation in the so-called 'Experimental' part to give spectroscopic data in an interpreted form, *i.e.* to every infrared peak is assigned a bonding or functional group, to every H-NMR peak a proton. That suggests that confirmation was achieved by a complete structural interpretation of the spectra rather than by mere analogy. When configuration became important in the late 1980s, the type of explicit argumentation changed. I can only assume the constitution was still determined in the traditional manner. There are still a few cases (mainly in natural products chemistry), where authors explicitly claim to have determined the (relative) configuration by purely chemical means. However, the typical argument for a certain configuration gradually became that it is 'proved' by spectroscopic analysis (mainly NMR).

In other words, spectroscopic methods gradually received more weight in structure determination along with the shift from constitutional to configurational analysis. Contrary to a common perception, it is only in the past 10 to15 years that spectroscopy has been accepted by ordinary synthetic organic chemists as an *independent* standard means for the determination of the structure of a new substance, and as such also to determine substance identity.

From Substance Identity to Molecular Species Identity

The fingerprint approach and the structure determination approach (in both its classical chemical and spectroscopic forms) are means to determine chemical substance identity. From a logical point of view, molecular structure, as the outcome of a sophisticated theoretical interpretation of chemical and spectroscopic properties, is a complex theoretical property of material samples and, consequently, of a chemical substance. Insofar as molecular structure serves to determine substance identity, we may consider it an outstanding (but not the only!) essential property of chemical substances. It is important to stress the ontological status: a molecular structure is a *property* of a chemical substance. Speaking in general philosophical terms, substances are ontologically prior to properties: no property without a substance.

I would argue, however, that spectroscopic instrumentation has radically undermined this ontological status. During the final decades of the twentieth century, chemists' ontological attitudes changed radically, a shift unparalleled since the change from the metaphysical to experimental concepts of species identity in late eighteenth century chemistry. Molecular structures are no longer considered *properties* of substances; they are now the species whose identity is to be determined and which are subject to chemical classification. In this situation, any discussion about material substances appears to have become obsolete.[30]

Of course, there was always a metaphysical tradition which preferred to talk about geometrical structures rather than material substances, a tradition which goes back to Plato's *Timaeus*. However, most chemists have remained essentially anti-metaphysical in their attitudes since the late eighteenth century. This lack of interest in philosophy, sometimes even antipathy, allowed chemists to use concepts such as molecular structure in a pragmatic way, as conceptual tools without worrying too much about the ontological implications.[31] Moreover, there is still a certain ambiguity in ordinary chemical language that conceals the ontological issues at stake.[32] Chemists typically use the same terms and names to refer to both material substances and molecular structures. Sometimes they carry this ambiguity to extremes. For instance, after defining a symbol '**11a**' for a suggested configurational structure in the main part of a chemical paper, the authors tell us in the experimental section that they have put 6 g of **11a** in a vessel and then heated it for 2 hours!

What evidence do we have then for an ontological change? And why should spectroscopic methods have had any effect on any such change? The first indirect clue is the parallel decline of the canonical substance characterisation. Nowadays, *material* characterisation of new products is reduced to the utmost minimum for the reproducibility of the preparation; sometimes even if that appears to be a problem, for instance, if no data for the corresponding purification fraction (boiling point or retention value) is provided. However, sophisticated instrumentation now permits spectroscopic characterisation at a microscopic level, where classical methods of substance characterisation fail. If the spectroscopic analysis is directly coupled with chromatographic separation, one can determine molecular structures in-line, so to speak, without any intermediary step of material characterisation. Unlike chemical substances, molecular structures have no boiling points, melting points and colours. If the latter have become the objects of chemical investigation and classification, there is no longer any need for such data.

Another indirect clue is the recent interest in structure analysis by X-ray diffraction. In contrast to most spectroscopic methods, X-ray diffraction does not fit the traditional chemical method of structure determination but is rather a completely independent method that provides structures of quite a different type.[33] Conformational analysis, as one of the major purposes of X-ray diffraction today, was formerly of little interest, since one could rarely put different conformational forms in different bottles, let alone separate reaction vessels. Two conformational forms were considered to belong to the same chemical substance. Thus, even the relatively liberal *Chemical Abstracts* (in a sense, the professional guide to ontological issues) did not register different conformational forms until 1996 at the earliest.[34] The recent interest in conformational analysis suggests a change of perspective in that structural differences at the conformational level of the same substance become constitutive for making a distinction between different species.

There are even more direct proofs of an ontological change or crisis. Spectroscopic methods frequently allow determining molecular structures in mixtures. Hence, the object under empirical investigation need not any longer be a pure chemical substance. One can determine the identity of a molecular structure spectroscopically without having a corresponding chemical substance. Since spectroscopic analysis of mixtures is sometimes subject to failure, chemistry journals usually demand purification including purification tests by elemental analysis. However, there are cases where purification is impossible, simply because a chemical substance corresponding to the molecular structure cannot exist.

Three examples spring to mind. The first one is the so-called 'matrix isolation spectroscopy'.[35] A gaseous reaction mixture is mixed with argon and then rapidly frozen so that intermediary states, which cannot be isolated by purification, are trapped in the solid argon matrix. This matrix is then subjected to spectroscopic investigation. The second one is high-resolution mass spectroscopy, which provides a series of mass peaks which can be interpreted as signals of certain molecular fragment species. The third example is the application of high-speed spectroscopy, usually promoted as 'femtosecond spectroscopy', which uses short laser pulses and Fourier transform analysis. This method enables the determination of molecular structures of intermediates and so-called van-der-Waals complexes,[36] either in solutions or in molecular beam collision experiments, with lifetimes in the range of 10^{-13} s.

All these methods provide data for structural characterisation of molecular or quasi-molecular species, for which we have no corresponding chemical substances.[37] Thus, structure determination no longer serves to determine substance identity alone. Instead, it has become an independent means for determining identity of a new kind of chemical species, molecular or quasi-molecular structures. Once spectroscopy was established as an independent method of structure determination, a fundamental ontological change in chemistry became possible.

Most chemists knew that *Liebigs Annalen* was one of the most conservative journals. Even in the late 1980s, it stubbornly rejected papers about these new species, whereas most other journals, even *Chemische Berichte*, were much more liberal. Only after *Liebigs Annalen* merged with other European journals to form the new *European Journal of Organic Chemistry* in 1998, did reports on the new chemical species begin to appear in its pages. Overall, the number of these quasi-molecular species reported in chemistry journals is still very low, much less than 1%. However, the mere fact that a few became accepted as chemical species by the chemical community, and have even been assigned registry numbers by *Chemical Abstracts*, proves that the fundamental ontological change has already taken place. Officially, the conservative reservations of *Liebigs Annalen* were justified on the grounds of avoiding errors in data interpretation. The journal demanded elemental analysis of isolated chemical substances in order to prove purity, to leave the spectroscopic analysis free of possible errors. However, I suspect that there were tacitly also deep reservations about the ontological change at issue.

Conclusion

Spectroscopic instruments are tools for various purposes. Analytical chemists use them for qualitative and quantitative analysis, *i.e.* for collecting information about what kind and how much of an already known species is present in a given sample. In addition, spectroscopic instruments are used nowadays for the first characterisation of new species, on which analytical chemistry depends. Species characterisation intends to fix the identity of a species by providing sufficiently many properties considered as essential properties. Thereby, a concept of species identity is presupposed which is the topic of the present paper.

We have seen that classical modern chemistry, discarding both metaphysical principles and observational properties, considered experimental properties (mainly chemical properties) as essential for determining the identity of pure chemical substances that were thereby considered the basic species of chemistry. Since that concept of species identity drew on an open and potentially infinite set of essential properties, practical substance identification by comparing properties of samples was, of necessity, provisional and vague. Classical chemical structure theory provided a solution by selecting those chemical properties as essential properties which are sufficient for chemical structures elucidation. This is tantamount to saying that chemical structure, as a complex theoretical property of a chemical substance, is its essential property. However, owing to the hypothetical character of chemical structure elucidation, many chemists remained reluctant to base substance identification on structural identity alone. Instead, in all crucial cases, chemists avoided both the vagueness of the empirical approach and the hypotheticality of the theoretical approach and relied on an operational concept of species identity by mixing samples and investigating mixing effects. Such was the situation in the mid-twentieth century, before the adoption of spectroscopic instruments in chemical laboratories.

At first, spectroscopic methods supplemented and improved traditional approaches of determining substance identity. As to the empirical approach, spectroscopic plots or data sets provided sufficiently rich information to be used as characteristic fingerprints of millions of substances, and thus replaced the operational approach even in crucial cases. As to the theoretical approach, particularly infrared, NMR and mass spectral data provided specific information to be used in chemical structure elucidation, and thus put chemical structures on a more secure basis than before. In the course of instrumental refinement, spectroscopic methods slowly became acknowledged as *independent* and reliable standard means for structure elucidation on both the constitutional and configurational level in the 1980s, which brought about a change in the ontological attitude of chemists in two steps.

Originally, the basic species of chemistry were chemical substances whose identity were determined first by experimental properties and then by molecular structures as complex theoretical properties. In classical chemical structure theory, molecular structures were hypothetical entities whose ontological status each depended on the hypothesis of structure elucidation of the corresponding substance. The more this was supplemented by spectroscopic means on independent grounds, the more chemists conceived of molecular structures as real entities.[38] Thus, chemists no longer considered molecular structures simply as properties of chemical substances; instead, molecular species became ontologically on par with chemical substances. The doubling of chemical species was not reflected by chemists as they used the same terms and names for both kinds; nor had it an impact at first on the practice of species identification, since classical chemical structure theory presupposed as its basic theorem a strict one-to-one relationship between chemical substances and molecular structures (at the configurational level). Nonetheless, there were two different kinds of species waiting for a decision on which should count as the basic one in chemical classification.

The most important impact of spectroscopic methods was that it finally made chemists decide in favour of molecular species. Once established as independent means of structure determination, spectroscopic methods were also used to characterise quasi-molecular species for which there exist neither a corresponding chemical substance nor a classical approach of chemical structure elucidation, such as conformational states, intermediary states in solution, van-der-Waals complexes and molecular fragments in mass spectroscopy. The fact that chemists, in very recent times, have come to consider these quasi-molecular species to be on a par with ordinary molecular species proves that they have ceased to regard chemical substances as the basic chemical species.

This decision has many far-reaching consequences. Chemical classification has become more complex and allows more differentiated concepts than hitherto. However, this change also implies many new ontological and conceptual problems that have probably escaped the attention of most chemists.[39] Species that are defined by spectroscopic properties alone, *i.e.* by their physical properties, cease to be relational entities as chemical substances and classical structural formulas and thus lose chemical information about reactivities. Furthermore, there is a lack of well-defined identity criteria for the new quasi-molecular species owing to the ambiguity of the term 'molecular structure'.[40] For instance, should we consider all possible conformations or all rotational states as different chemical species? How many quasi-molecular species should we recognise in pure water? And should we really re-adopt a change of aggregation state as a change of species identity? Without reasonably selected criteria for species identity of quasi-molecular structures, chemical species classification will clearly collapse. Spectroscopic instrumentation does not provide such criteria. It is simply a tool that is going to challenge chemists to examine their ontological attitudes.

Acknowledgements

At the earliest state of this paper, my brother, organic chemist Dietmar Schummer, helped me to clarify possible misunderstandings. I am grateful also for the many useful comments I received from historians of chemistry at the conference, particularly from the official commentators Arnold Thackray and James Bennett. The final version greatly benefited from detailed and extremely helpful comments from chemists, historians and philosophers: Guiseppe Del Re, Joseph E. Earley, Robin F. Hendry, Roald Hoffmann, Peter Morris, Jaap van Brakel, Stephen J. Weininger and Guy Woolley.

Notes and References

1 *Cf.* R. Hooykaas, 'The Discrimination Between "Natural" and "Artificial" Substances and the Development of Corpuscular Theory', *Archives Internationales d'Histoire des Sciences*, 1947/48, **1**, 640–651; W. Newman, 'Technology and Alchemical Debate in the Late Middle Ages', *Isis*, 1989, **80**, 423–445; B. Obrist, 'Art et Nature dans l'Alchimie Médiévale', *Revue d'Histoire des Sciences*, 1996, **49**, 215–286; J. Schummer, 'The Notion of Nature in Chemistry', in *Proceedings of the Fourth Summer Symposium on the Philosophy of Chemistry and Biochemistry, Poznan, 7–10 August, 2000*, E. Zielonacka-Lis and P. Zeidler (eds.), Poznan, forthcoming. For a reinterpretation of Aristotle's distinction, see J. Schummer, 'Aristotle on Technology and Nature', *Philosophia Naturalis*, 2001, **38**, 105–120.

2 *Cf.* C. A. Russell, 'The Changing Rôle of Synthesis in Organic Chemistry', *Ambix*, 1987, **34**, 169–180.

3 *Cf.* J. W. Llana, 'A Contribution of Natural History to the Chemical Revolution in France', *Ambix*, 1985, **32**, 71–91. Of course, the critical view on the received notion of elements was already prepared by Boyle.

4 There is a wealth of other prominent historical examples of identity issues in chemistry. To mention just three more: (1) In the late sixteenth century, the ontological status of fire, as one of the Aristotelian elements, was called into question and remained so for centuries. The ontological issue was whether fire is a substance at all, identical with itself, or an attribute or a state of other substances. (2) Another example is the ontological status of phlogiston as discussed in the eighteenth century. Different experiments proved either an earth-like (Stahl), or an air-like (Cavendish), or a fire-like (Rouelle) nature, all of which could hardly be reconciled with one self-identical material substance. The pre-nineteenth century history of the chemical elements or principles is particularly rich in variants of ontological status, from mere 'explanatory entities', to property-conferring principles and reified material substances; for more details see J. Schummer, *Realismus und Chemie*, Königshausen & Neumann, Würzburg, 1996, Chapter 4.3. (3) The early nineteenth century debate between Berthellot and Proust (and others) about the difference between homogeneous substances with definite and with varying composition depended on the question whether definite composition is considered essential for chemical species identity or not. On this and related issues in mineralogy see R. Hooykaas 'The Concept of "Individual" and "Species" in Chemistry', *Centaurus*, 1958, **5**, 307–322.

5 I intentionally exclude natural products because their origin provides additional links for establishing species identity.

6 T. Bergman, *Opuscula physica et chemica*, (1779–80); quoted from Llana, 'A Contribution of Natural History', p. 78.

7 R. Siegfried and B. J. Dobbs, 'Composition: A Neglected Aspect of the Chemical Revolution', *Annals of Science*, 1968, **24**, 275–93.

8 *Cf.* F. Szabadváry, *History of Analytical Chemistry*, Pergamon, London, 1996, Chapter IX. For a more detailed study of analytical chemistry of that period with emphasis on the German laboratory practice and education system see E. Homburg, 'The Rise of Analytical Chemistry and its Consequences for the Development of the German Chemical Profession (1780–1860)', *Ambix*, 1999, **46**, 1–32.

9 *Cf.* W. H. Brock, *The Fontana History of Chemistry*, Fontana, London, 1992, pp. 173–176 and p. 688. Szabadváry (*History of Analytical Chemistry*, p. 150 ff.) mentions an early textbook of analytical chemistry from Labadius (1801), with 'the earliest records of standards methods used for testing the purity of analytical grade reagents […] in many cases very similar to present day methods'.

10 Brock, *The Fontana History of Chemistry*, p. 174.

11 *Cf.* J. Schummer, 'The Chemical Core of Chemistry I: A Conceptual Approach', *Hyle*, 1998, **4**, 129–162.

12 It should be noted that this is how chemists actually characterised their new substances and thereby provided the only information for substance identification to colleagues. For a more enlarged list with stronger emphasis on chemical properties, see Marcelin Berthelot's suggestion on how to identify natural with artificial substances in his *Chimie Organique fondée sur la Synthèse*, Mallet-Bachelier, Paris, 1860, Vol. II, pp. 778–86 (reprinted: Bruxelles, Culture et Civilisation, 1966).

13 *Cf.* J. Schummer, 'Towards a Philosophy of Chemistry', *Journal for General Philosophy of Science*, 1997, **28**, 307–336. Since all material properties are context dependent, variation of the context generates new properties; *e.g.* variation of pressure conditions generates different boiling points. While these points may be summarised in a function, as one complex property, chemical properties resists such mathematical strategy. By variation of reaction conditions including reaction partners, combinations of reaction partners and so on, one can generate indefinitely many properties.

14 'Wenn zwei Stoff bezüglich einiger Eigenschaften übereinstimmen, so thun sie es auch bezüglich aller anderen Eigenschaften' (W. Ostwald, *Grundriß der Allgemeinen Chemie*, 3rd ed., Engelmann, Leipzig, 1899, p. 1).

15 For a discussion of the status of that assumption and the problems arising from the metaphysical misinterpretation by dialectical materialists with regard to the 'hot topic' of resonance structures, see Schummer, *Realismus und Chemie*, pp. 253, 279 ff.

16 *Cf. e.g.* the classic paper of R. K. Merton, 'Priorities in Scientific Discovery: A Chapter in the Sociology of Science', *American Sociological Review*, 1957, **22**, 635–659.

17 Quoted from the *Chemical Abstracts Index Guide*.

18 *Cf.* Schummer, 'The Chemical Core of Chemistry'.

19 J. Schummer, 'Scientometric Studies on Chemistry I: The Exponential Growth of Chemical Substances, 1800–1995', *Scientometrics*, 1977, **39**, 107–123.

20 *Cf.* J. Schummer, 'Coping with the Growth of Chemical Knowledge: Challenges for Chemistry Documentation, Education, and Working Chemists', *Educación Química*, 1999, **10**, 92–101.

21 J. Schummer 'Scientometric Studies on Chemistry II: Aims and Methods of Producing New Chemical Substances', *Scientometrics*, 1997, **39**, 125–140.

22 *Cf.* J. K. Taylor, 'The Impact of Instrumentation on Analytical Chemistry', in *The History and Preservation of Chemical Instrumentation*, J. T. Stock and M. V. Orna (eds.), Reidel, Dordrecht, 1986, pp. 1–10; and in much more detail, D. Baird, 'Analyti-

cal Chemistry and the "Big" Scientific Instrumentation', *Annals of Science*, 1993, **50**, 267–290.

23 Here and in the following I refer to the 'on-stream dates for several analytical techniques' collected in Taylor, 'The Impact of Instrumentation', p. 8.

24 It is probably interesting to note that these are the only instrumental methods used in organic chemistry for substance characterisation that have played a rôle worth mentioning (> 10%). For instance, given the complementary character of information from IR and Raman spectroscopy, it is surprising to find nothing of the latter. Organic chemists' confinement to a few standard techniques does obviously not depend only on the cost of instrumental equipment. Just compare the cost of, say, a 400 MHz H-NMR spectrometer with the relatively simple technology needed for measuring circular dichroism, of which I found only two instances in the whole period. Electrochemical methods are nearly absent. Only very recently, some organic chemists have been applying cyclic voltammetry (10% found in 2000). For the rôle of chromatographic characterisation, see below.

25 According to *A History of Analytical Chemistry*, H. A. Laitinen and G. W. Ewing (eds.), American Chemical Society, Washington DC, 1977, p. 220. Unfortunately, the book is not very reliable, mainly because it remains unclear whether the authors are telling only a US story or not.

26 Note that the present analysis is based on research papers reflecting the average view of working chemists, whereas some prominent individuals including textbook authors already had rather 'futuristic' views in the same period.

27 Laitinen and Ewing, *A History of Analytical Chemistry*, p. 158.

28 Since the 1980s, ultraviolet/visible spectroscopic characterisation has been increasing again (Figure 1) because the method can conveniently be used in-line with HPLC (high pressure liquid chromatography).

29 I should add that chemists occasionally include also quantum chemical calculations in their argumentation and thus completely ignore what philosophers have taught about the difference between theory and experiment.

30 *Cf.* also P. Laszlo, 'Chemical Analysis as Dematerialization', *Hyle*, 1998, **4**, 29–38.

31 *Cf. e.g.* P. J. Ramberg, 'Pragmatism, Belief, and Reduction. Stereoformulas and Atomic Models in Early Stereochemistry', *Hyle*, 2000, **6**, 35–61.

32 In a recent paper, Emily R. Grosholz and Roald Hoffmann have argued for the productivity of such ambiguities in the forefront of research ('How Symbolic and Iconic Languages Bridge the Two Worlds of the Chemist: A Case Study From Contemporary Bioorganic Chemistry', in *Of Minds and Molecules*, N. Bushan and St. Rosenfeld (eds.), Oxford University Press, New York, 2000, pp. 230–247).

33 *Cf.* Schummer, 'The Chemical Core of Chemistry', pp. 149–151.

34 *Cf. Registration Policy for the CAS Chemical Registry System*, Columbus, OH, p. 6; I refer to the version valid at least until 1996, which was not changed over the past 30 years (private communication).

35 The first infrared spectroscopic analysis of 'matrix-isolated' molecular species was reported by George C. Pimentel and co-workers in 1954. For brief, more or less historical notes on that technique see W. J. Orville-Thomas, 'The History of Matrix Isolation Spectroscopy' in *Matrix Isolation Spectroscopy*, A. J. Barnes *et al.* (eds.), Reidel, Dordrecht, 1981, pp. 1–11. Note that 'matrix-isolated' is a nonsense term, *i.e.* a *contradictio in adjectum*, since the molecular species are not isolated but frozen up in a solid *solution*. Using the term 'isolation' was probably part of rhetoric to make the new approach acceptable by conservative chemists.

36 *Cf.* also J. E. Earley, 'Modes of Chemical Becoming', *Hyle*, 1998, **4**, 105–115.

37 Interestingly, the *CA Index Guide* of 1996 still says, 'Intermediates that are not isolated and characterised are not indexed'. A brief look into a 1990 volume of *Chemische Berichte* reveals, however, that also 'non-isolated' molecular species, *e.g.* carbocations, received CAS registry numbers if they were spectroscopically characterised. To be sure, the rhetorical use of 'matrix isolation' has helped here.

38 See also L. Slater, 'Organic Chemistry and Instrumentation: R. B. Woodward and the Reification of Chemical Structures', in this volume.

39 For more details on these conceptual and ontological problems, see Schummer, 'The Chemical Core of Chemistry', pp. 139–143. A rare exception among chemists is R. Hoffmann, *The Same and Not the Same*, Columbia University Press, New York, 1995, Part I.

40 Here, all the arguments against microstructural essentialism apply, see J. van Brakel, 'Chemistry', in *Handbook of Metaphysics and Ontology*, H. Burkhardt and B. Smith (eds.), Philosophia, München, 1991, Vol. I, pp. 146–147.

Organic Chemistry and Instrumentation: R. B. Woodward and the Reification of Chemical Structures

LEO SLATER

Chemical Heritage Foundation, 315 Chestnut Street, Philadelphia, Pennsylvania 19106-2702, USA, Email: leobslater@yahoo.com

Introduction

In the course of the nineteenth and twentieth centuries, the field of organic chemistry has undergone two major revolutions. The first was the emergence of structure theory in the nineteenth century.[1] The second occurred during the middle decades of the twentieth century, as new modes of thought and newly available analytical instruments – new cognitive and physical tools – transformed scientific practice.[2] This historical transition has been termed the 'Instrumental Revolution' by Dean Stanley Tarbell and Ann Tracy Tarbell:

> [It] gave the organic chemist by 1955 powerful methods for purification and structure proof; some of these derived from techniques known for many years, which now became routinely available with relatively trouble-free instruments.... The new instrumentation allowed completely new ways of determining structures and of studying reactions of organic compounds. In days or weeks chemists could now solve problems which would have required months or years of careful work, or which might have been completely insoluble by classical methods. Many reaction rates and mechanisms, impossible to study before, now became accessible to attack.[3]

Thirty years before the Tarbells published their account, Robert Burns Woodward (1917–1979) had termed this change in the pace and practice of organic chemistry his discipline's 'second great revolution'.[4] For Woodward and his

Robert Burns Woodward (left) and Bernhard Witkop, 1956. Gift of Bernhard Witkop, Chemical Heritage Foundation Image Archives, Othmer Library of Chemical History, Philadelphia, Pennsylvania

contemporaries, the 1940s and 1950s were truly revolutionary, a period marked by new concepts of molecular shape and interaction and by instruments that gave new access to the molecular realm.[5] The increased use of instruments by organic chemists was crucial to changes in theory, mechanism and representation during this period. Indeed, Joachim Schummer has argued elsewhere in this volume that the meaning and ontological status of chemical structures was transformed in the course of this second revolution.

My own work focuses on the career of one chemist, Robert Burns Woodward. While Woodward, who was at Harvard University from 1937 until his death in 1979, does not represent the experience of his whole profession, he was both exemplary and exceptional. Certainly other chemists and other institutions, notably the University of Illinois, participated in the growth and transformation of organic chemistry in the United States, and Woodward was by no means solely responsible for the changes in theory and practice that took place in organic chemistry during his career. Nevertheless, he was most assuredly a powerful and influential member of the community that welcomed organic chemistry's 'second great revolution'. While Woodward's career was in some ways unique and his beliefs not universally shared, he was successful and outspoken, and he aggressively pursued a research agenda that embraced the instrumental revolution.[6]

For the historian, the practices of chemists are critical to an understanding of the second revolution. The physical methods of ultraviolet and infrared absorption spectroscopy, mass spectrometry and nuclear magnetic resonance spectroscopy (NMR)[7] gave organic chemists rapid and detailed knowledge of numerous properties pertaining to the mixtures and pure materials with which they worked.[8] As Harry Wasserman, one of Woodward's first graduate students in the 1940s, wrote on the occasion of Woodward's 60th birthday in 1977:

> In particular, Woodward was the first to take advantage of the modern
> electronic theory of reaction mechanisms to guide him through the ob-
> stacle-strewn pathways of organic reactions. At the same time he was one
> of the first to recognise the power of methods such as ultraviolet, infrared,
> and nuclear magnetic resonance spectroscopy for the determination of
> structure. These approaches, combined with his rigorous intellectual
> analysis of the entire synthetic route, including close attention to
> stereochemical factors, led to an astonishing series of synthetic accomplish-
> ments.[9]

These words captured the essence of what was to Woodward and his cohort a
two-fold revolution. Alongside mechanistic thinking, Wasserman emphasised
the utilisation of instruments for structure determination as characteristic of the
Woodwardian method. Wasserman pointed to how Woodward's innovations
proved successful both for himself and for organic chemistry as a discipline.

Woodward's devotion to these new tools persisted throughout his career. In
the early 1940s, Woodward published a series of papers correlating the ultraviolet spectra of α,β-unsaturated ketones (and dienes) with their structures. These
correlations of structure and spectra became known as the Woodward rules (also
referred to as the Woodward–Fieser rules in acknowledgement of Louis and
Mary Fieser's recasting of them).[10] In particular the earlier work on ultraviolet
spectra, and Arnold Beckman's introduction of a commercial ultraviolet instrument in 1941, pushed organic chemists to take note of advances in physical
chemistry and instrumentation. Similarly, in 1961, Woodward and co-authors
William Moffitt, Albert Moscowitz, William Klyne and Carl Djerassi published
their work on the 'Octant Rule', which correlated the optical rotatory dispersion
– often abbreviated to ORD – spectra of saturated ketones with their structures.[11] Both of these generalisations about spectra and structure created new
utility for spectroscopic measurements. These correlation techniques typified
theory in organic chemistry. They were theory in the practical sense of allowing
for the ready analysis of sets of facts, spectroscopic data in this instance, in
relation to one another; they were not, in most instances, theory by a more
abstract or mathematical definition. The Woodward rules foreshadowed Woodward's later work with Roald Hoffmann, the Woodward–Hoffmann rules, for
which Hoffmann shared the 1981 Nobel Prize with Kenichi Fukui.

Woodward was also well known for his synthetic and structural work in the
natural-products area. Other highlights of his career included his wartime work
on the structural elucidation of penicillin and synthetic approaches to quinine.
During the 1950s, Woodward collaborated with Pfizer on the structural analysis
of a new series of antibiotics: terramycin, aureomycin, and magnamycin. His
other achievements in the field of structure determination include many milestones in organic chemistry: strychnine (1947), ferrocene (1952), cevine (1954),
ellipticine (1959), oleandomycin (1960), and tetrodotoxin (1964). His synthetic
achievements included cortisone (1951), strychnine (1954), chlorophyll (1960),
tetracycline (1962), colchicine (1963), cephalosporin C (1965) and vitamin B_{12}

(1972). In 1963, the Swiss drug firm CIBA (later Ciba–Geigy, now Novartis) set up the Woodward Research Institute for him in Basel. This yielded dual appointments as the Donner Professor of Science at Harvard and as the Director of the Institute. In 1965 Woodward won the Nobel Prize 'for his outstanding achievements in the art of organic chemistry'. Between Cambridge, Massachusetts, and Basel, more than 400 graduate and postdoctoral students trained in his laboratories. Let us now turn to Woodward and instrumentation.

The Second Revolution

Speaking in New York in 1956, Woodward commented at length on the nature of the second revolution and in particular about the conceptual and technological changes:

> In emphasising the growth of the general recognition of the uses of theoretical considerations in the practical problems, I do not wish to neglect an almost equally important aspect of the advances of the past two decades. I refer to the marvelous phys[ical] tools – led by infrared measurements – which we now possess. Again, the examples which we have been able to provide of their use have given overwhelming evidence of their utility.[12]

Woodward hoped that the example of his success would inspire others to adopt his revolutionary *modus operandi*. Woodward's reference to 'the advances of the past two decades' was an acknowledgement of the commercial instruments that synthetic organic chemists could use with their own hands. The principles behind these advances were older than two decades. Indeed, on the other side of the Atlantic in 1936, almost exactly twenty years before Woodward's talk, Nevil V. Sidgwick had commented on the impact of these new physical principles on nineteenth-century structure theory. In his Presidential Address to the Chemical Society of London, he said:

> [O]ur knowledge of the meaning of these structures has developed, especially in the last 20 years, to an enormous extent. We have applied to their investigation a whole series of physical methods, based on the examination of the absorption spectra in the infra-red, the visible and the ultra-violet, and of the Raman spectra: on the measurement of specific heats and heats of combustion, of the dielectric properties, and of the scattering of X-rays and electron waves, as well as on the study of chemical dynamics: to mention only the most important.[13]

The distinction between Sidgwick's and Woodward's twenty years of progress was that the period of 1916–1936 saw the broad application of these principles and techniques in chemistry and physics, while Woodward's subsequent period, 1936–1956, saw some of these principles and techniques (particularly those of absorption spectroscopy) black-boxed and made directly available to Woodward's sub-discipline, organic chemistry.

By the adoption of analytical instrumentation in the middle part of the twentieth century, these organic chemists dramatically altered their practices.

The difficult process of structure determination became routine, at least for small molecules. Instrumentation underpinned theory, which in organic chemistry took the form of generalisations or rules relating structure to chemical and physical properties (such as the Woodward rules or the Octant Rule) and, as analysis became increasingly routine, the new practitioners were freed to focus on synthesis. This instrumental revolution established what molecules and reactions could be represented and how.[14] Thus, chemists set limits on what claims could be made about nature and reified chemical structures and reaction mechanisms[15] by the regular use of these instruments. The former distinction made by nineteenth-century chemists[16] between the conventional arrangement of atoms in a molecule and the actual structure of the molecule disappeared from the professional discourse of organic chemists in the wake of these new instruments. The conception of 'actual' molecular structure is itself problematic, but to understand what the chemists were doing, one must accept it as a possibility at the very least. From the inception of the structure theory, these two beliefs, that of structure either as the true form of nature or as a shorthand for experimental findings, were possible and were the source of much debate within the chemical community. As Alan Rocke has noted: 'Kekulé did not deny in principle the determinability of actual arrangements, [though some did, LS] stressing that this question was *independent* of his treatment of rational (resolved) formulas as relative, conventional, and based solely on chemical reactions'.[17] This position is also supported by Schummer's analysis of *Liebigs Annalen*, up to and including the 1950s.

Work done in chemical physics and physical chemistry during the 1920s and 1930s transformed beliefs about how molecules were held together, of how bonds were formed and broken, and of how reactions occurred. More mathematical models of bonding and reactivity, particularly in the wake of quantum mechanics, gave new theoretical grounding to structure theory,[18] as Sidgwick had already suggested in 1936:

> To Kekulé the links [between atoms] had no properties beyond that of linking; but we now know their lengths, their heats of formation, [*etc.*]... Throughout all this work the starting point has always been the structural formula in the ordinary organic sense.[19]

Yet the search for definitive structures for natural products, for example, strychnine, that had begun in the nineteenth century continued unabated and largely unchanged in spite of the revolution in chemical bonding. Even as the status of structural formulae changed, they maintained their 'ordinary organic sense' (Sidgwick's term). Nevertheless, the middle portion of the twentieth century saw chemical structures revealed by the use of physical instruments. These new practices led to changes in the epistemological and ontological status of chemical structures within organic chemistry. Ultraviolet, infrared and NMR all employed the measured interaction of electromagnetic radiation – light or radio waves – with the materials under study. These instrumental practices built on both the structure theory of the late nineteenth century and the electron theories that emerged in the 1920s: quantum mechanics, Lewis structures and the dynamic models of chemical mechanisms.[20] These theories of atoms, molecules and electrons gave rise to a vast realm of molecular possibilities. It was the instrumental revolution and its concomitant mechanistic approach (*mechanistic* in the sense of *reaction mechanisms*) that gave boundaries and control to this unruly territory. Organic chemists used instruments and rules to delimit their cognitive and laboratory practices.

In his 1956 manifesto entitled 'Synthesis', Woodward described the situation for theory in organic chemistry, particularly with regard to the reaction mechanisms:

> The great advance of the recent past has been the recognition of the entities responsible for the maintenance of ... nearest-neighbour relationships, and a description in simple general terms of wide applicability and precision, of their fluid nature, and of the laws to which they are subject. The resulting edifice of organic chemical theory enables us...to assert that the outcome of very few organic reactions is unexpected, and fewer inexplicable. Of course, this assertion must be qualified, in the sense that not infrequently several outcomes are possible; in the present day most or all of these will be disclosed to the chemist willing to examine the problem with assiduity, but the choice of that one which will in fact obtain is often difficult. So, much art remains; there is ample room for further progress, and much is being made by the theorists in this essentially quantitative direction. None the less, modern theory permits synthetic planning, and reduction of plans to practice, on a scale which was hitherto quite as impossible as were the simpler syntheses of the last century before the elaboration of the structure theory.[21]

For Woodward this theory, the understanding of molecules as having shapes and reactivities governed by 'nearest-neighbour relationships', was practical and fluid. These relationships allowed chemists to understand the capability of a part of a molecule to undergo a transformation in terms of the shapes and locations of the rest of the molecule. As Woodward put it in his Nichols Lecture, the edifice of organic chemical theory that resulted from 'the recognition of the entities responsible for the maintenance of nearest-neighbour relationships, and a description in simple, general terms, of wide applicability and precision, of their fluid nature, and of the laws to which they are subject' permitted accomplishments which had not been possible before.[22] This structural understanding allowed the thoughtful practitioner to reliably predict the possible outcomes of chemical reactions. Nevertheless, Woodward's assiduous chemists must be thoughtful without being fanciful.

The representation of structures in organic chemistry allowed for the depiction of a myriad of possible chemical compounds and intermediates. Woodward demanded a distinction between constructive supposition and idle speculation. From his own work, he discussed illustrations of the 'interplay of between synthesis and theory...'..[23] One of his examples, in which an optically active ketone (**I**) was transformed into a racemic (having no optical activity) phenol (**II**), displays where the place of models lay in regard to truth, at least for Woodward.

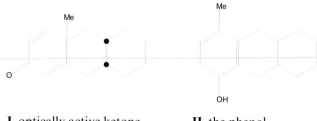

I, optically active ketone   **II**, the phenol

He wrote:

> It is our feeling that the fugitive nature of intermediates in chemical
> reactions does not confer upon them license with respect to molecular
> dimensions and characteristics. Rather it is likely that a given intermediate
> is as much an assemblage of atoms characterised by precise bond distances
> and bond angles as are the molecules from which it is obtained, and into
> which it is transformed. Consequently, we deplore the tendency to accept
> as an intermediate any species solely on the ground that it can be concep-
> tualised, and that it is not implausible.[24]

Rules of bonding and molecular geometry, *theory*, were not sufficient to allow for
the acceptance of a reaction intermediate. These reaction intermediates were
forms of matter that were presumed to exist somewhere during the course of an
individual reaction, not to be confused with synthetic intermediates that were
distinct, stable, isolable materials produced at each stage of a multistep synthesis.
Though these intermediates were short-lived and conceptual by nature, Woodward wanted to restrict idle speculation. The postulated intermediates must
share all the essentials of their starting materials and their products.

Logic and, whenever possible, physical data were Woodward's tools for giving
credence to a proposed intermediate:

> In the case of the 1,3 rearrangement of dienones, for example, of [**I**] to the
> phenol [**II**], it is easy to imagine that the departing group proceeds directly
> from its point of initial attachment to its resting place, and that during the
> course of its progression, it is associated only in a general way with the
> unsaturated groupings of the molecule.

The unsaturated portion of the molecule was the right-hand section. The postulated 1,3 shift was the simplest explanation of the mechanism, but criteria must
be examined:

> This possibility can be readily understood, it can be easily symbolised [*cf.*
> **III**], and it is not implausible. But is it true? Our examination of the
> rearrangement of the optically active ketone [**I**] revealed that the change
> led to a racemic product. Clearly, the molecule must pass into a symmetri-
> cal condition; the only symmetrical intermediate which can be envisaged is
> [**IV**], and the change is a succession of 1,2 rearrangements, rather than a
> 1,3 rearrangement involving a generalized intermediate.

III   **IV**

Truth was paramount and deduction could reveal how the transformation took place. By Woodward's logic, certain claims *should not* have been made and others *could* have been made. To say more about the transformation required additional information, as Woodward wrote: 'Of course, the aspect of the system during the progress of the 1,2 shifts remains to be established, and in order to examine it, even more subtle methods of scrutiny must be brought to bear'.[25] Woodward focused this scrutiny with theory and instrumentation. He set strictures on the kind of truth claims that were allowable. In the case of this optically active ketone, representations set boundaries too broad to delimit a unique solution. Further, not everything that models plausibly allowed to be represented should be claimed just because no evidence existed to the contrary; positive evidence beyond the rules of representation was required. At the same time, representations – structural diagrams – limited the range of possible solutions. Woodward's demand for truth in the representations of even transitory forms of matter underscores the new status of structures as the objects of study, rather than as properties of the substances represented. Woodward's language of scrutiny and truth was informed by the new molecular access that instruments provided.

At the same time that Woodward strove to limit and define the molecular possibilities and structural representations, others were still struggling with older concerns about the ontological status of structures as true or unique. For instance, George Wheland of the University of Chicago, addressed the status of chemical structures in his textbook, *Advanced Organic Chemistry* in the following way:

> Indeed, the more fundamental question may be asked why it is supposed that the molecule can be adequately represented by any structure whatever. That is to say, no single experiment, or combination of a small number of experiments, can prove that the valences of the atoms must be those supposed, or even that the concept of structure itself is valid. The justification of these aspects of the structural theory is rather that they permit a self-consistent, logical and extremely useful interpretation of an enormous number of independent experiments of many different kinds. They do not permit an interpretation of *all* the experiments that can be performed.[26]

In Wheland's philosophical construction, what was most important for a proposed chemical structure was its qualities of self-consistency, logic and utility: they were all that one could claim for a structural solution. Wheland believed that chemical structures were just a shorthand, and an imperfect one at that. In this we see the fundamental values of the discipline, logic and utility, were employed on both sides of the status question, even as the new instrumental practices transformed beliefs. Yet even as Wheland wrote, times were changing. Woodward observed in 1956, two years after completing the synthesis of strychnine:

> A traditional task of synthesis has always been the verification of structure. To some extent this assignment is an arbitrary one, since a degradative reaction may quite as well partake of rigour as a synthetic one. But the value of checks and controls in any science can hardly be questioned, and it is certainly true that it is often one thing to construct an hypothesis which is consistent with all known facts, and quite another to demonstrate that the hypothesis forms a sound and realistic basis for future action. [In this case, the hypothesis was a chemical structure.] Furthermore, while analytical and degradative work must always be primary, it is often synthesis which provides the simplest, most rigorous and final proof. It is possible to argue, for example, that this function has been performed for strychnine by our recent synthesis, without in any way depreciating the importance of the structural investigations which led to the elaboration of the constitution of the alkaloid.[27]

This gives us the traditional rôle of synthesis and an intimation of the rising importance of synthesis relative to analysis. These comments were a nod to the position of an older generation of chemists for whom the chemical work as opposed to instrumental work, was foremost.

As Wasserman observed, Woodward's vision of the revolution in organic chemistry was two-fold. Novel ways of thinking about chemical processes and the use of new analytical instruments mutually reinforced one another as parts of chemical practice. In Woodward's words:

> To have placed primary emphasis on the importance of theory ... is not to derogate a second factor of very [great] moment – the development and application of physical methods. No new general principle is involved here. Chemists should always have been willing, nay eager, to utilise to the full any experimental method or tool which would give them greater and more precise insight into the operations which they were conducting. But for a long time organic chemists, for some reason – perhaps a too great compartmentalisation of chemistry, perhaps a too insular pride – were loath to make use of even the relatively simple physical tools which were available. Happily, this recalcitrance has now largely disappeared.[28]

According to Woodward, the previous generation of chemists was too narrowly focused or too proud to make use of the new generation tools. And what precisely were these tools? Two techniques were already in widespread use by 1956. As Woodward wrote:

> Ultraviolet spectra have long been with us, and during the last decade have

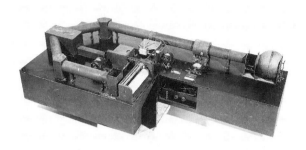

General Electric's recording ultraviolet spectrometer, designed by A. C. Hardy, 1930s. Photograph courtesy of NMAH

moved into the prominent use which they deserve. But no single tool has had a more dramatic impact upon organic chemistry than infrared measurements. The development, just after the second Great War, of sturdy and simply operated machines for the determination of infrared spectra has permitted a degree of immediate and continuous analytical and structural control in synthetic organic work which was literally unimaginable fifteen years ago. The power of the method grows with each day, and further progress may be expected for a long time to come.[29]

Ultraviolet spectroscopy exploited the variable absorption of light by sample compounds in the region of the spectrum from about 210 nm through the visible (approximately 380 nm to 800 nm). Typically, molecules containing multiple double bonds, conjugated or aromatic systems, absorbed in this region. Infrared was a major advance because many modes of vibration for common chemical structures were found at the lower energies above 800 nm, in the infrared. Among these common constituents were those of alcohols, amines, and unconjugated double bonds.

Instruments were a key component of Woodward's professional development. His publications on the spectra and structure of α,β-unsaturated ketones (the Woodward rules) were his first major chemical achievement. These rules were well noted in the chemical community. Lord Todd, an eminent British chemist, former President of the Royal Society, and long-time friend of Woodward, wrote in his autobiography:

He was one of those very rare people who possessed that elusive quality of genius, and was certainly the greatest organic chemist of his generation, and possibly of this century ... my attention was first drawn to him by a paper he published in 1941 (at the age of twenty-four) on the ultraviolet absorption spectra of α,β-unsaturated ketones; it seemed to me to herald a breakthrough in the use of spectroscopy in the study of molecular structure. In this I was right, and Woodward went on from strength to strength.[30]

Though Woodward relied heavily on physical, that is, spectroscopic, data in developing his rules for determining the structures of these compounds from their ultraviolet spectra, this was an application for *organic*, not physical, chemists. Indeed, as W. David Ollis, one of Woodward's collaborators on strychnine, observed: '... Woodward did not receive encouragement from physical chemists in his adoption of physical methods as structural aids. They did not understand why the Woodward rules (1941–42) for correlating the ultraviolet spectra of unsaturated ketones actually worked'.[31] Although not unusual for this period, the Woodward rules were among the first of this type of generalisation, and were quite quantitatively precise (in the sense of attaching standardised values to empirical results) and comprehensive in scope.

In the wake of these new instruments, Woodward pleaded for two changes in practice:

> The organic chemist properly has great respect for pure compounds, and the result has been that much present practice involves infrared measurements only on individual substances of assured purity. To adopt this protocol is to throw away one of the larger advantages of the method. The routine examination of virtually every reaction mixture, however crude, or lacking in tangible prospect of yielding a desired product, often provides a clue to important developments which could not otherwise be made.

Instruments that could increase the productivity of the organic chemist were to remove, at least some of the time, the painstaking task of isolating every intermediate in a synthesis. In Schummer's analysis, this identification of molecular structures within mixtures was evidence of an ontological shift in the status of these structures, as they no longer need be associated with a pure, isolated substance. Instruments could also provide insight into the reaction process by revealing by-products. Woodward continued:

> Second, the tendency to allocate to specialists the responsibility for the making, and in particular the interpretation, of physical measurements, is to be deplored. The capacity of the physical specialist to place his results properly in the context of an organic chemical investigation is often narrow and unrealistic, and the organic chemist will find himself magnificently rewarded, who takes pains to be himself in a position to understand and interpret the physical aids he wishes to use. In any event, physical methods, *and the principle that they should be used whenever possible*, are now part of our armamentarium, and we expect no surcease of further developments in this direction.[32]

Organic chemists should take charge of their own physical methods and make their use and interpretation part of their day to day practice. Woodward was claiming for synthetic organic chemists territory that once had belonged to their physical collaborators.[33] By the end of the 1950s, a new and still more powerful tool was emerging: NMR. Woodward had worked with chemists at Pfizer for a

During its product life, the Varian A-60 became a workhorse in over 1000 chemical laboratories in universities and chemical companies around the world. In common with many other NMR machines, the A-60 shown here has been modified during its active life, as shown by the box above the console. Science Museum Collections, inventory number: 1975–423. Science Museum/Science and Society Picture Library

number of years on a project to determine the structures of a number of tetracycline antibiotics. In a letter to Vladimir Prelog in the summer of 1958, Woodward mentioned that:

> [W]e regarded the structure of oleandomycin as virtually established. Our structure ... does not contain many features of special interest, but we found the work fascinating for two reasons. For one, it was the first case in which we made extensive use of nuclear magnetic resonance spectra. Indeed, these measurements were the major factor in our deduction of the structure..., and I have become convinced that the tool is one of the very first importance in structural work....[34]

Woodward recognised early on that NMR was a new and powerful instrumental technique that would do away with the arduous tasks of structural determination. Not just instruments were necessary for modern practice, but also the principle that they should be used whenever possible.

But not all were pleased with the turn of events. One such was Sir Robert Robinson, writing in 1973:

> I wonder whether chemistry would have been richer or the poorer if Baeyer had been able to put indigo into a machine and get the structure of it right away; we should probably know nothing about the isatins and numerous important synthetic reactions, and so on. It would have been a serious loss, and that is what will happen in the future if the disciples of the physical methods of end-all and be-all organic chemistry get their way.[35]

To this attitude Woodward had previously penned a response:

This short history [of strychnine] should give pause to those whose talent for despair is lavished upon organic chemistry ornamented and supplemented – or as they fancy, burdened – by magnificent new tools which permit the establishment in days or weeks of enlightenments which once would have required months or years. While it is undeniable that organic chemistry will be deprived of one special and highly satisfying kind of opportunity for the exercise of intellectual *élan* and experimental skill when the tradition of purely chemical structure elucidation declines, it is true too that the not infrequent dross of such investigation will also be shed; nor is there any reason to suppose that the challenges for the hand and the intellect must be less, or the fruits less tantalising, when chemistry begins at the advanced vantage point of an established structure.

Of course, men make much use of excuses for activities which lead to discovery, and the lure of unknown structures has in the past yielded a huge dividend of unsought fact, which has been of major importance in building organic chemistry as a science. Should a surrogate now be needed, we do not hesitate to advocate the case for synthesis.[36]

Woodward viewed the new instrumental practices and the newly stabilised structure theory as an opportunity that allowed synthetic organic chemistry to blossom.

Conclusion

I would make two points in conclusion. One, to reiterate Woodward's remarks that instruments allowed synthesis to flower as a topic of chemical inquiry, and two, that chemical structures were, as never before, equated by chemists with the materials they represented. By which I meant that we have on the one hand the materials of chemistry, the pure compounds, liquids, crystals, *etc.*, each with measurable properties including chemical composition, and each capable of undergoing chemical transformations, or reactions. And on the other hand we have a set of conventional representations of these materials, chemical structure theory, which allowed mapping the organisation of these properties and transformations. I believe that in the 1860s these two things, the materials and the structure theory, were, even for the proponents of the structure theory, unmistakably distinct. As a result of the second great revolution, these two were indistinguishable for organic chemists by the 1960s. In fact the reification of these *theoretical* chemical structures, which had emerged in the second half of the nineteenth century, became the foundation of chemical practice for organic chemists. By this I mean that the move from structures to the materials and back again became almost unconscious and effortless. This can in large part be attributed to the adoption of instrumental techniques which allowed organic chemists in their own laboratories and with their own hands to gather detailed physical data on compounds.[37] The increasing immediacy and volume of this *physical* data eventually removed the time-consuming process of *chemical* analysis. Therefore, the decreasing time between the isolation of a material and its analysis – the extraction of its properties – promoted the collapse of the distinction between structural representation and material. Furthermore, the advent of NMR allowed for a total one-to-one mapping of physical data to chemical structure. In the way that UV absorption had allowed for the determination of substitution patterns in α,β-unsaturated systems, NMR potentially allowed the individual detection of every hydrogen atom and every carbon atom in a molecule and could with further refinement give information about how these atoms were connected and arranged. Through the mediation of instruments, structures became increasingly stable concepts, no longer merely a kind of shorthand.

For chemists, the fundamental understanding of chemical compounds and their properties rested on the determination of three-dimensional arrays of atoms in space. With instrumental techniques, chemists concluded a project begun with the atomic theory and continued through the 1850s, 1860s, and 1870s with the structure theory of Kekulé and Archibald Scott Couper and the tetrahedral carbon atom of Jacobus Hendricus van't Hoff and Joseph Achille Le Bel. According to Johannes Wislicenus of Leipzig, at the close of the nineteenth century, van't Hoff's 'doctrine of atomic arrangement in three dimensions' was 'a logical and necessary stage, perhaps the final stage, in the chemical theory of atoms'.[38] What van't Hoff believed the object of his study to be was 'nothing else than the spatial – *i.e.* the real – positions of these points, the atoms'.[39] I would claim that, by the 1960s, instruments had made this belief a reality. The instrumental revolution established what molecules and reactions could be represented and how. Indeed, Schummer suggests that, nowadays, molecular structures are no longer considered properties of substances; they are species, objects of study, in their own right.

Notes and References

1 For more on the structure theory in the nineteenth century see Alan J. Rocke, *The Quiet Revolution: Hermann Kolbe and the Science of Organic Chemistry*, University of California Press, Berkeley, CA, 1993, especially Chapters 6, 7 and 10; A. J. Rocke, 'Subatomic Speculations and the Origin of Structure Theory', *Ambix*, 1983, **30**, 1–18; A. J. Rocke, 'Kekulé, Butlerov and the Historiography of the Theory of Chemical Structure', *BJHS*, 1981, **14**, 27–57; George E. Hein, 'Kekulé and the Architecture of Molecules' and Henry M. Leicester, 'Kekulé, Butlerov, Markovnikov: Controversies on Chemical Structure from 1860 to 1870', both in *Kekulé Centennial*, O. Theodor Benfey, Symposium Chairman, American Chemical Society, Washington, DC, 1966, Chapters 1 and 2; and Benfey, *From Vital Force to Structural Formulas*, Beckman Center for the History of Chemistry, Philadelphia, PA, [1964], 1992.

2 For more on nineteenth century theory and practice see, for example: Ursula Klein, 'Origin of the Concept of Chemical Compound', *Science in Context*, 1994, **7**, 163–204; and Alan J. Rocke, 'Subatomic Speculations and the Origin of Structure Theory', *Ambix*, 1983, **30**, 1–18.

3 Dean Stanley Tarbell and Ann Tracy Tarbell, *Essays on the History of Organic Chemistry in the United States, 1875–1955*, Folio Publishers, Nashville TN, 1986, p. 336.

4 R. B. Woodward, (hereafter, RBW), 'Synthesis', in *Perspectives in Organic Chemistry*, Alexander Todd (ed.), Interscience Publishers, Inc, New York, 1956, 155–184, p. 156.

5 For more on Woodward, see: *Robert Burns Woodward: Architect and Artist in the World of Molecules*, Theodor Benfey and Peter J. T. Morris (eds.), Chemical Heritage Foundation, Philadelphia, 2001; and L. B. Slater, 'Industry and Academy: The Synthesis of Steroids', *Historical Studies in the Physical and Biological Sciences*, 2000, **30**, 443–480.

6 Carl Djerassi, in his thoughtful and generous comments on this paper, correctly pointed out that Woodward, though a powerful *advocate* of instrumental methods, was not the most active *user* within the discipline.

7 For more on the history of NMR, see: D. M. Grant and R. K. Harris (eds-in-chief), *Historical Perspectives*, Vol. 1 in *Encyclopedia of Nuclear Magnetic Resonance*, John Wiley, New York, 1996; Timothy Lenoir and Christophe Lécuyer, 'Instrument Makers and Discipline Builders: The Case of Nuclear Magnetic Resonance', *Perspectives On Science*, 1995, **3**, 276–345; Felix W. Wehrli, 'The Origins and Future of Nuclear Magnetic Resonance Imaging', *Physics Today*, June 1992, 34–42; Peter G. Morris, 'Present and Future Application of NMR to Medicine and Materials Science', in R. Bud and S. Cozzens (eds.), *Invisible Connections: Instruments, Institutions and Science*, Bellingham, SPIE Optical Engineering Press, WA, 1991, 217–233; K. Marsden and Ian D. Rae, 'Nuclear Magnetic Resonance in Australia 1952–1986', *Historical Records of Australian Science*, 1990, **8**, 120–50; and Geoffrey Bodenhausen, 'A Retrospection on Ten Years of Two-dimensional Nuclear Magnetic Resonance Spectroscopy', *Chimia*, June 1984, **38**, 215–8.

8 I will largely pass over mass spectrometry in this paper, though one should note Carl Djerassi's major contributions in structural elucidation using this technique. See Carsten Reinhardt's paper in this volume.

9 H. H. Wasserman, 'Profile and Scientific Contributions of Professor R. B. Woodward', *Heterocycles*, 1977, **7**, 1–17, p. 2.

10 R. B. Woodward, 'Structure and the Absorption Spectra of α,β-Unsaturated Ketones', *Journal of the American Chemical Society*, 1941, **63**, 1123–1126; RBW, 'Structure and Absorption Spectra, III. Normal Conjugated Dienes', *Journal of the American Chemical Society*, 1942, **64**, 72–75; RBW, 'Structure and Absorption Spectra, IV. Further Observations on the α,β-Unsaturated Ketones', *Journal of the American Chemical Society*, 1942, **64**, 76–77. All reprinted in Benfey and Morris, *Robert Burns Woodward*, pp. 47–56. Also see Louis F. Fieser, Mary Fieser and Srinivasa Rajagopalan, 'Absorption Spectroscopy and the Structures of the Diosterols', *Journal of Organic Chemistry*, 1948, **13**, 800–806.

11 R. B. Woodward, W. Moffitt, A. Moscowitz, W. Klyne and C. Djerassi, 'Structure and the Optical Rotatory Dispersion of Saturated Ketones', *Journal of the American Chemical Society*, 1961, **83**, 4013–4018. Reprinted in Benfey and Morris, *Robert Burns Woodward*, pp. 244–249.

12 Harvard University Group, Faculty Papers, by permission of the Harvard University Archives and Dr Crystal Woodward, hereafter HA:HUG(FP), 68.10, Box 24, Folder 'Nichols Award [1955–1976]'.

13 N. V. Sidgwick, 'Presidential Address: Structural Chemistry', *Journal of the Chemical Society*, 1936, 533–538, p. 533.

14 A number of authors have dealt with the historical role of instruments in twentieth-century science: Peter J. T. Morris and A. S. Travis, 'The Rôle of Physical Instrumentation in Structural Organic Chemistry', in *Science in the Twentieth Century*, J. Krige and D. Pestre (eds.), Harwood Academic Publishers, Amsterdam, 1997, 715–739;

Nicolas Rasmussen, *Picture Control: The Electron Microscope and the Transformation of Biology in America, 1940–1960*, Stanford University Press, Stanford, 1997; Davis Baird, 'Analytical Chemistry and the 'Big' Scientific Instrumentation Revolution', *Annals of Science*, 1993, **50**, 267–290; Yakov M. Rabkin, 'Uses and Images of Instruments in Chemistry', in *Chemical Sciences in the Modern World*, S. Mauskopf (ed.), Univesity of Pennsylvania Press, Philadelphia PA, 1993, 25–42; Robert E. Kohler, *Partners in Science: Foundations and Natural Scientists, 1900–1945*, University of Chicago Press, Chicago, 1991, especially pp. 358–391; J. B. Willis, 'Spectroscopic Research in the CSIRO Division of Chemical Physics 1944–1986', *Historical Records of Australian Science*, 1990, **8**, 151–181; Lily E. Kay, 'Laboratory Technology and Biological Knowledge: The Tiselius Electrophoresis Apparatus, 1930–1945', *History and Philosophy of the Life Sciences*, 1988, **10**, 51–72; S. Nunziante Cesaro and E. Torracca, 'Early Applications of Infrared Spectroscopy to Chemistry', *Ambix*, 1988, **35**, 39–47; Y. M. Rabkin, 'Technological Innovation: The Adoption of Infrared Spectroscopy by Chemists', *Isis*, 1987, **78**, 31–54; and E. Bright Wilson and John Ross, 'Physical Chemistry in Cambridge, Massachusetts', *Annual Review of Physical Chemistry*, H. Erying (ed.), 1973, **24**, 1–27.

15 For more on the study of reaction mechanisms by physical organic chemists, see Mary Jo Nye, *From Chemical Philosophy to Theoretical Chemistry: Dynamics of Matter and Dynamics of Disciplines, 1800–1950*, University of California Press, Berkeley, CA, 1993, especially Chapters 6, 7 and 8.

16 M. J. Nye, 'Philosophies of Chemistry', pp. 6–14; and 'Kekulé's Benzene Theory and the Appraisal of Scientific Theories', in *Scrutinizing Science: Empirical Studies of Scientific Change*, Arthur Donovan, Larry Laudan and Rachel Laudan, (eds.), Kluwer Academic Publishers, Boston, 1988, 145–161, pp. 154–159.

17 Alan J. Rocke, *Chemical Atomism in the Nineteenth Century: From Dalton to Cannizzaro*, Ohio State University Press, Columbus, OH, 1984, p. 269.

18 For more on the impact of quantum mechanics on chemistry and the rise of quantum chemistry, see Kostas Gavroglu and Ana Simoes, 'The Americans, the Germans, and the Beginnings of Quantum Chemistry: The Confluence of Diverging Traditions', *Historical Studies in the Physical and Biological Sciences*, 1994, **25**, pp. 47–110.

19 Sidgwick, 'Presidential Address', p. 533–4.

20 See for example, Andrea I. Woody, *Early 20th Century Theories of Chemical Bonding: Explanation, Representation and Theory Development*, PhD Dissertation, University of Pittsburgh, 1997; Kenneth T. Leffek, *Sir Christopher Ingold, a Major Prophet of Organic Chemistry*, Nova Lion Press, Victoria, BC, 1996, Chapter 6; Trevor I. Williams, *Robert Robinson: Chemist Extraordinary*, Clarendon Press, Oxford, 1990, Chapter 7; Martin D. Saltzman, 'The Development of Physical Organic Chemistry in the United States and the United Kingdom, 1919–1939: Parallels and Contrasts', *Journal of Chemical Education*, 1986, **63**, 588–593; Robert E. Kohler, 'G. N. Lewis's Views on Bond Theory 1900–1916', *British Journal for the History of Science*, 1975, **8**, 233–239; Kohler, 'The Lewis–Langmuir Theory of Valence and the Chemical Community', *Historical Studies in the Physical Sciences*, 1975, **6**, 431–468; Kohler, 'Irving Langmuir and the Octet Theory of Valence', *HSPS*, 1972, **4**, 39–87; Kohler, 'The Origins of G. N. Lewis's Theory of the Shared Pair Bond', *HSPS*, 1971, **3**, 343–376.

21 Woodward, 'Synthesis', p. 156–157.

22 HA:HUG(FP), 68.10, Box 24, Folder 'Nichols Award [1955–76]'.

23 Woodward, 'Synthesis', p. 177.

24 Woodward, 'Synthesis', p. 178.

25 Woodward, 'Synthesis', p. 178–179.

26 G. W. Wheland, *Advanced Organic Chemistry*, second edition, J. Wiley, New York, 1949, p. 94.

27 Woodward, 'Synthesis', p. 164–165.

28 Woodward, 'Synthesis', p. 157.

29 Woodward, 'Synthesis', p. 157–158.

30 Alexander Todd, *A Time to Remember*, Cambridge University Press, New York, 1983, p. 114.

31 W. D. Ollis, 'Robert Burns Woodward – An Appreciation', *Chemistry in Britain*, 1980, **16**, 210–216, p. 210.

32 Woodward, 'Synthesis', p. 158.

33 In significant ways, this territorial expansion parallels that of Wendell Stanley. See Angela N. H. Creager, 'Wendell Stanley's Dream of a Free-Standing Biochemistry Department at the University of California, Berkeley', *Journal of the History Biology*, 1996, **29**, 331–360.

34 RBW to Prelog, August 8, 1958. HA:HUG(FP), 68.10, Box 34, Folder 'Prelog [1950–77]'.

35 Quoted in Trevor I. Williams, *Robert Robinson: Chemist Extraordinary*, Clarendon Press, Oxford, 1990, p. 184.

36 R. B. Woodward, M. P. Cava, W. D. Ollis, A. Hunger, H. U. Daeniker and K. Schenker, 'The Total Synthesis of Strychnine', *Tetrahedron*, 1963, **19**, 247–288, p. 248. Reprinted in Benfey and Morris, *Robert Burns Woodward*, pp. 136–177.

37 Rabkin has commented on this in regard to infrared spectroscopy, see his 'Technological Innovation in Science', p. 52.

38 Johannes Wislicenus, 'Preface to the Second Edition', 1894, in J. H. van't Hoff, *The Arrangement of Atoms in Space*, translated by Alfred Werner, second ed., Longman, Green, London, 1898, p. vii.

39 J. H. van't Hoff, *The Arrangement of Atoms in Space*, second ed., p. 5.

The Chemistry of an Instrument: Mass Spectrometry and Structural Organic Chemistry

CARSTEN REINHARDT

Lehrstuhl für Wissenschaftsgeschichte, Universität Regensburg, D 93040 Regensburg, Germany, Email: carsten.reinhardt@psk.uni-regensburg.de

Introduction

During the two decades after 1945, the chemical sciences and technologies experienced a major transformation that is sometimes called the 'second chemical revolution'. The crux of this transformation was the introduction of instrumental methods and techniques both in chemical research and routine analysis. Prior to this period, chemists determined the chemical structure of an unknown compound mainly through known reactions: identifying the products by chemical means and establishing quantitative data through gravimetric and volumetric methods. After the 1960s, chemists have generally obtained their results using a variety of instruments that permit the analysis of chemical substances in terms of their physical properties.[1]

From the very beginning of chemistry, specific tools and techniques have been used for the purification, isolation and synthesis of compounds, and physical methods have been applied for the determination of their properties. In the early development of structural organic chemistry two physical methods, molecular weight determination and detection of molecular asymmetry by optical rotation, were important. Others, for example melting and boiling points, gave no direct information about the structure of organic molecules but were of use in general identification. For structural elucidation, chemists relied almost exclusively on the investigation of chemical reactions. This situation changed slowly in the first half of the twentieth century with the development of spectroscopic techniques and X-ray and electron diffraction. A textbook published in 1945/6 surveyed an impressive array of methods already in use, ranging from the determination of melting and freezing points to ultraviolet, infrared and visible light absorption spectroscopy and, among many others, mass spectrometry.[2] But most of these techniques were still in their infancy, at least with regard to their use in the structural elucidation of complex organic molecules. It appears that physical instrumentation in structural organic chemistry only took off in the late 1950s. Thus, the 1955 edition of *Determination of Organic Structures by Physical Methods*[3] contained nearly the same contents as the handbook published ten years earlier. But six years later, in 1961, 'advances have been made in approaches to the structures of molecules at an unprecedented rate', and the editors made no apology for a dramatic expansion of the chapters on nuclear magnetic resonance, optical rotatory dispersion and mass spectrometry.[4] In 1965, Peter Schwarz, editor of *Physical Methods in Organic Chemistry*, discussed the reasons for this tremendous increase. Military needs during the war stimulated advances in instrumentation, and made possible the use of physical methods as routine tools. At the same time, natural products (Schwarz gave steroid hormones as an example) provided challenges of a new order of complexity to the organic chemist. Furthermore, the increasing interest in physical organic chemistry drew attention to physical methods of investigation. Economic factors also played an important rôle: the post-war boom in government funding of scientific research stimulated the development and use of scientific instruments at a fast growing pace.[5]

In dealing exclusively with one of these many physical techniques, mass spectrometry, I intend to supplement Schwarz's arguments. In particular, I shall identify critical moments and experiments, that, borrowing a phrase from Carl Djerassi's book on optical rotatory dispersion (another important physical technique from the same era), became the 'prince who awoke the sleeping beauty' of mass spectrometry.[6] From the outset, it is important to state that the development of mass spectrometry is not another example of (instrumental) technology-push or (scientific) demand-pull, since both preconditions already existed in 1951, almost a decade before mass spectrometric techniques were used in structural elucidation of unknown compounds. The 'prince' was the rationalisation of the fragmentation processes of organic molecules in the mass spectrometer, which allowed a semi-empirical correlation of spectra and structure. This endeavour began seriously in the mid-1950s, and it was accomplished with the help of the concepts and terminology of physical organic chemistry.[7] The characteristics of most physical techniques in chemistry can be seen from two different, but related, points of view, the 'theoretical' and the 'empirical' approaches. With the theoretical approach, the sought for information is obtained by interpreting the data in terms of physical theories, and it provides detailed information about molecular geometry and energetics, as well as other parameters. In general, this theoretical approach is limited to rather simple molecules, and its use belongs to the physicists and physical chemists. The second approach is based on the empirical correlation of physical properties with structural features. Its main underlying assumptions are that the measured physical property can be localised in the molecule and, to a lesser or greater degree, changes in other parts of the compound do not affect this property. There are two reasons why most of the use of physical methods in organic chemistry was based on the empirical approach: the molecules under consideration are usually too complex for a fundamental solution, and the organic chemist has to use empirical correlations. Moreover, this type of reasoning corresponded to arguments based on evidence already familiar to organic chemists. This argument has been proposed by Nicolas Rasmussen for biological research and the electron microscope.[8] For structure determination in organic chemistry, Joachim Schummer suggests that the reasoning in classical chemistry and in research based on spectroscopic techniques is basically the same, which allowed for a smooth incorporation of new instrumental techniques in chemical practice.[9]

The development of mass spectrometry shows some remarkable parallels to infrared spectroscopy.[10] Both techniques originated in physics, grew to instrumental maturity during World War II, mainly in the chemical and petroleum industries, and were extensively applied in chemical research after the end of the war, with a lead-in time for infrared. In both cases the input of industry was crucial, *via* the development of reliable instrumentation and the data collections necessary for the interpretation of spectra. A difference, it seems, is that mass spectrometry was applied mainly to the analysis of hydrocarbon mixtures and that the random fragmentations and rearrangements shown by this class of compounds delayed the application of mass spectrometry in structural research.

Use of Mass Spectrometry in the Petroleum and Chemical Industries

In a mass spectrometer, positively charged particles (and some negative ions) are formed by electron bombardment of a compound, normally present in the vapour state. These positive ions and their fragments are then accelerated in an electric field and deflected by a magnetic one. Variation of one of these fields focuses the ions with the same mass to charge ratio at a detector, the resulting mass spectrum is a recording of the abundance of the positive ions *versus* their mass. The technique can be traced back to the work of J. J. Thomson, who in 1912 first observed the existence of stable isotopes, ^{20}Ne and ^{22}Ne. Important contributions during the inter-war period were made by Francis W. Aston at Cambridge University, Arthur J. Dempster of the University of Chicago, Alfred O. Nier of the University of Minnesota, and many others. Most of their work dealt with the determination of isotopes and their natural abundances.[11]

The application of mass spectrometers to organic chemical analysis was initiated by the war-time needs of the petroleum and synthetic rubber industries after 1940, and was made possible mostly through the work of Harold W. Washburn of Consolidated Engineering Corporation (later Consolidated Electrodynamics Corporation of Pasadena, Ca., CEC), John A. Hipple of Westinghouse, and their collaborators.[12] Before the war, hydrocarbon gas mixtures from the petroleum industry's new catalytic crackers were analysed by distillation using Podbielniak fractionating columns together with infrared spectroscopy. The use of infrared spectroscopy had already resulted in a considerable saving of time, but mass spectrometry improved on that greatly. In 1943 it was shown that a nine-component mixture of C_5 and C_6 hydrocarbons could be analysed in four hours by mass spectrometry, compared with ten days using the traditional method of refraction index measurements of fractions from the fractionating columns.[13] Nevertheless, complete analysis required distillation into narrow-boiling fractions and the solution of sets of large numbers of simultaneous equations. In addition, for the butadiene production programme which supplied the feedstock for synthetic rubber, mass spectrometry competed in terms of accuracy and precision with infrared, ultraviolet and distillation methods, as is shown by a report in which more than 70 laboratories in the US were evaluated for performance in analytical procedures.[14] When applied in a highly standardised manner (and using mainly instruments of one manufacturer, CEC), mass spectrometry contributed to the exact determination of the light hydrocarbons of gas mixtures in the C_3 to C_5 regions generally found in industry. Though there were problems with the reliability of the instruments, especially with the electronic circuits of the detectors (in the beginning Faraday cups with vacuum tube electrometers), the use of mass spectrometers in the petroleum and chemical industry provided the basis for its later application in structural organic chemistry in a threefold way. The instrument manufacturers, for example CEC, produced relatively large numbers of instruments and developed the mass spectrometer as a highly reliable and precise instrument suitable for organic analysis. Many spectra and fragmentation patterns were obtained and published, mainly through the efforts of the Hydrocarbon Research Group of the American Petroleum Institute (Research Project 44). Some of the chemists who were trained in the use of the instrument in this way later applied it to research in more complex organic compounds, the most renowned being Seymour Meyerson of Standard Oil of Indiana (Indiana Standard) and Fred W. McLafferty of the Dow Chemical Co in Midland, Michigan.

When in 1951, Milburn J. O'Neal and Thomas P. Wier at the Houston Manufacturing Research Laboratory of the Shell Oil Company described a heated inlet system for the analysis of the high boiling fractions of petroleum, the use of mass spectrometry for structural elucidation of complex molecules seemed to be close at hand.[15] Though they investigated compounds up to a molecular weight of 600, and excellent instruments were commercially available at the time, the extension of mass spectrometry to research in structural organic chemistry was not achieved immediately. Klaus Biemann, one of the pioneers of the structural approach, later used the success of the mass spectrometer in the petroleum industry to explain this delay. As an outcome, most organic chemists regarded mass spectrometers as very expensive and elaborate instruments, which were tricky to handle. This was due entirely to the quantitative industrial approach, involving the analysis of mixtures of compounds of known structure.[16] Furthermore, the fragmentation patterns of hydrocarbons did not show any promise for correlation with structures. This had to await systematic investigations in molecules with an aromatic moiety or other functional groups, an endeavour that began in the petrochemical industry, notably, with John H. Beynon at ICI, Meyerson at Indiana Standard and McLafferty at Dow.

Fred W. McLafferty and Herbert Woodcock at Dow Chemical Co, Midland, Michigan with an early 90° sector mass spectrometer, c. 1951. The instrument was built by Victor Caldecourt from Westinghouse parts, supplied by Russ Fox and John Hipple. Courtesy of Fred W. McLafferty

McLafferty first came into contact with mass spectrometry in 1950 at the Dow Corporation's Spectroscopy Laboratory, where forays had already been made into the development of spark source spectra and infrared spectroscopy. Dow bought a 90° Nier-type mass spectrometer from Westinghouse, and Victor Caldecourt introduced McLafferty to mass spectrometry.[17] The spectroscopy laboratory dealt mainly with analysis connected to manufacturing processes and McLafferty was able to establish mass spectrometry as an important analytical instrument, often in competition with infrared spectroscopy. Additionally, McLafferty made an effort to alter the 'terrible' reputation of mass spectrometry for analysis of organic compounds, owing to 'random rearrangements' of hydrocarbons, to use his phrase.[18] By contrast, the mechanism of the so-called 'specific rearrangements' could be worked out, thereby assisting the elucidation of the structure of the molecules under consideration. The yield of specific ions can be increased by the formation of sterically favourable transition states or energetically more stable products. In the case of hydrocarbon molecules, there may be several decomposition paths with comparable energy requirements and for this reason several rearrangements may occur in a random manner. If the molecule contains a functional group, this group may control the fragmentation process and lead to a specific rearrangement. As Sir Cyril Hinshelwood put it in 1959:

There is one aspect of mass spectrometry where the cleavage into laymen and experts does not yet apply. This is, I should like to emphasise, a good thing, because the most live [*sic*] and truly scientific parts of any subject are those which excite the attention of the layman and the expert equally. I am referring now to the fundamental physics and chemistry of the processes by which the ions are formed, and in which they fragment to give the characteristic pattern shown in the mass spectrum of a complex molecule. The whole sequence of events is still in many ways rather a mysterious one, but it is of fundamental interest, and, indeed the behaviour of molecules in this fragmentation is a new and self-contained chapter of chemical kinetics.[19]

One of the most famous rearrangements is the so-called McLafferty rearrangement which involves the migration of a hydrogen atom to a double bond. This reaction type had already been mentioned in a 1952 publication,[20] but its generalisation and the proof of its great utility for the interpretation of mass spectra had to await the systematic studies of McLafferty at Dow.[21] The mechanism involves a six-membered cyclic transition state for the hydrogen rearrangement and includes the elimination of an olefin molecule from unsaturated molecules such as alkyl esters or ketones. Investigations with deuterated compounds by Biemann's group at MIT and Einar Stenhagen's in Sweden in the early 1960s showed clearly that the proposed mechanism is correct.[22] Moreover, the McLafferty rearrangement turned out to be a general reaction, not limited to ketones and esters, but occurring in compounds with carbon, nitrogen or oxygen content, and the double bond could be part of an olefinic or even an aromatic system. The reasoning in terms of physical organic chemistry was that the hydrogen rearrangement yields a stable olefinic neutral product, not a radical species; although the cation now contains an unpaired electron, this is resonance stabilised; the more favourable enthalpy can more than compensate for the unfavourable entropy. One example is the cleavage of a C–O bond in benzyl-methyl ether, in which the rearrangement of the McLafferty type is in competition with the formation of the tropylium cation.

The tropylium ion, $C_7H_7^+$, first described by William von Eggers Doering and Lawrence H. Knox in 1954,[23] was a very interesting species, because of its aromatic behaviour. Meyerson and his colleagues Paul Rylander and Henry Grubb from the Research Department of Indiana Standard gave a considerable boost to mass spectrometry by showing that similar compounds could be formed by a rearrangement in a mass spectrometer.[24] In the mid-1950s, the decomposition of alkylbenzenes was thought to be a straightforward cleavage of a carbon–hydrogen or carbon–carbon bond in the side-chain. But spectra of deuterated toluenes and inconsistencies in appearance potentials indicated that the origin of the ion with the mass 91 ($C_7H_7^+$) was not that simple. The reasoning of Meyerson, Rylander and Grubb is a good example of how rigorous experiments were able to contribute to the establishment of specific rearrangement mechanisms. Deuteration of toluene either in the side-chain or the ring positions and careful examination of the mass spectra obtained proved that the seven hydrogen atoms in $C_7H_7^+$ were equivalent. This finding led to the proposal that the toluene molecule underwent a rearrangement to a seven-membered ring, the tropylium ion. Though this ion is energetically more stable than the benzyl ion, such a rearrangement had not been observed in solution. But in the low pressure gas phase of the mass spectrometer, where neither solvation nor bimolecular processes were available, the tropylium ion was formed. The dissociation of alkylbenzenes through the tropylium ion and not the benzyl ions allowed the reinterpretation of some hitherto anomalous data, *e.g.* the cleavage of the phenyl–methyl bond in polymethylbenzene molecules, the absence of directive effects in isomeric dialkylbenzenes, and the strength of the benzyl–hydrogen bond.

Though these two examples show the power of chemically-based reasoning, there have been more fundamental, physical approaches. In 1963, McLafferty compared these two approaches with a situation familiar to all chemists:[25]

> There have been two general approaches to the physical, or mechanistic, explanations in organic chemistry – the approach from physics through the development of quantum mechanics, molecular orbital theory, *etc.*, on the one hand, as compared to the more intuitive or empirical approach of physical-organic chemistry with such effects as inductive effect, resonance, *etc.*, on the other hand.

The equivalent of the quantum mechanical approach in mass spectrometry was the quasi-equilibrium (or statistical) theory. But only the spectra of very simple molecules could be calculated rigorously, making difficult the application of this theoretical approach to more complex organic molecules. It was the semi-empirical approach which was put to use in the field of natural products, at first mainly alkaloids and steroids.

Klaus Biemann at MIT

In the late 1950s, most mass spectrometrists were still concerned with mass spectra of compounds of known structure. Exceptions included the group around Stenhagen at the University of Gothenburg who used the mass spectrometers designed and constructed by Ragnar Ryhage from the Karolinska Institute in Stockholm, mostly for the elucidation of the structures of long-chain fatty acids and esters. Another early group that engaged in systematic studies of the structures of natural products was headed by Klaus Biemann at MIT, with a focus on alkaloids, amino acids and peptides.[26] Biemann had a strong background in synthetic organic chemistry thanks to his studies with Hermann Bretschneider at the University of Innsbruck (Austria) and during a post-doctoral fellowship at MIT under George Büchi he worked on the synthesis and structure of natural products. This work was sponsored by the Swiss company Firmenich & Cie., a manufacturer of food and perfume flavours, and it was through this cooperation that Biemann became interested in the use of mass spectrometry. In 1956, he heard a conference talk on the identification of fruit flavour components with the help of mass spectrometry, and, having experience with ultraviolet and infrared spectra, he considered the possibility of applying mass spectrometry to structure determination in a similar manner. He was appointed instructor at the analytical division of the department of chemistry at MIT in 1957. As he wanted to retain his interests in the chemistry of natural products, Biemann decided to investigate this promising technique. Mass spectrometry offered considerable potential for bringing his new post and principal

Klaus Biemann (back, left) discussing a photographic plate used to record high resolution mass spectra in the CEC 21-110 high resolution mass-spectrometer shown partially on the left, with Walter McMurray (standing on the right) and Peter Bommer (seated, front centre), 1963. Photograph, courtesy of Professor Klaus Biemann

interest together, and he persuaded the head of the department, the famous chemist Arthur C. Cope, to spend the then enormous sum of $60 000 (which Cope took from the William Barton and Emma Rogers Fund, made available to him just a year earlier) for the purchase of a CEC 21-103C mass spectrometer. Seven years later, Cope stated that this was one of the best uses made of the fund, because Biemann 'has become recognised as the foremost person in the world in the application of mass spectrometry to the determination of complex organic compounds', and that this first instrument had been supplemented by three additional ones, purchased with government funds.[27] Biemann's project was also supported with a $10 000 grant from Firmenich & Cie. for the purchase of the instrument, and a grant for a post-doctoral fellow to work with Biemann. In return, Biemann acted as a mass spectrometry consultant for the Swiss company, analysing samples (at an average of 15 spectra a month),[28] and assisting with the setting up of an in-house mass spectrometry unit at Geneva. Together with a grant from the National Institutes of Health, which paid for a second post-doc, this sparked off the programme, initially planned to deal with 'analysis of amino acids, structure of small peptides, origin of rearrangement peaks in mass spectra (using isotopic labelling), correlation of mass spectra and structure, *etc*'.[29] For the latter, Biemann's intention was to start out with alicyclic compounds, such as cyclohexane derivatives and mono- and bicyclic terpenes, and because he knew of the interest of Firmenich & Cie. in this field, he suggested that they send him samples. Later, Biemann's research was stimulated by new and interesting compounds supplied by Firmenich. An example is *cis* rose oxide, a monoterpene ether discovered in 1959 by Casimir F. Seidel and Max Stoll at Firmenich, and found to be an important ingredient of Bulgarian rose oil. Its structure was elucidated through a collaboration between the chemists at Firmenich, Albert

Eschenmoser's group at ETH Zurich, and Klaus Biemann's group at MIT.[30]

Biemann's interest in indole alkaloids originated during his work with Büchi.[31] Alkaloids being of considerable interest to both pharmacologists (because of their biological activity) and chemists (because of their complicated structures), were an ideal target for mass spectrometry.[32] Many alkaloids were only available in minute amounts, but usually in series with closely related structures. Hitherto, chemists had elucidated the structure of a specific alkaloid either by its chemical degradation into smaller identifiable molecules or by its conversion into other alkaloids of known structure. This required laborious chemical preparations and often considerable amounts of starting material. The identity of these degradation products with the samples was then established by infrared spectroscopy or melting point measurements.

Although they are very complex, alkaloids possess certain features that eased the burden of unravelling their structures. They usually contain aromatic and alicyclic rings, and in a structurally-related group of alkaloids, many members differ only in the substituents in the aromatic ring. Hence, the mass spectra for the fragments containing the alicyclic rings yield peaks with identical mass numbers and similar intensities. By contrast, the fragments that retain the aromatic ring show peaks that differ by the difference in the mass of their substituents. Known as the 'mass spectrometric shift', this technique was developed by Biemann and put to good use by his group and then others. Biemann, for example, in his investigations of indole and dihydroindole alkaloids, compared the conversion products of ajmaline and sarpagine, whose mass spectra were very similar but differed by sixteen mass units. Because the structure of ajmaline was already known, Biemann and his co-workers were able to confirm a proposed structure of sarpagine by mass spectrometry. This took place just before the much more laborious chemical methods allowed a direct comparison of the conversion productions of ajmaline and sarpagine by melting point and infrared measurements.[33] Biemann, in collaboration with the group of William I. Taylor at CIBA, also investigated a number of related alkaloids from *Tabernanthe iboga*, namely ibogamine, ibogaine and ibogaline. This investigation began, it seems, as a comparative study to the sarpagine research. Biemann looked for alkaloids whose carbon skeleton was isomeric with the equivalent structure in sarpagine.[34] The chemists at CIBA Pharmaceutical Products at Summit, New Jersey, were able to provide Biemann with the alkaloids he needed. Though the structures of the first two alkaloids, ibogaine and ibogamine, had been established, the small amount available for ibogaline prevented the proof of its structure. UV and IR spectra showed that ibogaline contained a 5,6-dimethoxy-indole moiety. Biemann simply used the mass spectrometric shift technique to demonstrate the validity of the proposed structure.

Another technique, deuterium labelling, enabled structural elucidation of iboxygaine, a hydroxyl derivative of ibogaine, and indicated a fragmentation scheme for ibogaine that allowed a rational correlation of many of the peaks with fragments of known structure. This approach was based on the assumption that fragments which are stabilised by electrons of an adjacent heteroatom or a π-bond show stronger peaks than fragments which are not. Moreover, Biemann proposed that multiple cleavages of bonds 'should occur in a concerted fashion and lead wherever possible to neutral molecules rather than fragments with a number of unpaired electrons'.[35] This followed a generalisation of the so-called 'Stevenson's rule',[36] according to which the positive charge after cleavage will reside on that fragment which has the lower ionisation potential. In most cases this is brought about by delocalisation of the charge by resonance or the inductive effects of substituents. In 1965, Biemann and his collaborator Peter Bommer compared the state of the art in organic mass spectrometry, with the vast amount of data available in the form of qualitative observations, with the situation:

> prevailing in organic chemistry in general that one in principle knows all the factors which concertedly govern the behaviour of an organic molecule, and should thus be able to predict by calculation the best synthesis of a given molecule, but any synthesis devised today is almost exclusively based upon empirical observations and experience and not on quantum mechanical calculations. Seemingly empirical rules based on at least qualitative, in many instances even quantitative, understanding of reaction mechanisms are providing the organic chemist today with a much better basis for the design of a workable synthetic route, and it is to a large extent the physical chemist who has produced this basis. In the much younger field of organic mass spectrometry it would seem that here again the physical chemist may be able to provide the much-needed fundamental data to put the empiricism of the organic chemist on better grounds.[37]

It is characteristic of the period that Bommer and Biemann added a cautious remark. Because most of the underlying experimental data had been found by measurements in solution, their application to mass spectrometry, where unimolecular fragmentations and rearrangements occur in the vapour phase, had to be taken with a pinch of salt. They argued that in order to make a choice when only a few structures were possible or under consideration, it was possible to choose between them on the basis of mass spectrometry.

This argument was reminiscent of a discussion that Biemann had in 1962 with Robert L. Autrey, assistant editor of the *Journal of the American Chemical Society*, about the reliability of results obtained with mass spectrometry. Autrey thought that mass spectrometric methods were not sufficiently reliable to enable the structural elucidation of hitherto unknown compounds.[38] On the basis of two referee reports, he responded to a manuscript submitted by Biemann and his collaborators Margot Friedmann-Spiteller and Gerhard Spiteller:

> We realise that mass spectrometry is a powerful method, and we are confident of the correctness of your conclusions. However, its use is still in its infancy, and the samples to which you have applied it are open, perhaps, to question as to their homogeneity. You have detected 20 compounds, but we question that you have 'isolated' 16 of them. There is no demonstration, other than by gas chromatography, of their purity. None of the usual criteria of purity have been applied; none of the compounds have been

characterised by any of the classical means: there are no melting points, no analyses, few ultraviolet and no infrared spectra. The chemist or chemical taxonomist discovering one of these compounds in another plant may be quite unable to ascertain that he has in hand a known compound, as few institutions have available a mass spectrometer of the calibre to obtain these data.[39]

Biemann replied that he was himself concerned about the possibility that 'other workers in the field of alkaloid chemistry may use these mass spectrometric techniques too freely or may draw unjustified conclusions'. He explained that in order to make clear the applicability of mass spectrometry he had written a book, which had just appeared in print.[40] Furthermore, he questioned the critique concerning the isolation of these alkaloids, claiming that this would be a matter of definition. But he altered the text of the relevant table in the final manuscript, now stating only that the alkaloids are listed in the order of elution from an alumina column. From the beginning, and in contrast to the criticism that no melting points were reported, Biemann and co-authors Friedmann-Spiteller and Spiteller had given these data in cases where these compounds had been obtained in crystalline form. In reference to the lack of elemental analyses, he stated in the final publication that this:

> is balanced by the presentation of mass spectrometric molecular weights (and the entire spectrum), which are very reliable for this type of compound ... and leave as little ambiguity (albeit of a different kind) regarding the true empirical formula as does a conventional elemental analysis. Furthermore, we feel that the mass spectrum is at least as good a method for the characterisation of organic compounds as other spectra at present accepted for this purpose.[41]

Two years earlier, and in correspondence with an editor of the same journal, Biemann had indicated the criteria for justifying the use of mass spectrometric determination of molecular weights rather than elemental analysis. Conceding that the use of the new technique 'might lead to situations where a negligent investigator never bothers to prepare the new compounds in a pure state', he suggested the acceptance of mass spectrometric molecular weights if:

(a) there is an indication of the purity of the material associated with the description of the experiment;
(b) the compound belongs to a class expected to give reasonably intense peaks for the molecular weight;
(c) the type of instrument used (including operating conditions such as temperature and ionising potential), and perhaps even the laboratory in which the spectrum was determined, is indicated; and
(d) the compound was derived in a reasonably clear-cut reaction from another completely characterised substance.[42]

But even this cautious proposal, and the knowledge of the possible inaccuracy of a conventional analysis of carbon and hydrogen, did not cause more than a lukewarm response. Marshall Gates, the editor of the *Journal of the American Chemical Society*, only offered to consider the acceptance of mass spectrometric data in a few straightforward cases.[43]

Carl Djerassi at Stanford University

With McLafferty and Biemann, Carl Djerassi was one of the early proponents of the use of mass spectrometry in structural organic chemistry. He was well aware of the possible pitfalls of assigning specific 'mechanisms' to fragmentations of molecules. At the end of the 1960s, Djerassi's investigations, which were based mainly on high resolution work and isotopic labelling techniques, have either strengthened or revised many of the earlier proposed fragmentation mechanisms. But Djerassi and his first collaborators in mass spectrometry, Herbert Budzikiewicz and Dudley H. Williams, stated in 1967:

> The quotation marks around the term 'mechanism' are still well deserved and the term is hardly used throughout our present book; 'rationalisation' is a much better substitute. Since the fragment ions are not isolated, only indirect support can be presented to describe their nature and the evidence is by no means as rigorous as in many other organic chemical reaction mechanisms. Nevertheless, the circumstantial evidence is now overwhelmingly in favour of the approach used in our first book – namely that much of organic mass spectrometry can be discussed in terms of the standard and really oversimplified language of the organic chemist. It is largely for that reason that mass spectrometry has found such rapid acceptance by organic chemists during the past few years, and it is precisely through the use of such oversimplified concepts and generalisations that the more detailed and refined knowledge of the future will be derived.[44]

Djerassi and his group at Stanford did much to establish mass spectrometry as a tool for the organic chemists, with the assumption that most of them would not measure the spectra by themselves but would only interpret them. Therefore they regarded the 'mechanistic' approach as the best from a pedagogical perspective. To achieve their goal they published a series of books, commencing with relatively simple organic molecules and then dealing with natural products such as alkaloids, steroids, terpenoids and sugars.[45]

As early as 1957, Djerassi, then at Wayne State University in Detroit, had used mass spectrometry for the determination of the empirical formula of the cactus sterol lophenol, which made possible its structure elucidation. The mass spectrometric measurements had been done in the laboratory of Rowland Reed in Glasgow. After taking leave of absence at Syntex in Mexico City, Djerassi joined the faculty of Stanford University in 1959/60, when a new chemistry department was created under the skilful leadership of William S. Johnson, formerly professor of chemistry at the University of Wisconsin, Djerassi's *alma mater*.

According to Djerassi, chemists active in the field of natural products had to use every possible assistance in the elucidation of their unknown compounds, very often available only in very small amounts. For this reason, physical methods entered organic chemistry invariably through the efforts of the natural product chemists, and the success of these methods was such that they literally obliterated the traditional methods in this field.[46] While still at Wayne State University, Djerassi had made an important foray into this field with optical rotatory dispersion (ORD), which he first used to investigate steroid ketones.[47] His interest in mass spectrometry was stimulated by the work of Paul de Mayo and Reed at Glasgow, Stenhagen and Ryhage in Sweden, but above all by the investigations on alkaloids by Biemann. Biemann and Djerassi met in Australia at the IUPAC International Symposium on the chemistry of natural products in August 1960, and Biemann agreed in principle to spend a few months as visiting lecturer on mass spectrometry at Stanford to introduce the new technique there. Djerassi chose the same type of instrument as Biemann had used, a CEC 21-103C with heated inlet system, regarded as the best instrument available at the time. Already in December 1960, Herbert Budzikiewicz, an Austrian post-doctoral fellow, joined Djerassi's group and he subsequently became the main operator and investigator of mass spectrometry in Djerassi's laboratory at Stanford.

Djerassi was interested in the structural elucidation of alkaloids by mass spectrometry, as was Biemann. However, together with Budzikiewicz, he chose the steroid ketones, the same class of compounds which had been of great use in the ORD work, for the first systematic investigation with mass spectrometry. In this endeavour, they were 'motivated by the belief that a semi-empirical study of the mass spectrometric fragmentation patterns of a group of closely related substances (together with deuterated analogues) would lead to generalisations, which might prove very fruitful in structural and stereochemical investigations of natural products'.[48] For this particular class of compounds, their view proved to be too optimistic. Extensive correlations between the mass spectra of saturated steroidal ketones and their structures were not feasible; but labelling them with deuterium made it possible to investigate dissociations and rearrangements.[49] Moreover, they were able to show the influence of interatomic distances for the McLafferty rearrangement. Other compound classes, especially the steroidal ketals, proved to be more suitable for charge localisation upon electron impact. In general, Djerassi's strategy indeed proved to be a very successful one.

With his colleagues, Djerassi shared the assumption that the relatively high ionising energy in the ion source of the mass spectrometer (normally 70 eV) is sufficient to remove an electron from any bond in the molecule, and thus results in an electronically excited positive ion, which does not decompose immediately. Its excess electronic energy can be transferred *via* vibrations/oscillations to yield lower-lying electronic states, and in cases where the molecules contain hetero-atoms or π-bonds, it is possible to localise the charge prior to further decomposition. Therefore, most compounds of interest for the organic chemist do not decompose statistically, but it is possible to identify the molecular ion according to the canonical rules of physical organic chemistry, and, on this basis, to rationalise the mechanism of its fragmentation. This could be shown for example through the introduction of functional groups in compounds which had given mass spectra with many fragments (and therefore many peaks), which simplified the spectrum in favourable cases to such an extent that only one or two fragment ions carried the bulk of the ion current.[50] Reasoning along these lines allowed the prediction of bond fissions, which were assumed to be homolytic, but was not only based on analogies to the accepted rules of modern organic chemistry. It was assisted by the use of additional techniques, such as the measurement of the appearance potential of a molecular ion, which is equal to the dissociation energy of a compound plus the ionisation potential. Thus, the appearance potential can provide hints to predict which bonds are broken, and therefore which ions are formed. Equally important was the investigation of so-called metastable peaks, which originated when positive ions further decomposed after their acceleration but before they reached the analyser region of the mass spectrometer. With both methods, and together with kinetic studies, it was possible to indicate the structures of the ions formed.

To establish unequivocally the composition of a fragment ion, a suitable method was already at hand: organic high resolution mass spectrometry. This technique had originated in the mid-1950s with the work of Beynon at ICI in Manchester, and made use of the double-focusing mass spectrometers of the Nier–Johnson or the Mattauch–Herzog types (originally developed for the measurement of isotope abundances). With their help it was possible to determine the mass of an ion to within one part in 10^6, for example to differentiate between the species $C_{16}H_{22}O_2$ (246.1620 amu) and $C_{17}H_{26}O$ (246.1984 amu), a difference of 0.0364 mass units. Until 1964, high resolution mass spectrometry was used to check only a few of the more interesting peaks, or, according to Biemann, to do low-resolution mass spectrometry with occasional accurate mass measurements. To make use of the smaller peaks, it was necessary to use a computer to assemble and chart all the information. Biemann, in collaboration with Walter McMurray, Peter Bommer and Dominic Desiderio, developed the element mapping technique to present the data in a concise form.[51] This is actually a print of the mass-to-charge values in vertical order (increasing from top to bottom), with a separate column for each set of ions with a given heteroatom content (increasing from left to right). Other forms of presentation were later developed by Al Burlingame (which preserved the original bar graph of the low resolution spectra) and McLafferty, who introduced a three-dimensional graph presentation.[52]

High-resolution techniques helped tremendously with the assignment of structures to fragmentation peaks, and in the following years the computer played a more and more dominant rôle. An interesting example of the expectations raised by the use of computers in the 1970s is the application of artificial intelligence for the interpretation of data assembled by mass spectrometry, and later other physical methods as well.[53] Djerassi, who linked up with Joshua Lederberg and Edward Feigenbaum at Stanford for this purpose, used one of the programmes developed in the mid-1970s to test the unambiguity of structure elucidation published in the chemical literature and based on the chemical and spectral data alone. With the help of CONGEN (CONstrained structure GENeration), graduate students found out that in no instance was the published structure the only one consistent with the data given. Since additional possible structural assignments existed, the published structures were not as certain as the authors of the

*The News Gazette, Champaign-Urbana, April 28, 1965 offered the following account of the
occasion shown in this photograph: 'Experts on analysis technique. Authorities on mass
spectrometry, scientific analysis technique, inspect one of Purdue University's mass
spectrometers during a break in a Purdue short course on using the instruments to
determine molecular structure. From left are guest lecturers Prof. K. L. Rinehart,
University of Illinois, and Prof. A. L. Burlingame, University of California, and Prof. F. W.
McLafferty, Purdue, course chairman. Representatives of industrial and government
laboratories and universities from 18 states and two Canadian provinces are taking the
course, sponsored by the Purdue chemistry department. The course is the first in the nation
offered on the subject.' Courtesy of Fred W. McLafferty*

papers had indicated, or they had a basis in data not mentioned in the paper. Though Djerassi, only partly tongue-in-cheek, as he recalled, proposed to establish the programme as an automated journal referee, this approach demonstrated the limitations rather than the potential for the use of artificial intelligence for determination of chemical structures.

The success of mass spectrometry in chemical research of the 1960s was bound to chemically oriented research programmes, and was formulated in the language of physical organic chemistry. Even though this was a very successful approach in the end, it was received with some reservations, as seen, for example, in the editorial policy of the *Journal of the American Chemical Society*. The orientation towards physical organic chemistry had weighty cognitive and pedagogical reasons, but (perhaps) was partially influenced by the relatively marginal rôle of natural product chemistry in the United States (when compared with the synthetic approach and especially physical organic chemistry). Thus also for reasons of status, the use of concepts and terminology of physical organic chemistry was a legitimate decision. The adoption of mass spectrometry was promoted by the extensive publishing of textbooks by its advocates, kept alive by the innovation of new techniques, often in close collaboration with the instrument manufacturers, and boosted by a generous funding policy of the government. To give an often quoted sentence of the philosopher Ian Hacking in its full version: 'Experimentation has a life of its own, interacting with speculation, calculation, model building, invention and technology in numerous ways'.[54]

Acknowledgements

I must thank Klaus Biemann, Carl Djerassi and Fred W. McLafferty for access to and permission to quote from their personal papers, and their comments on this paper. The helpful feedback of Peter J. T. Morris, Anthony S. Travis, and the participants of the CHMC conference in London is also gratefully acknowledged.

Notes and References

1 D. Baird, 'Analytical Chemistry and the "Big" Scientific Instrumentation Revolution', *Annals of Science*, 1993, **50**, 267–290; P. J. T. Morris and A. S. Travis, 'The Rôle of Physical Instrumentation in Structural Organic Chemistry', in *Science in the Twentieth Century*, J. Krige and D. Pestre (eds.), Harwood Academic Publishers, Amsterdam, 1997, pp. 715–739. Baird traces back the revolution (coined in terms of I. B. Cohen's and Ian Hacking's concepts) in analytical chemistry to the 1920s–1950s. Morris and Travis give an excellent overview of their topic.

2 A. Weissberger (ed.), *Physical Methods of Organic Chemistry*, two vols., Interscience Publishers, New York, 1945–1946 (Techniques of Organic Chemistry).

3 E. A. Braude and F. C. Nachod (eds.), *Determination of Organic Structures by Physical Methods*, Academic Press, New York, 1955.

4 F. C. Nachod and W. D. Phillips, *Determination of Organic Structures by Physical Methods*, Vol. 2, Academic Press, New York, 1962, p. vii.

5 J. C. P. Schwarz (ed.), *Physical Methods in Organic Chemistry*, Holden-Day, San Francisco, 1965, pp. 2–3.

6 C. Djerassi, *Optical Rotatory Dispersion: Applications in Organic Chemistry*, McGraw-Hill, New York, 1960, p. 19. The authors of this chapter were Anthony N. James and Berndt Sjöberg, both of Djerassi's laboratory.

7 For an analysis of the interplay of instrumental development and the mechanistic approaches of physical organic chemistry see the contribution of Leo Slater in this volume.

8 N. Rasmussen, *Picture Control. The Electron Microscope and the Transformation of Biology in America, 1940–1960*, Stanford University Press, Stanford, 1997, pp. 13–14.

9 See Schummer's contribution in this volume.

10 Y. M. Rabkin, 'Technological Innovation in Science. The Adoption of Infrared Spectroscopy by Chemists', *Isis*, 1987, **78**, 31–54.

11 J. H. Beynon and R. P. Morgan, 'The Development of Mass Spectrometry. An Historical Account', *International Journal of Mass Spectrometry and Ion Physics*, 1978, **27**, 1–30; H. Remane, 'Zur Entwicklung der Massenspektroskopie von den Anfängen bis zur Strukturaufklärung Organischer Verbindungen', *NTM*, 1987, **24**(2), 93–106; H. A. Laitinen and G. W. Ewing (eds.), *A History of Analytical Chemistry*, Division of Analytical Chemistry, American Chemical Society, Washington DC, 1977, pp. 216–229.

12 Beynon and Morgan, 'The Development of Mass Spectrometry. An Historical Account', p. 28.

13 H. W. Washburn, H. F. Wiley and S. M. Rock, 'The Mass Spectrometer as an Analytical Tool', *Industrial and Engineering Chemistry* (*Analytical Edition*), 1943, **15**,

541–547, p. 546.

14 C. E. Starr, Jr. and Trent Lane, 'Accuracy and Precision of Analysis of Light Hydrocarbon Mixtures', *Analytical Chemistry*, 1949, **21**, 572–582.

15 M. J. O'Neal and T. P. Wier, 'Mass Spectrometry of Heavy Hydrocarbons', *Analytical Chemistry*, 1951, **23**, 830–843.

16 K. Biemann, 'Applications of Mass Spectrometry' in *Elucidation of Structures by Physical and Chemical Methods*, K.W. Bentley (ed.), Part One, Interscience Publishers, New York 1963, pp. 259–316, on p. 260.

17 Interview of Fred W. McLafferty by the author, 16 and 17 December 1998.

18 Fred McLafferty with reference to the 1951 Pittsburgh Analytical Conference: 'This was really the beginning of applying mass spectrometry to other than hydrocarbon molecules. Something that I didn't realise [until] some years later, mass spectrometry was having a terrible reputation for being good for quantitative analysis but no good for qualitative analysis and the reason was that almost all spectra that were run were hydrocarbons, and hydrocarbons are the worst for mass spectrometry, because of what I've called in my book random rearrangements'. Interview by the author of Fred W. McLafferty, 16 and 17 December 1998.

19 C. Hinshelwood, 'Opening Remarks', in *Advances in Mass Spectrometry*, J. D. Waldron (ed.), Pergamon Press, New York, 1959, xiii–xiv, on p. xiv.

20 G. P. Happ and D. W. Stewart, 'Rearrangement Peaks in the Mass Spectra of Certain Aliphatic Acids', *Journal of the American Chemical Society*, 1952, **74**, 4404–4408.

21 F. W. McLafferty, 'Mass Spectrometric Analysis. Broad Applicability to Chemical Research', *Analytical Chemistry*, 1956, **28**, 306–316, pp. 312–313; F. W. McLafferty, 'Mechanism of Rearrangements in Mass Spectra', *Chemistry and Industry*, 1958, 1366–1367; F. W. McLafferty, 'Mass Spectrometric Analysis. Molecular Rearrangements', *Analytical Chemistry*, 1959, **31**, 82–87.

22 K. Biemann, *Mass Spectrometry. Organic Chemical Applications*, McGraw-Hill, New York, 1962, pp. 119–120.

23 W. Doering and L. H. Knox, 'The Cycloheptatrienylium (Tropylium) Ion', *Journal of the American Chemical Society*, 1954, **76**, 3203–3206.

24 P. N. Rylander, S. Meyerson and H. M. Grubb, 'Organic Ions in the Gas Phase, II. The Tropylium Ion', *Journal of the American Chemical Society*, 1957, **79**, 842–846.

25 F. W. McLafferty, 'Decomposition and Rearrangements of Organic Ions' in *Mass Spectrometry of Organic Ions*, F. W. McLafferty (ed.), Academic Press, New York, 1963, p. 311.

26 K. Biemann, 'The Massachusetts Institute of Technology Mass Spectrometry School', *Journal of the American Society for Mass Spectrometry*, 1994, **5**, 332–338.

27 Arthur C. Cope to J. A. Stratton (president of MIT), July 22, 1964. MIT Archives, AC 134, Box 33, Folder 9.

28 Memorandum 'Organisation of MS-Research', April 9, 1959. Biemann Papers, private collection, folder Firmenich et Cie.

29 Biemann to Max Stoll (head of research of Firmenich et Cie.), April 22, 1958. Biemann Papers, folder Firmenich et Cie.

30 C. F. Seidel, D. Felix, A. Eschenmoser, K. Biemann, E. Palluy and M. Stoll, 'Zur Kenntnis des Rosenöls. II. Die Konstitution des Oxyds $C_{10}H_{18}O$ aus bulgarischem Rosenöl', *Helvetica Chimica Acta*, 1961, 44, 598–606.

31 Interview of Klaus Biemann by the author, 10 December 1998.

32 K. Biemann, 'Determination of the Structure of Alkaloids by Mass Spectrometry', in *Advances in Mass Spectrometry*, R. M. Elliott (ed.), Vol. 2, Macmillan, New York, 1963, 408–415.

33 M. F. Bartlett, R. Sklar and W. I. Taylor, 'Rauwolfia Alkaloids XXXIII. The Structure and Stereochemistry of Sarpagine', *Journal of the American Chemical Society*, 1960, **82**, 3790.

34 Biemann to William I. Taylor, March 21, 1960. Biemann Papers, folder CIBA.

35 K. Biemann, 'Determination of the Structure of Alkaloids by Mass Spectrometry', p. 412.

36 D. P. Stevenson, 'Ionization and Dissociation by Electronic Impact. Ionization Potentials and Energies of Formation of *sec*-Propyl and *tert*-Butyl Radicals. Some Limitations of the Method', *Discussions of the Faraday Society*, 1951, **10**, 35–45.

37 P. Bommer and K. Biemann, 'Mass Spectrometry', in *Annual Review of Physical Chemistry*, H. Eyring, C. J. Christensen and H. S. Johnston (eds.), 1965, **16**, 481–502, pp. 487–488.

38 Compare this with the account of Joachim Schummer in his contribution to this volume.

39 Robert L. Autrey to Klaus Biemann, October 9, 1962. Biemann Papers, folder Aspidosp. The publication referred to is K. Biemann, M. Friedmann-Spiteller and G. Spiteller, 'Application of Mass Spectrometry to Structure Problems, X. Alkaloids of the Bark of Aspidosperma Quebracho Blanco', *Journal of the American Chemical Society*, 1963, **85**, 631–638.

40 Biemann to Robert L. Autrey, October 18, 1962, Biemann Papers, folder Aspidosp. The book is *Mass Spectrometry. Organic Chemical Applications*, McGraw-Hill, New York, 1962.

41 Biemann *et al.* 'Application of Mass Spectrometry to Structure Problems, X', p. 637.

42 Biemann to Marshall Gates, November 4, 1960. Biemann Papers.

43 Marshall Gates to Biemann, November 16, 1960. Biemann Papers.

44 H. Budzikiewicz, C. Djerassi and D. H. Williams, *Mass Spectrometry of Organic Compounds*, Holden-Day, San Francisco, 1967, p. vi.

45 H. Budzikiewicz, C. Djerassi and D. H. Williams, *Interpretation of Mass Spectra of Organic Compounds*, Holden-Day, San Francisco, 1964 (second printing 1965); *Structure Elucidation of Natural Products by Mass Spectrometry*, two vols, Holden-Day, San Francisco, 1964.

46 Interview of Carl Djerassi by Jeffrey L. Sturchio and Arnold Thackray, 31 July 1985. Chemical Heritage Foundation, Philadelphia, PA. Quoted with kind permission of Carl Djerassi.

47 C. Djerassi, *Steroids Made it Possible*, American Chemical Society, Washington DC, 1990, pp. 53–56.

48 H. Budzikiewicz and C. Djerassi, 'Mass Spectrometry in Structural and Stereochemical Problems, I. Steroid Ketones', *Journal of the American Chemical Society*, 1962, **84**, 1430–1439, p. 1431.

49 See Budzikiewicz, Djerassi and Williams, *Structure Elucidation of Natural Products by Mass Spectrometry*, Vol. 2, p. 64.

50 Budzikiewicz, Djerassi and Williams, *Mass Spectrometry of Organic Compounds*, pp. 9–12.

51 K. Biemann, P. Bommer and D. Desiderio, 'Element-mapping. A New Approach to the Interpretation of High Resolution Mass Spectra', *Tetrahedron Letters*, 1964, **26**, 1725; K. Biemann, 'High Resolution Mass Spectrometry of Natural Products', *Journal of Pure and Applied Chemistry*, 1964, **9**, 95–118.

52 For an overview see Budzikiewicz, Djerassi and Williams, *Mass Spectrometry of Organic Compounds*, pp. 37–43.

53 Djerassi, *Steroids Made it Possible*, pp. 102–114; R. K. Lindsay, B. G. Buchanan, E. A.

Feigenbaum and J. Lederberg, *Applications of Artificial Intelligence for Organic Chemistry: The Dendral Project*, McGraw-Hill, New York, 1980.

54 I. Hacking, *Representing and Intervening. Introductory Topics in the Philosophy of Natural Science*, Cambridge University Press, Cambridge, 1983, p. xiii.

Impact of Instrumentation on Biomedical and Environmental Sciences

Innovation in Chemical Separation and Detection Instruments: Reflections on the Rôle of Research-technology in the History of Science

NICOLAS RASMUSSEN

School of Science and Technology Studies, LG Morven Brown Bldg., University of New South Wales, Sydney 2052, Australia, Email: Nicolas.Rasmussen@unsw.edu.au

Introduction

The empirically rich essays of Morris, Travis and Cerutti each discuss major developments in chemical instrumentation, and together this section represents an important new resource for the history of organic chemistry, biochemistry, and related fields – particularly in the post-war period. The 1950s and 1960s were by any account a period of extraordinarily rapid scientific progress, in terms both of the knowledge of biomolecules and the behaviour of organic chemicals in the environment, and in the novel instrumentation enabling this progress. There are many questions that can be raised about this flourishing of science, but one especially large question that intrigues me here (as elsewhere) is whether knowledge or technology was in the driver's seat.[1] That is, were developments such as gas chromatography (GC) and mass spectroscopy (MS), and ultraviolet and infrared spectroscopy before them, called into being to meet an already perceived need? Or on the other hand, did these technologies effectively create new fields of inquiry by making certain types of question tractable?

This is a question about technological determinism in scientific change. New research-technologies can shape disciplines and institutions. New research-technologies can make certain directions of inquiry particularly easy or attractive, boosting the fortunes of some fields at the expense of others. And beyond these merely 'permissive' rôles of research-technology (to borrow a distinction from developmental biology) in enabling particular investigations, perhaps research-technologies can even play an 'instructive' rôle, leaving their impression on the content of scientific knowledge.[2] This possibility has epistemologically[3] radical potential.

Is our knowledge a contingent by-product of the technologies we happen to employ? Or would our knowledge be much the same (perhaps with different gaps and strengths) given different paths of overall technological development? The more that new instrumentation appears to arise from the work of, and to fill the perceived needs of, those scientific investigators that actually employ it in their research programs, the less likely the first, contingentist, alternative seems. On the other hand, the more that new instrumentation is developed within the investigative context of its use, the more it is likely to reinforce the existing ways of life and patterns of authority prevailing in the disciplines of the users, and the less likely it is to reveal radically new phenomena, since the new instrumentation will be calibrated against what is already known by the users before they accept it. Thus the more we consider new instrumentation and technique to be a powerful factor in scientific change, the more we open the way to a view of the content of scientific knowledge as contingent and technologically determined. And conversely, the more we resist the idea that technology can shape knowledge, the more leeway we give conservative social and intellectual forces within established disciplines to play a decisive rôle.

The Invention of the ECD and its Dissemination to New Contexts of Use

It is clear which horn of the dilemma concerning technology the story of the Electron Capture Detector (ECD) pushes us toward. Lovelock's background was in chemistry and physics, but as Peter Morris explains, when Lovelock invented the ECD he was working on a biochemical problem – distinguishing components of complex lipid mixtures – within a medical research institution. Thus Lovelock was acting as a life scientist, and thoroughly immersed in this investigative-user context when he developed the new instrument. Used as a detector of substances at the end of a GC apparatus that separated the bi-omolecules, it was sensitive enough to pick up each of the molecular species in the samples of the size, complexity, and constitution routinely studied in this sort of biochemistry. Indeed from the life science point of view, the instrument was the GC/ECD combination, as neither separation nor detection alone filled the biochemist's purpose of distinguishing the different components. Packaged thus it was plainly a user's invention, and though extremely useful among biochemists, revealed nothing earth-shattering to them in the short run. Indeed, a nice example of the way a new instrument may be rejected by conservative forces within disciplines is presented by Morris, with his anecdote about the early trial of GC/ECD on 'pure' methyl caproate by Lovelock, James and Martin. Because the detector was so sensitive as to pick up traces of contaminants under the trial conditions, and no peak at the expected retention time, the ECD was judged to be 'useless and wholly anomalous'.

The ECD enabled incremental progress and was accepted fairly readily by biochemists for certain purposes, but not so in other investigative contexts such as petroleum chemistry, where there may have been a similar 'pent-up demand' (as Morris nicely puts it) for distinguishing volatile compounds in complex mixtures. Here, unlike in biochemistry, the ECD made few inroads as one of many different detectors to be used with GC. The reason for the difference, Morris argues, was that in petroleum chemistry, rapidity of fraction identification was paramount and quantities were not limiting. In biochemical contexts, on the other hand, sensitivity was paramount, because purification procedures from biological sources often yielded only minute quantities for analysis. Neither revolutionary change nor even much in the way of incremental advance was offered by the ECD to petroleum chemistry. However, a new field of investigation tangential to biochemistry was to emerge in which the ECD would work profound changes for the petroleum chemist, a field that arose as a by-product of petroleum chemistry – or more precisely as a consequence of the environmental release of the products of the petrochemical industry. I am referring of course to pesticides. By the beginning of the 1960s, especially after the alarm raised by Rachel Carson in her book *Silent Spring* (1962) began to be felt in the form of environmental regulation and legal action, a new demand arose for characterisation and quantification of organic chemical pollutants (and identification of polluters) diluted to the extreme in the environment.

Because of the ECD's extreme sensitivity and ability to distinguish different compounds with similar biological effects, the ECD displaced many bioassays that had been used as pollution detectors and quickly became the essential instrument for environmental monitoring. Morris even goes so far as to suggest that increasingly stringent limits on pesticide contaminants through the later 1960s and 1970s were not only enabled by the increasing sensitivity of the ECD, but actually motivated by the technology – together with the attitude that any level of contamination was a threat. Thus outside the original use/context, in environmental toxicology, the new technology had a profoundly revolutionary effect. Without questioning Morris's analysis or this conclusion of localised technological determinism that fits so well with the considerations outlined above, I would like to suggest that another major part of the story is missing. Bioassays also were developing, and the Ames bacterial assay for mutagenicity played a crucial role along with the ECD in making the detection and prevention of contamination in the sub-ppb domain crucial. The Ames test, which was (and still often is) taken to have proved that there is no safe concentration threshold for carcinogenic compounds has a fascinating and politically fraught history that needs further elaboration before we can fully understand the events under consideration here. So do events outside the laboratory which belong more properly to cultural history, such as the manner in which fears of nuclear weapons and radioactive contamination influenced political stances toward environmental chemicals.[4]

Instrumentation in Environmental Analysis

Peter Morris's tale of ECD is effectively a detailed chapter from the end of the overarching story of environmental analysis woven together by Anthony Travis. Focusing on the demands of those charged with industrial hygiene in chemical firms, Travis details the linked histories of detection methods and detection limits for organic chemical contaminants. (There is also a parallel, untold history of chemical engineers needing the same kinds of instruments in quality and process control, as Travis notes.) Starting in the 1930s with the generation of instruments founded on spectrographic principles, Travis convincingly links the development of infrared and ultraviolet spectrometers to the chemical firms requiring the instrumentation. These devices were sometimes built in-house like Eastman–Kodak's mass spectrograph, and sometimes by 'spin-off' instrument firms feeding on the demand and expertise of the chemical manufacturers, as with Perkin–Elmer's symbiotic growth in the shadow of American Cyanamid's Bound Brook dye plant. But in either case, the instruments can be described as inventions from within the context of use – that is, inventions by chemists for use on chemical plant premises in monitoring processes and wastes. The gas chromatographic and mass spectrometric instruments that succeeded them were also home-grown products of the investigative contexts in which they were first employed (mainly biochemical contexts, as Morris and Cerutti both indicate).

After their initial invention, however, the further perfection and (especially) standardisation of these MS and GC instruments was driven by external forces in the 1960s, particularly environmental regulation and pressures raised by litigation, according to Travis. So far as this goes, the conclusion seems meshed with the technological determinism supported by Morris's account: lower detection limits drove lower acceptable limits. But there is another side, a case to be made for continuity and conservatism. Apart from the matter of increasing sensitivity, the practices of the new environmental analysts in contexts like the Environmental Protection Agency (EPA) were essentially the same as those developed by the industrial hygiene personnel of the chemical firms that helped create these instruments. Such continuity seems unsurprising given that instruments are seldom (probably never) truly generic; that is, instruments require compatible 'software' and tend to carry with them many features of the scientific culture in which they were first developed.[5]

Thus a set of techniques developed by the industry was adopted by their watchdogs and enemies and turned against it. Predictable as it may seem in hindsight, this turn of events probably surprised the chemical industry, and it strikes me as ironic. One wonders how many of these instruments would have been developed by the chemical industry in the 1930s through 1950s if environmental regulations and litigation pressures such as emerged in the 1960s had been present. After all, what cannot be detected cannot be entered in evidence. But such speculations ignore the role of biochemists developing separation and detection methods within their own contexts of investigation; perhaps all the necessary chromatographic or spectroscopic technology would, like the ECD in Morris's account, have been invented by biochemists and supplied to the EPA, even if the petrochemical industry had discouraged such sensitive detection methods.

Instruments of Separation in the Life Sciences

Luigi Cerruti's careful account of the development of methods for separation of large molecules in the period from the late 1940s to the early 1960s beautifully illustrates how (as others previously have noted) these new technologies opened an entirely new domain to investigation: the specific sequence of monomers composing proteins and other polymeric biomolecules.[6] Though never stated in unambiguous terms, one thing stands out from the many instances of (sometimes competing) new techniques he describes, from paper, thin-layer, and gel chromatography to starch electrophoresis to PAGE: these methods overwhelmingly were developed by biochemists working on life science problems in life science institutions. Even the Wool Industries Research Association's laboratory in Leeds fits this description, as it was supported in its protein structure work during the 1930s by the medically ambitious Rockefeller Foundation program in 'molecular biology'.[7] However sophisticated in physics and chemistry the biochemist-inventors named by Cerutti were, their inventions were not the product of carefully calculated physical chemistry theory simply 'applied' to someone else's biological problems and materials. They were the product of tinkering by investigators doing life science, investigators who needed to use the new inventions to solve their immediate, biochemical problems.

As discussed above, this is the typical mode of origin of successful new instrumentation, in the conservative, incremental mode. It is developed by users, calibrated and domesticated by them, and thus is unlikely to drive any sudden revolution either in the social order or the knowledge-base of the disciplines in which it is developed. But this view seems counterintuitive in the case at hand: did not these methods for analysing amino acid and (later) nucleic acid sequences in biomolecules lead to a revolution in biochemistry? I would argue that they did not revolutionise biochemistry *per se*, which already knew that proteins and sugars were polymers; biochemistry was able incrementally to solve the problems it had posed itself, and together with the adaptation of physical chemistry by Pauling to the problem of polymer folding, the ability to sequence helped the field to grow along established lines. The revolution worked by the power to learn biological polymer sequences came, instead, in neighbouring fields of life science.

Once could point to the co-linearity established between protein sequences and the nucleic acid sequences that encode them (the discoveries associated with the names of Watson, Crick, Pauling and Sanger) as the revolution that chromatography and electrophoresis brought to molecular genetics in the late 1950s. Pauling's discovery that sickle cell and 'normal' haemoglobins vary at only one amino acid residue was a part of that revolution, and the starting point of a related revolution in human genetics that is finally reaching its logical conclusion today with the Human Genome Project. And as Cerruti also aptly tells, electrophoretic methods for separating variants of a given protein opened new vistas of inquiry for population biologists in the 1960s, since the new methodology allowed them to study variant genes without breeding experiments, and thus to study undomesticated, wild species. But beyond the bonanza of data on formerly unstudied creatures brought by the 'find-'em-and-grind-'em' (as some population geneticists wryly put it) heyday of the isozyme, this case also provides an example of how the adoption of an instrument from outside can intellectually revolutionise a discipline. Before the electrophoresis era nobody in evolutionary biology could coherently raise the issue or pose the question of how much genetic variation is 'neutral' in the sense of having no discernible phenotype. But ever since this issue has been central to population genetics.

Theoretical population genetics, by the way, as the most theoretically developed and mathematically sophisticated discipline in life science, represents a powerful counter-instance to Cerutti's concluding off-hand comment that the primacy of practice is out of place with the hierarchy of disciplines. The thrust of this comment would seem to be that empirical fields suffer from lower status than theoretical fields, despite ongoing remarkable contributions from the empirical side, such as all the many innovations in chromatography developed by the empirical biochemists included in Cerutti's account. The raw fact is that among the basic life sciences, theoretical population genetics has perhaps the lowest status of all. Not only does it suffer from paltry funding compared to experimental, empirical fields (which might be attributed to the inherently more costly nature of the latter), but it is looked down upon by the molecular biologists, immunologists, biochemists, cell biologists, and even by its sister fields of genetics precisely for the degree to which it deals in elegant mathematical models based on ingenious simplifying assumptions. Life scientists understand from their tradition that life's complexity has always escaped the oversimplifications of theorists, and therefore they do not prize the theoretical in the same way as physical scientists.[8] Thus I believe that Cerutti looks at the history of his biochemists from the perspective of other chemists; in this perspective it may indeed seem surprising that macromolecule separation methods owe so little to physical chemical theory. But this may be the wrong perspective for understanding what is essentially a field of life science.

Conclusion

In considering these three fine research articles, I could not forget a story once told me by my grandfather, Harry Sootin. As a young and idealistic chemist in the 1920s, newly minted by City College of New York, he had taken a job in quality control at a synthetic fibre plant (Du Pont, according to the family's uncertain collective memory). Not long after he began in this post, his routine analyses identified a large batch of monomer as contaminated to the extent that it fell well outside quality standards. He repeated the analysis and arrived at a nearly identical result. So he conscientiously reported the bad batch to his department chief. The chief told him that he must be mistaken, and that he had to repeat his analysis. Reaching the same conclusion a third time, he was foolish or principled enough (a bit of both, I believe) to insist on the accuracy of his analysis, and to record the batch as faulty. He promptly was fired, and replaced by another chemist who reached the 'correct' conclusion. Harry went on to become a high school chemistry teacher and a historian of science, which was in any case a better career choice for a man of his character.

This story gives us a nice parable of the fate that can befall a new instrument when it produces surprising findings. Thus we can well understand what Eric von Hippel has observed, that most innovations in technique that are widely adopted arise from within the context of use.[9] Those introduced from outside are less likely to win acceptance, and more likely to meet resistance. Hence the special importance in the history of science of those rare instruments that escape the contexts of their initial invention and use, that enter a field from outside, and that therefore preserve their revolutionary potential.

Notes and References

1 This issue has motivated much of my work. See *e.g.* Nicolas Rasmussen, *Picture Control: The Electron Microscope and the Transformation of Biology in America, 1940–1960*, Stanford University Press, Stanford, CA, 1997; *idem*, 'Making a Machine Instrumental: RCA and the Wartime Beginnings of Biological Electron Microscopy', *Studies in the History and Philosophy of Science*, 1996, **27**, 311–349; *idem*, 'Fact, Artifacts and Mesosomes: Practicing Epistemology with the Electron Microscope', *Studies in the History and Philosophy of Science*, 1993, **24**, 227–265; *idem*, 'Freund's Adjuvant and the Realization of Questions in Postwar Immunology', *Historical Studies in the Physical and Biological Sciences*, 1993, **23**, 337–366.

2 I have explored the utility of the embryological 'permissive/instructive' distinction in Nicolas Rasmussen, 'Instruments, Scientists, Industrialists, and the Specificity of "Influence": The Case of RCA and Biological Electron Microscopy' in I. Loewy and J.-P. Gaudillere (eds.), *The Invisible Industrialist*, MacMillan, London, 1998, pp. 173–208.

3 Epistemology is the field of classical philosophy devoted to the nature of knowledge. I use the term here to refer to the methodologies of scientists and the theories of knowledge, and how best to obtain it, implicit in these methodologies.

4 The classic source for this insight is Ralph Lutts, 'Chemical Fallout: Rachel Carson's *Silent Spring*, Radioactive Fallout and the Environmental Movement', *Environmental Review*, 1985, **9**, 211–225.

5 Nicolas Rasmussen, 'What Moves When Technologies Migrate?: "Software" and Hardware in the Transfer of Biological Electron Microscopy to Postwar Australia', *Technology and Culture*, 1999, **40**, 47–73.

6 Robert Olby, *The Path to the Double Helix*, University of Washington Press, Seattle, 1974; Soraya DeChadarevian, 'Sequences, Confirmation, Information: Biochemists and Molecular Biologists in the 1950s', *Journal of the History of Biology*, 1996, **29**, 361–383; Joseph Fruton, *Proteins, Genes, Enzymes; The Interplay of Chemistry and Biology*, Yale University Press, New Haven, 1999.

7 For a small sample of the voluminous literature on the Rockefeller Foundation program in molecular biology, see Olby, *Path* (note 6 above); Robert Kohler, *Partners in Science: Foundations and Natural Scientists, 1900–1945*, University of Chicago Press, Chicago, 1991; Pnina Abir-Am, 'The Discourse of Physical Power and Biological Knowledge in the 1930s: A Reappraisal of the Rockefeller Foundation's "Policy"

in Molecular Biology', *Social Studies of Science*, 1982, **12**, 341–382.

8 Like Cerutti, I base these assertions merely on anecdote and personal experience. I can claim over ten years as an active researcher in laboratory biology, in several different fields, as the grounds for my description of the attitudes of life scientists.

9 Eric von Hippel, *The Sources of Innovation*, Oxford University Press, New York, 1988, Chapter 2 *et passim*. As Yakov Rabkin has shown, the case of infrared spectroscopy fits the pattern, in that academic chemists after the war were none too eager to adopt the device, despite its demonstrated utility in certain chemical industries; as in the case of the electron microscope, these scientists had to be bribed and coaxed by instrument firms to make a place for it in their academic programs; see Rabkin, 'Technological Innovation in Science: The Adoption of Infrared Spectroscopy by Chemists', *Isis*, 1987, **78**, 31–54; and Rasmussen, 'Making a Machine Instrumental', (note 1 above).

'Parts per Trillion is a Fairy Tale': The Development of the Electron Capture Detector and its Impact on the Monitoring of DDT

PETER J. T. MORRIS

Introduction

In the late 1950s, James Lovelock, a chemist turned medical researcher, working at the National Institute for Medical Research at Mill Hill on the northern outskirts of London, developed a highly sensitive device called an electron capture detector (ECD).[1] This device transformed the monitoring of the pesticide DDT during the 1960s. This is a story which can be told in different ways and at various levels. For instance, it would be possible to analyse the evolution of the ECD in detail, showing how it arose and how it was shaped by its social and technological environment. As the infant ECD had a rival in the field of pesticide residue analysis, the microcoulometric approach, a comparative analysis would also be possible and I intend to combine these two approaches elsewhere. Here, however, I will show: how both the place and the person fit Terry Shinn's concept of 'interstitiality'; examine the impact of these new analytical techniques on the monitoring of DDT (rather than the better known case of the halocarbons[2]); and finally, briefly survey the unexpected consequences of its invention.

The Person

James Lovelock's interest in science was fostered by walks with his father, who was knowledgeable about natural history, reading science fiction, and visits to the Science Museum and the Natural History Museum in London.[3] After he left

James Lovelock, 1950s. Photograph courtesy of NIMR

school, Lovelock joined the chemical consultants, Murray, Bull and Spencer, and studied chemistry at evening classes. While working there, Lovelock learnt the importance of accuracy in analysis:

> I grew to regard accuracy in measurement as almost sacred. As Humphrey Murray frequently told me, lives and jobs can depend on the right answer.... The hands-on experience I gained as an apprentice to this wise man was a priceless gift which has served [me] throughout my life as a scientist.[4]

When the Second World War broke out, the evening classes stopped, but Lovelock was able to obtain a scholarship to study chemistry at Manchester University, which then had one of the best chemistry departments in the country. Lovelock later recalled working under Alexander Todd:

> He was a young man in his thirties, full of idealism and a fine professor to his students. He also treated us like apprentices and would spend a great deal of time talking to us personally as we worked in the laboratory.[5]

The Place

After he graduated in 1941, Lovelock began working for the Medical Research Council at the National Institute for Medical Research in Hampstead. Its founding director, Sir Henry Dale, was a pioneering pharmacologist, and the father-in-law of Alexander Todd.[6] In 1942, Dale was succeeded by (Sir) Charles Harington, a brilliant biochemist who directed the institute for two decades, longer than any other director. Lovelock has remarked that Harington 'was uncompromising in his pursuit of excellence'.[7] The 1950s were the heyday of the NIMR, when it was staffed by Fellows of the Royal Society, and current and

National Institute for Medical Research, Mill Hill, 1950s. Photograph courtesy of NIMR

Sir Charles Harington. Photograph courtesy of NIMR

future Nobel Laureates. It was very different from the British universities. Short-term contracts were the rule, in stark contrast to the tenure most university scientists then enjoyed. According to Lovelock, Harington:

> would tell worried scientists who sought security 'If you are good, you don't need tenure; if not, you should not be here'. He would then grin and say: 'Don't worry; such is our reputation, that you will find no difficulty in getting a job elsewhere'.[8]

In addition, there was an informal atmosphere at the NIMR, which engendered co-operation and interaction across disciplinary boundaries.[9] 'Physicists, chemists, biologists and medical scientists talked and planned together in the coffee room or the cafeteria',[10] and Lovelock has recently argued that:

> [these] casual encounters between at most five scientists, usually one to one, are worth at least as much as the think tanks or 'brainstorming' sessions so beloved of administrators.[11]

There was also a strong empirical tradition at the NIMR:

> it was the tradition of those days never to read the literature, especially textbooks. Senior scientists asserted that our job was to make the literature, not read it. This recipe worked well for me. Had I read the literature of ionisation phenomena in gases before doing my experiments, I would have been hopelessly discouraged and confused. Instead, I just experimented.[12]

A key feature of British research in this period was the construction of scientific apparatus in-house, often from odd bits and pieces, and the NIMR was no exception:

> We were among the best paid scientists in the country, but there was very little money available for equipment. Instead, we had either to make our own or have it made in the Institute workshops.[13]

Feeling the Draught

Between 1946 and 1951, Lovelock worked at the Medical Research Council's Common Cold Research Unit at Harvard Hospital in Salisbury, Wiltshire. This state-sponsored research on the common cold in the 1940s and 1950s was a combination of medical optimism following the success of penicillin in the Second World War and the desire of the post-war Labour government to adopt a more egalitarian approach to health. As Lovelock has wryly noted, the Medical Research Council was 'not entirely unaffected by public opinion and political pressure'.[14]

When little was known about the common cold, it made sense to begin with the folklore that you caught colds in the winter because you were cold. As Lovelock later recalled:

> My job was to determine the extent of chilling objectively, then compare it with clinical data on the frequency of colds. The three factors important in chilling are temperature, humidity and air movement. The first two are easy to measure, but the air movements in a closed room ... were so slight as to be undetectable by the simple anemometers then available.[15]

He constructed two very sensitive anemometers and on the basis of his research concluded that the common cold was not caused by being cold. This may seem self-evident nowadays when every newspaper article on the subject starts with the assertion that you cannot catch a cold from a chilly draught, but it appears that Lovelock was the first to demolish this myth scientifically.

Saving the Hamster

By the time he returned in 1951 to the NIMR (which had been relocated to Mill Hill a year earlier), Lovelock had become a medical researcher with a chemical background, rather than a chemist. Furthermore, at the Common Cold Unit, he had developed his skills as a craftsman-engineer, able to make instruments and other apparatus from whatever materials lay at hand. Thus, Lovelock stood squarely within Shinn's 'research-technology'. If Lovelock was well qualified to operate within this field of 'research-technology', his working environment also played a crucial rôle. Although the NIMR was formally dedicated to biomedical research, its directors, particularly Harington, believed in freedom of research and individual researchers were encouraged to follow their lines of research to their logical conclusions, even when they lay outside medicine. As we have seen, there was a strong empirical tradition and a 'can-do' atmosphere, which may have stemmed from the wartime experiences of many of the NIMR's scientists. The scope for innovation was widened even further by close and frequent interaction between researchers on different floors.

Even if we accept that Lovelock was a scientist capable of making a major technical breakthrough and that the NIMR was certainly the place to make one, there has to be a starting point. Curiously, the path to the ECD began with the reanimation of frozen hamsters. When Lovelock returned to the NIMR, Harington asked him to assist three biologists Alan Parkes, Christopher Polge and Audrey Smith with their research on freezing biological materials, initially spermatozoa and blood. Having successfully completed this part of their research, the biologists then switched to freezing and then reanimating small animals.[16]

The reanimation process was quite difficult, because the warming surface of the body drew vital blood away from the heart and the animal died. By 1953, the biologists had got round this by pressing a hot spoon against the chest of a golden hamster, thereby successfully reviving it at the cost of a severe chest burn. Lovelock was troubled by the unnecessary pain caused by this procedure. With the help of a continuous wave magnetron obtained from the Royal Navy, he was able by 1955 to revive the hamsters using radio frequencies, essentially a primitive microwave oven.

Using his earlier work on the effect of freezing on red blood cells, Lovelock studied the damage caused to body cells by the freezing process and he realised that it was connected to the fatty acid composition of the lipids in the cell membrane. How could these acids be measured in the minute concentrations found in laboratory animals? Help was quite literally close at hand.

In the summer of 1951, two of Lovelock's colleagues at the NIMR, Archer Martin and Tony James, introduced the new technique of gas-liquid chromatography.[17] This was arguably the most important advance in chemical analysis since Bunsen and Kirchhoff developed spectral analysis almost a century earlier. Some important techniques take a decade or more to establish themselves (NMR

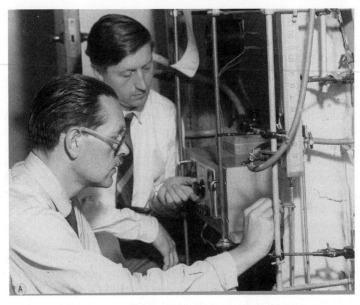

Archer Martin and Tony James, 1950s. Photograph courtesy of NIMR

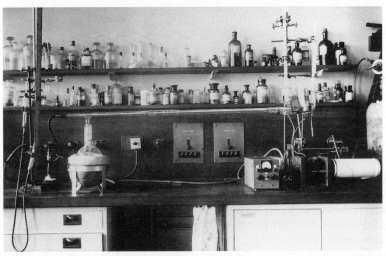

Early gas chromatography apparatus used by Tony James. Science Museum/Science and Society Picture Library

is a case in point), but despite its complexities, gas-liquid chromatography spread like wildfire in 1952 and 1953. The reaction of Denis Desty at BP was typical:

> One hot afternoon in the early summer of 1952 there arrived in the mail a small envelope which contained a chromatogram from a simple pen recorder and requesting samples of pure hydrocarbons for evaluation of a new separation technique, gas-liquid chromatography. The letter was sig-

Martin's laboratory at the NIMR, 1950s. Photograph courtesy of NIMR

ned by A. J. P. Martin at the National Institute for Medical Research at Mill Hill, London. In order to whet our appetite he indicated that the recording was that of the separation of a milligram or so of gasoline from the car of his collaborator A. T. James. It had been obtained in about 30 minutes and, even from a superficial view, a separation had been achieved similar to that obtained on a 100 plate distillation column over perhaps a month. As you can imagine the impact on both myself and my old boss S. F. Birch was fairly spectacular. We immediately rang the Institute at Mill Hill and it soon emerged that Martin had very recently designed a very sensitive omniverous [*sic*] vapour detector, which allowed the detection of hydrocarbons for the first time.... Having hastily fixed an appointment for the next day, we dashed over to Mill Hill with suppressed excitement and met Martin and James, who seemed somewhat surprised that we should arrive so quickly after the despatch of their letter.[18]

There had been a pent-up demand from chemists in specific areas, such as petroleum refining and biochemists, for a technique that could separate tiny amounts of relatively volatile compounds quickly and effectively. The reasons for this demand varied. In the rapidly expanding petrochemical industry, the key factors were the rapid separation of complex mixtures to permit the monitoring of production lines. By contrast, biochemists and analytical chemists were anxious to separate and thus analyse mixtures of similar compounds which were available only in tiny quantities.

The latter was exactly the problem faced by Lovelock in 1955. He recalled that Martin and James's first paper had been on the separation of fatty acid esters and went upstairs to ask them if gas chromatography would be able to advance his research. At first, they were enthusiastic, but when they saw the tiny amounts involved, they realised that their existing detector (a gas density balance) would be too insensitive. Martin suggested Lovelock should try to isolate larger quantities or if that failed, he could always try to invent a more sensitive detector.[19] While this remark was probably intended as a joke, it illustrates the 'can-do' attitude that existed at the NIMR.

Finding the Peaks

However it is carried out, chromatography (literally colour-writing) can only succeed if there is some practicable means of identifying the different fractions as they emerge from the column. The solid chromatographic column demonstrated by Mikhail Tswett in 1907 effectively only worked with coloured compounds, notably chlorophyll. When Martin introduced paper chromatography in 1944, his proposal to use ninhydrin as a marker for amino acids was crucial to the success of the technique. In the case of gas-liquid chromatography, Martin and James used titration initially, but as well as being rather slow, it could only be applied to acids and bases such as amines. Two other detectors were soon introduced, the katharometer, which had already been used for gas absorption chromatography by Courtney Phillips at Oxford in 1949, and the gas density balance by Martin and James.

According to Arnaldo Liberti, the Italian pioneer of gas chromatography, the latter had its pros and cons:

> A gas density balance was indeed an outstanding detector and one of its most attractive features was the possibility of determining the molecular weight of an eluted compound… This most interesting aspect, however, has not found a wide application because of the care needed to operate this detector.[20]

James has remarked that the gas density balance was 'unfortunately too difficult for many instrument firms to construct'[21] and Raymond Scott found it a frustrating instrument to construct:

> The gas density balance consists of a Wheatstone network of capillary tubes drilled out of a solid block of high conductivity copper…. We filled three copper blocks before we eventually made a working gas density bridge. It worked very well, but it also stimulated me to look for an alternative and simpler method of detection. My colleagues and I did not want to face the difficulties of making another density balance.[22]

The early katharometers also had their problems, as David Grant who was working at the Coal Tar Research Association in 1954 later recalled:

> An exasperating routine job was the fitting of these sensors when they burnt-out, as they frequently did. The circuit consisted of an elementary Wheatstone bridge fed by wet acid accumulators, which of course, required

regular recharging. The bridge output was monitored by a rather ungainly commercial recorder with a five-second response and a frustrating propensity to 'go dead' at critical times (as for instance when a peak was due!).[23]

Addressing the 1956 London Symposium on Gas Chromatography as Chairman, Tony James was critical of existing detectors and concluded:

> The field is open for the development of more sensitive and simpler detectors, and it is to be hoped that workers in this field will not rest content with the instruments at present available.[24]

Creating the Electron Capture Detector

Certainly Lovelock had no intention of being content. Lovelock has claimed[25] that he preferred to spend his time inventing rather than carrying out experiments to collect larger amounts of fatty acids, but he was no doubt aware of the potential value of a more sensitive detector, especially for biochemical work. At this point, he recalled one of the new anemometers he had constructed at the Common Cold Unit. Made from radium scraped from old aircraft gauges, it had worked on the principle of ion drift; the slow positive ions were easily disturbed by small air currents. Unfortunately, field trials in the Arctic showed that it was also very sensitive to cigarette smoke. Ironically, current health and safety rules would have prevented both the construction of the anemometer and the cigarette smoke that showed its potential as a detector. Lovelock has very strong opinions on this issue:

> Fortunately we were not hampered, like now, by a well-intentioned but hindering health and safety bureaucracy. Scientists who used dangerous chemicals or radioactive materials were expected to be personally responsible. There was some risk but I doubt if, under the stifling restrictions of today, I would have had the persistence to carry on with so uncertain a project as the infant Electron Capture Detector.[26]

John Otvos and David Stevenson at the Shell Oil research centre at Emeryville in California had developed a beta-ray ionisation chamber based on strontium-90 to analyse gases. The Emeryville team and their Dutch colleague Hendrik Boer soon realised the value of this device for gas chromatography and they developed their quite different detectors by 1955.[27] The Boer detector (now called a cross-section detector) was a dual chamber apparatus, in which the effluent gas from the chromatograph flowed through one cell and pure carrier gas through the other. The difference between the ionisation currents in the two chambers was converted into a potential, which was measured by an electrometer. The voltage applied to the ionisation chambers was between 100 and 300 volts and the carrier gas was usually nitrogen or hydrogen. The results were comparable to those achieved with a gas density balance or a katharometer, but the device could not be described as sensitive.

Lovelock combined their work with his anemometer to create the first generation ECD, which had a single chamber, thus removing the need to have a

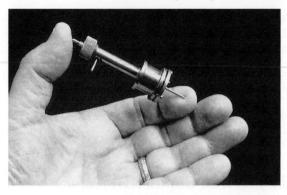

Modern electron capture detector (held by James Lovelock). © J. E. Lovelock, 2000.
Reprinted from Homage to Gaia: The Life of an Independent Scientist *by James*
Lovelock (2000) by permission of Oxford University Press

parallel flow of pure carrier gas.[28] Lovelock later recalled making his first
prototype and, characteristically, he emphasised its home-made character:

> My first detector was a simple cylindrical ion chamber, about two mil-
> lilitres in volume and contained a one billion Bequerel strontium 90 source
> of beta radiation. I remember bending the stiff and fiercely radioactive foil
> behind a sheet of thick glass until it fitted the detector cavity... In the
> middle of the cavity was a small collecting electrode, connected to a
> home-made electrometer. The chamber was polarised by connecting the
> outer case to a voltage source ... and we used an automobile spark-plug as
> the insulator that held the anode. The electrometer was quite literally
> home-made – it used a pair of vacuum tubes in a balanced cathode follower
> circuit and I made it on our kitchen table.[29]

There was a problem with the carrier gas. Light carrier gases such as hydrogen
and helium would have worked best. However, helium was too expensive and
hydrogen was too dangerous for overnight operation at high temperatures, so
nitrogen was the usual carrier gas. Lovelock was able to reproduce Boer's results,
but it was much less sensitive than the gas density balance and of no use for the
delicate work he had in mind. He then decided to experiment with a lower
potential, of the order of 30 volts. This seemed promising and James gave him a
test mixture of fatty acid esters to try out. When Lovelock carried out the
separation running the detector at 100 volts, he obtained four small peaks.
Surprisingly, when he lowered the potential of the detector to 10 volts, he
observed numerous peaks running off the scale. Lovelock was eager to demonstrate his new detector to Martin and James, but matters did not turn out as he
expected:

> I thought that the search was over and we now had a truly sensitive
> detector. I asked James and Martin to come try it, which they did, bringing
> with them an allegedly pure sample of methyl caproate. I shall never forget
> the look of amazement on Tony James's face as peak after peak was drawn

from a small sample of this substance. Worse, none of them had the retention time of methyl caproate or of any other fatty acid ester. We now know that what was seen were traces of electron absorbing impurities in the sample, but at that time it seemed to be a useless and wholly anomalous device.[30]

Later experiments revealed that the ECD was extremely sensitive to the presence of traces of certain impurities. A tiny amount of carbon tetrachloride on the silicone seal, for instance, was sufficient to sabotage any subsequent attempts to use the detector.

Frustrated with the ECD, Lovelock turned to the investigation of other ionisation processes. At this point, he started to be concerned that his research was drifting away from its biomedical origins:

The physics of thermal energy electrons was hardly what the physicians running the Medical Research Council expected of me. I recall asking Sir Charles Harington, 'Can I spend some of my time finding out how the Electron Capture Detector works?' I added, 'There is no certainty that it will be of practical use but to me it is fascinating science'. He replied, 'I am happy to leave it entirely to your own judgement. This is a scientific Institute and so long as what you do is good science I am not much concerned about whether or not it has an immediate medical value'.[31]

At this point, in late 1956, fortune smiled on Lovelock. He was trying to use a potential in the detector chamber that was greater than the ionisation potential of most organic compounds but much less than that of the carrier gas. One day, the stores were out of nitrogen and the storekeeper asked if argon would do. By accidental good fortune, Lovelock's technician agreed. Since argon happens to have a similar ionisation potential to nitrogen, Lovelock then ran his experiment using argon and the same chamber he had used for his prototype ECD, at a very high potential of 700 volts. This set-up gave excellent results with fatty acid esters and he thought the problem was solved. Alas, when the argon ran out and was replaced with the usual nitrogen, the sensitivity sank back to its former mediocre level.

Further investigation revealed that the improved performance was a result of the Penning effect. In 1934, Frans Michel Penning at Philips had discovered that a metastable state of argon, formed under these conditions, transfers its energy to another gas molecule on collision provided that the ionisation potential of the other molecule is less than the energy level of the metastable argon.[32] The organic compound is ionised by the argon atom rather than by direct electron capture. The number of molecules ionised in this way is fairly low, but still sufficient to allow effective detection of the eluates as they pass out of the chromatograph.

Around this time, Boer paid a visit to Mill Hill and was introduced by Martin to Lovelock. Boer later recalled that Archer Martin:

in a most gentlemanly manner ... asked me if I had ever considered means other than β-rays for the ionisation of organic molecules, such as excited argon atoms. I answered, truthfully, that being an organic chemist, such odd things had never occurred to me...[33]

As there was a dearth of suitable detectors in the late 1950s, requests for an argon detector from laboratories flooded in from around the world. Typically, Raymond Scott visited Martin in 1957 and was introduced to Lovelock, who gave him all the information required for the construction of an argon detector. Scott later recalled the hair-raising construction of his own detector:

> The argon detector required a 10 [millicurie] strontium 90 source which was indeed a 'hot source'. It arrived in a silver sheet fairly rigid and contained between lead blocks. It had to be bent into a tube in order to insert it into the cavity of the detector ... two of my colleagues, [held] (with long crucible tongs) a sheet of glass in front of my face to protect me from beta rays and [I wrapped] my hands in lead sheet obtained from the wrappings of several two-ounce packets of pipe tobacco. I managed to bend the silver sheet into a cylinder and inserted it into the argon detector ... I took far less risks than Madame Curie and with any luck I hope I will live as long. Incidentally, the micro-argon detectors that were developed subsequently by Jim Lovelock used a 50 [microcurie] radium source and these we were quite happy to cut up with a pair of scissors as at that time they only represented the activity in half a dozen luminous dials from a wristwatch.[34]

It is doubtful that the argon detector could be disseminated in the same way today!

When Lovelock went to New York in the spring of 1958 to give a paper about the argon detector, he met Seymour 'Sandy' Lipsky, who was working on the metabolism of fatty acids in human plasma at Yale Medical School.[35] They found they had a shared interest in the analysis of fatty acid esters and in ionisation detectors. Lipsky had built his own version of the Shell detector, but used tritium as his beta radiation source to get round government rules. He was having problems with it, but Lovelock pointed out he was using an excessively weak source. In gratitude, Lipsky invited Lovelock to work at Yale for several months in the academic year 1958–9.

As soon as Lovelock arrived, the two scientists decided to combine the argon detector with Marcel Golay's novel capillary column. With this combination, they were able to separate methyl oleate from its *trans* isomer methyl elaidate. This created great excitement among the users of gas chromatography and Lovelock was immediately invited to work with Al Zlatkis at the University of Houston to improve the chromatographic separation of petroleum products. This trip was to have a major influence on Lovelock. When he left the NIMR in 1961, he became Professor of Chemistry at the College of Medicine at Baylor University in Houston, Texas.

Lovelock also used his sabbatical at Yale to transform the ECD. The original ECD was very troublesome to use and produced erratic results. After meeting Ken McAffee of Bell Telephone Laboratories, who had used pulses of high

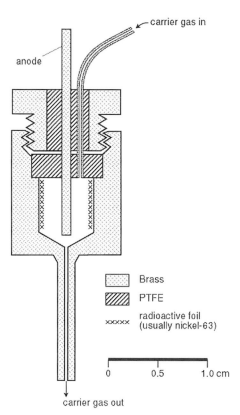

Schematic diagram of the early electron capture detector (after Lovelock)

How the ECD Works

The radioactive source generates a population of electrons in nitrogen gas and these electrons are mopped up by electron-absorbing elutes (notably chlorinated hydrocarbons) from a gas chromatograph. The fall in the number of electrons is detected by a drop in the potential across the walls of the ECD or, in the modern version, by sampling of the electron population by potential pulses.[36]

---

potential to study electron drift, Lovelock realised he could solve many of the problems with the original ECD by replacing its continuous low potential with high potential pulses. Eventually, he refined the concept even further by creating a constant population of electrons. These steps converted an infuriating device into a reliable, stable and relatively inexpensive detector.

As Lovelock has subsequently pointed out, the final version of the ECD was also highly sensitive:

It is not easy to describe the exquisite sensitivity of the ECD. One way is to imagine that you had a wine bottle full of a rare perfluorocarbon liquid somewhere in Japan and that you poured this liquid onto a blanket and left it to dry in the air by itself. With a little effort, we could detect the vapour that had evaporated into the air from that blanket here in Devon a few weeks later. Within two years, it would be detectable by the ECD anywhere in the world.[37]

Nevertheless, it was only of value for compounds with a strong affinity for electrons. In all other cases, the use of radioactive sources was considered a major drawback (even in the early 1960s) and other detectors were preferred, notably, the newly developed flame ionisation detector. For one embattled group, however, the ECD had arrived in the nick of time.

Finding a Molecule in a Billion

After the Second World War, numerous new pesticides were marketed, notably the chlorinated hydrocarbons and organophosphate compounds (see Table 1).[38] As they were very efficacious and comparatively cheap, they were soon used on a large scale. In 1948, Paul Müller was awarded the Nobel prize in medicine for his 'discovery of the strong contact insecticidal action of dichloro-diphenyl-trichloromethylmethane' (DDT), which was described as being 'of the greatest importance in the field of medicine'. DDT production in the United States more than tripled from 17 tonnes in 1953 to 56 tonnes in 1959.[39]

Methods of analysing the residues left by these pesticides were soon introduced, notably the Schechter-Haller method for DDT. Even in the late 1950s, 1 part per million (ppm) was an excellent result for DDT and routine tests were rarely better than 2.5 ppm.[40] Francis Gunther, the doyen of pesticide residue analysis, was very conservative in his assessment of analytical results. As late as 1973, he declared:

> I don't think anyone can determine parts per trillion of any chemical, let alone DDT. Our analytical capabilities are just not that reliable. In parts per billion there is a very high unreliability factor. Parts per trillion is a fairy tale.[41]

By contrast, the older inorganic pesticides, such as white arsenic, could be routinely measured to 0.2 parts per million.[42] The new organic pesticides, containing chlorine or phosphorus, were very different from the alkaloids that were the staple of the poisoner and the toxicologist.[43] Furthermore, unlike the inorganic pesticides, they were soluble in plant fats and waxes, thereby increasing the risk of food contamination.[44] Many of the approved methods for analysing organic pesticides, including the Schechter-Haller method, used colorimeters which had been around since the 1860s.[45] Paul Mills of the US Food and Drug Administration introduced a rapid semi-quantitative paper chromatography test for the detection of DDT residues in 1959.[46]

When the monitoring of marine wildlife began in the late 1950s, biologists were able to show that pesticides such as DDT and dieldrin had harmful effects

Table 1 *Insecticides introduced between 1942 and 1951*[48]

Insecticide	Type*	Introduced by	When
DDT	CHC	Geigy	1942
TEPP	OP	IG Farben	1943
lindane (gamma-BHC)	CHC	ICI	1945
chlordane	CHC	Velsicol	1945
DDD (TDE)	CHC	Rohm & Hass	1945
parathion	OP	American Cyanamid	1947
dieldrin	CHC	J. Hyman	1948
aldrin	CHC	J. Hyman	1948
E 838	OP	Bayer	1948
heptachlor	CHC	Velsicol	1948
dimefox	OP	Pest Control Ltd	1949
malathion	OP	American Cyanamid	1950
endrin	CHC	J. Hyman	1951
demeton	OP	Bayer	1951

*CHC = chlorinated hydrocarbon   OP = organophosphate

on shrimps and crayfish at levels below existing detection limits.[47] In a paper presented in 1966, Gunther claimed that the desired minimum detectability for pesticides in 1955, given the state of knowledge of the time, was 0.02 ppm, down from 1 ppm a decade earlier.[49] As we have seen, this was significantly below the sensitivity of existing chemical techniques. By 1965, the desired minimum detectability level had fallen further to 0.1 parts per [American] billion (ppb), a concentration Gunther humorously compared to:

> establishing that the extreme dryness of a particularly good martini is due to the fact that only a single drop of vermouth was added to 125 000 [US gallons] of quality gin.

Lovelock has claimed that the development of the ECD was crucial to the detection of DDT in the environment and indeed for the publication of *Silent Spring* itself.[50] Rachel Carson was a marine biologist[51] and – hardly surprisingly in a book aimed at the general public – she does not discuss the analysis of pesticide residues at all. Carson drew heavily on the work of her fellow biologists and naturalists. Her book contains only a few citations that refer to residue analysis, and they all predate the introduction of gas chromatography. Typically, a 1961 paper[52] by Bill Durham of the US Public Health Service on the DDT content of the body fat of Alaskan natives, he used a spectrophotometric variant of the 'classical' Schechter-Haller method, with a detection limit of about 0.1 ppm at best.

By 1960, official analysts were becoming seriously concerned about their inability to detect very low levels of DDT and other chlorinated hydrocarbons rapidly, reliably and consistently. This concern was increased by a growing number of incidents involving the monitoring of foodstuffs. The most famous case was the 'cranberry scare' of 1959, just before Thanksgiving in late November when cranberry consumption reached its annual peak. Although the US Department of Agriculture had approved aminotriazole weedkiller for use in cranberry bogs in 1957, subsequent trials showed that it might be carcinogenic. On 9 November, Arthur Flemming, then US Secretary of Health, Education and Welfare, announced that the Food and Drug Administration (FDA) was going to check the entire cranberry crop for aminotriazole contamination. Although the FDA was eventually able to clear the crop before Thanksgiving, the sales of cranberries (unsurprisingly) fell sharply. In a scenario that was later to be repeated in the case of BSE, leading politicians including Vice-President Richard Nixon consumed cranberries in public to show that they were safe to eat.[53] Similarly, evaporated milk and butter were seized *en route* to Hawaii in 1960 because they were allegedly contaminated with pesticide residues.[54]

In Britain, the Pesticides Residues in Foodstuffs Sub-committee of the Society for Analytical Chemistry in 1960 commissioned P. H. Needham of the Rothamsted Experimental Station to investigate bioassays as a means of rapidly testing for pesticide residues in foodstuffs.[55] Needham visited various laboratories that used bioassay methods in the UK and in Europe. The sensitivity of the method was high. The city council laboratory in Zurich could detect DDT to a limit ranging from 0.1 to 0.005 ppm using the larvæ of the yellow fever mosquito (*Aedes ægypti*). There were even attempts to distinguish different insecticides using bioassays. Clearly the method was useful but there were problems. Guinea pigs were needed to feed the mosquitoes and Needham recommended the fruit fly (*Drosophila melanogaster*) as the best insect to use. There were also variations between different foodstuffs: a level of pesticide that was toxic to mosquitoes in processed peaches was non-toxic in processed peas.

Unsatisfactory as biological methods were, their sensitivity was much higher than existing chemical methods. While Needham was compiling his report, the sensitivity of chemical analysis was on the brink of a dramatic change. In 1960, R. Goulden and his colleagues at the Shell Research Centre in Sittingbourne, Kent, published the first account[56] of the detection of chlorinated hydrocarbon pesticides using the argon detector sold by Shandon Instruments. Soon afterwards, J. O. Watts and A. K. Klein at the FDA developed their own electron-capture detector based on the paper published in 1960 by Lovelock and Lipsey.[57] A. D. Moore at the US Forest Service also independently conceived the idea of analysing insecticide residues using a Barber–Colman Model A-4071A (radium-226) argon detector in 1961.[58] All three groups showed that the electron-capture detector could detect DDT and other chlorinated hydrocarbon pesticides to at least 1 ppm and often down to 0.25 ppm. By 1963, Watts and Klein had reduced the level for DDT to 0.1 ppm, a level comparable with conventional chemical methods, with a device which was still in an early stage of development.[59]

In a 1963 paper[60] that compared the electron-capture detector with other methods of detecting pesticides in milk, Lloyd Henderson of Foremost Dairies, San Francisco, concluded:

> The electron-capture gas chromatographic procedure at this stage of its development yielded lower total residues with fewer pesticides identified than did the other two chromatographic procedures.[61] The procedure, however, may be developed into a rapid and accurate method for the

detection and evaluation of pesticide residues in milk and dairy products.

By 1964, a report[62] by the UK Ministry of Agriculture, Fisheries and Food was able to tabulate DDT residues in butter and milk to a limit of about 0.01 ppm using gas chromatography, and presumably, the ECD. A year later, chemists at the UK Laboratory of the Government Chemist declared that gas chromatography (and again by implication, the ECD) had allowed the detection limits for pesticides to be reduced to the hitherto unimaginable level of 0.0001 ppm (0.1 part per billion).[63]

It appears that it was an incident in the winter of 1963, well after the publication of *Silent Spring*, that first brought the ECD to the fore in environmental analysis.[64] An abnormal number of fish were dying in the lower reaches of the Mississippi, but there was no obvious cause. Biological experiments soon revealed that there was a toxin in the river mud and in the livers of the dead fish, but what was it? The gas chromatograph and the ECD revealed very low concentrations of two chemicals, which were later shown by analytical chemists at Shell to be associated with the production of the chlorinated hydrocarbon pesticide endrin. Eventually, the sources of these chemicals was traced to a dump used by a chemical plant at Memphis, Tennessee.

At this point, the ECD encountered competition from the microcoulometric detector,[65] introduced by Dale Coulson of the Stanford Research Institute in 1959. Coulson combined the gas chromatograph with the existing technique of combustion analysis, in which the pesticide sample is burnt and the oxidation products analysed for inorganic chlorine and sulfur. For his detection system, he adopted the tube furnace constructed for pesticide work by E. D. Peters and his group at Shell Research at Emeryville.[66] The chloride ions formed in this way were then analysed using an automatic chloride analyser which had been developed for this system by Coulson and his co-worker Leonard Cavanagh.[67] This microcoulometric titration method operated on the basis of a constant silver concentration in the titration cell.

Henderson reported[41] good results with the microcoulometric method, which could measure DDT residues accurately to around 0.3 ppm. In 1965, Coulson introduced his own commercial model, but both the model and his company appear to have been short-lived.[68] Randall Hall of Purdue University developed a very sensitive detector[69] in 1974 on the basis of Coulson's work, and this has become known as the Hall® electrolytic conductivity detector (confusingly called an HECD). The Hall® detector was never popular in the UK, but is still recognised by the Environmental Protection Agency in the United States.[70]

By the mid-1970s, the ECD and microelectrolytic methods had altered the situation to the extent that the analysis of DDT residues, a major topic in the early 1960s, had disappeared from the pages of the *Journal of the Association of Official Agricultural Chemists*. The current sensitivity of the ECD is stated to be less than 10 femtograms (1×10^{-14} g) of lindane which is about 10 part per *quadrillion*, whereas the latest version of the Hall® detector is said to have a sensitivity of 5 picograms (5×10^{-12} g) of heptachlor or around 5 parts per trillion.[71] In short, the chemical analysis of chlorinated hydrocarbon residues has improved by up to 100 million-fold in the last three decades.

Worrying about Pollution

The acronyms of the 'environmental apocalypse' – first of all DDT, then PCBs, CFCs and TCDD – would have taken much longer to penetrate our collective consciousness if it had not been for the ECD. At the same time, the extreme sensitivity of the ECD has increased our anxiety about them. Lovelock has quoted Paracelsus to the effect that 'the poison is the dose',[72] and noted that even nerve gases have a threshold limit. He went on to say that 'what was needed was common sense' but as Lovelock himself has since remarked,[73] this has sometimes been in short supply.

Before the 1950s, it was usually assumed that the problematic level of any pesticide was its toxic level. Threshold levels were originally called 'tolerances', and they were just that, the amount of a substance that could be tolerated by an adult without any immediate and obvious injury or discomfort.[74] If the chemical be present in food or in the environment at large at a lower level, there was no grounds for concern. In 1946, Trustham West and George Campbell in their influential DDT textbook stated that 'quite large amounts of DDT must be applied to the skin or given by the stomach to produce fatal results'.[75] The emphasis was on mammalian toxicity, so a compound like DDT, which could be eaten by the spoonful without apparent harmful consequences, was used freely without worrying too much about any residues it might leave.[76] It was also assumed that any likely threshold level could be readily measured by existing analytical techniques. Both these assumptions were based on the use of earlier inorganic pesticides. Even a potent organic pesticide such as nicotine could be estimated using colorimetric methods.[77] The validity of these comforting concepts broke down with the pesticides introduced after the Second World War. Furthermore, there was growing public concern about the increases in cancer cases. The presumption that any problems would be a result of direct poisoning was replaced by the view that any chemical capable of causing tumours would be potentially harmful, even at very low concentrations.[78]

Any attempt by industry or even government to assert that levels of a given pollutant was harmless was met by the assertion that there were no 'safe' threshold levels, not least by Carson herself who declared:

> In effect, then, to establish tolerances is to authorise contamination of public food supplies with poisonous chemicals in order that the farmer and the processor may enjoy the benefit of cheaper production.[79]

To be fair, though, industry had in the past called for threshold levels that with hindsight turned out to be too high.[80]

One way out of this impasse was the imposition of a 'zero tolerance'. The flagship of this concept was the 'Delaney Clause' in the Food Additives Amendment of 1958 to the Food, Drug and Cosmetic Act of 1938, which set a zero tolerance level for any substance that was capable of causing cancer in man or animal.[81] At a time when chemical methods could not reliably measure residues below 0.02 ppm at the very best, this may have been sensible. Once gas chromatographs could measure residues at a level of a few parts per trillion, pesticides were in danger of becoming victims of the ever-improving techniques of analytical chemistry.[82] In the event, the Delaney clause was given a restricted application in this area by the courts and the issue never arose directly.[83] Nonetheless, DDT was banned in the United States in 1972, partly in the belief that it was carcinogenic.[84] Chlordane and heptachlor were withdrawn from general use in the United States by Velsicol in 1978, because they caused cancers in animals.[85]

By the late 1970s, chemical analysts could show that pollutants could be detected pretty well everywhere at a level of parts per quadrillion ($1:10^{15}$), even in Greenland ice cores.[86] As Gunther quipped prophetically in 1973:

> Yesterday we looked for little bits of a few things in some things; today we are looking for less of more things in any thing; tomorrow we will look for nothing in everything.[87]

Sometimes such findings have had important implications, but more often than not they have simply increased public anxiety without having any practical impact on public health. The concept that unacceptable levels of chemicals could be beyond the reach of the chemical analysts captured the minds of political activists. The development of an environmentalism that was hostile towards pesticides and other chemicals was largely spearheaded by Barry Commoner, a professor of plant physiology at Washington University, St Louis, in the mid-1960s. Commoner was radicalised by his earlier efforts to ban atmospheric nuclear tests and it is clear that he turned to chemical pollution as a consequence of the partial test ban treaty of 1963. His main concern in the late 1960s was the eutrophication of the Great Lakes, but he also campaigned against DDT, drawing a parallel between DDT and strontium-90 from fallout as early as 1963. Based in St Louis, he was also clearly influenced by the great Mississippi fish-kill.[88] Furthermore, improvements in chemical analysis have heightened the psychological impact of pollution by showing its presence in emotional areas such as baby food, schoolyards and mothers' milk, albeit in very low concentrations.[89]

The analytical problems have been solved, but the political issue of finding safe threshold levels, at least to the satisfaction of an increasingly sceptical public-at-large, remains both controversial and unresolved.

Notes and References

1 Although James Lovelock published his autobiography, *Homage to Gaia*, Oxford University Press, Oxford, 2000, since this paper was first written, two earlier autobiographical essays are still useful: James E. Lovelock, 'Midwife to the Greens: The Electron Capture Detector', *Microbiologia*, March 1997, **13**, 11–22 and James Lovelock, 'Travels with an Electron Capture Detector', *Resurgence*, March/April 1998, **187**, 6–12. Also see James E. Lovelock, 'The Electron Capture Detector: A Personal Odyssey', *Chemtech*, September 1981, 531–537; Lovelock's essay in L. S. Ettre and A. Zlatkis (eds.), *75 Years of Chromatography*, Elsevier, Amsterdam, 1979, pp. 277–284; and James E. Lovelock, 'Tales of a Reluctant Instrument Maker', in Robert E. Sievers

(ed.), *Selective Detectors: Environmental, Industrial and Biomedical Applications*, Wiley, New York, 1995, pp. 241–249.

2 For the history of the ECD and the halocarbons (also called chlorofluorocarbons or CFCs), see Lovelock, 'The Ozone War', Chapter Eight of *Homage to Gaia* and Lydia Dotto and Harold Schiff, *The Ozone War*, Doubleday, New York, 1978.

3 'Childhood', Chapter One of Lovelock, *Homage to Gaia*, and Lovelock, 'Travels with an Electron Capture Detector', p. 6.

4 Lovelock, 'Travels with an Electron Capture Detector', p. 6.

5 Lovelock, 'Travels with an Electron Capture Detector', p. 6.

6 http://www.nimr.mrc.ac.uk/history.htm. An official history of the National Institute for Medical Research is in progress. Also see Sir Arthur Landsborough Thomson, *Half a Century of Medical Research*, two volumes, HMSO for the Medical Research Council, London, 1973 and 1975, *passim*. For Sir Henry Dale, see W. Feldberg, 'Henry Hallett Dale', *Biographical Memoirs of Fellows of the Royal Society*, 1970, **16**, 77–174.

7 Lovelock, 'Midwife to the Greens', p. 12. For Sir Charles Harington, see Sir Harold Himsworth and Rosalind Pitt-Rivers, 'Charles Robert Harington', *Biographical Memoirs of Fellows of the Royal Society*, 1972, **18**, 267–308.

8 Lovelock, 'Midwife to the Greens', p. 12.

9 Lovelock gives a good example of how he unexpectedly assisted an eminent biochemist in *Homage to Gaia*, pp. 284–285.

10 Lovelock, 'Midwife to the Greens', p. 12.

11 Lovelock, *Homage to Gaia*, p. 285.

12 Lovelock, *Homage to Gaia*, p. 184.

13 Lovelock, 'Midwife to the Greens', p. 12.

14 Lovelock, 'Midwife to the Greens', p. 13.

15 Lovelock, *Homage to Gaia*, p. 182.

16 Lovelock, *Homage to Gaia*, pp. 101–110.

17 For the history of gas chromatography see Leslie S. Ettre, 'Gas Chromatography' in Herbert A. Laitinen and Galen W. Ewing (eds.), *A History of Analytical Chemistry*, Division of Analytical Chemistry of the American Chemical Society, Washington DC, 1977, pp. 296–306, and L. S. Ettre and A. Zlatkis (eds.), *75 Years of Chromatography*, Elsevier, Amsterdam, 1979. Since this paper has been written, two new histories have been published: Leslie S. Ettre, *Milestones in the Evolution of Chromatography*, ChromSource, Brentwood, TN, 2002; C. W. Gehrke, R. L. Wixom and E. Bayer, *Chromatography: A Century of Discovery, 1900–2000*, Elsevier, Amsterdam, 2001. I am indebted to Edward Adlard for these references.

18 Denis H. Desty in Ettre and Zlatkis, *75 Years of Chromatography*, p. 33. The timing is corroborated by weather records, as there was a heatwave (82°F to 89°F) between 27 June and 2 July 1952. It has been suggested that Martin sent a chromatograph of the exhaust gases from James's car rather than his petrol, but petrol seems more likely. Perhaps other accounts of this episode used the American term 'gas' and this has been misunderstood.

19 Lovelock, 'Midwife to the Greens', p. 14.

20 Arnoldo [*sic*] Liberti in Ettre and Zlatkis, *75 Years of Chromatography*, p. 257.

21 Anthony T. James in Ettre and Zlatkis, *75 Years of Chromatography*, p. 170.

22 R. P. W. Scott in Ettre and Zlatkis, *75 Years of Chromatography*, p. 398.

23 David W. Grant in Ettre and Zlatkis, *75 Years of Chromatography*, p. 117–118.

24 A. I. M. Keulemans, *Gas Chromatography*, Second Edition, Reinhold, New York, 1959, p. 99.

25 Lovelock, *Homage to Gaia*, p. 183.

26 Lovelock, 'Travels with an Electron Capture Detector', p. 8.

27 Keulemans, *Gas Chromatography*, p. 80–81. Also see the essay by Boer in Ettre and Zlatkis, *75 Years of Chromatography*, pp. 14–15.

28 Also see 'The ECD', Chapter Seven of Lovelock, *Homage to Gaia*. For technical descriptions, see Keulemans, Gas Chromatography, pp. 100–1; J. E. Lovelock, 'Ionization Methods for the Analysis of Gases and Vapors', *Analytical Chemistry*, 1961, **33**, 162–178; J. E. Lovelock and N. L. Gregory, 'Electron Capture Ionization Detectors' in Nathaniel Brenner, Joseph E. Callen and Marvin D. Weiss, (eds.), *Gas Chromatography: Third International Symposium Held Under the Auspices of the Analysis Instrumentation Division of the Instrument Society of America, June 13–16, 1961*, Academic Press, New York and London, 1961, pp. 219–229; J. E. Lovelock, 'The Electron Capture Detector: Theory and Practice', *Journal of Chromatography*, 1974, **99**, 3–12; and J. E. Lovelock and A. J. Watson, 'The Electron Capture Detector: Theory and Practice II', *Journal of Chromatography*, 1978, **158**, 123–138.

29 Lovelock, 'Travels with an Electron Capture Detector', p. 8.

30 Lovelock, *Homage to Gaia*, p. 185.

31 Lovelock, 'Travels with an Electron Capture Detector', pp. 8–9.

32 F. M. Penning, 'The Starting Potential of the Glow Discharge in Neon Argon Mixtures Between Large Parallel Plates, II. Discussion of the Ionisation and Excitation of Electrons and Metastable Atoms', *Physica (Amsterdam)*, 1934, **1**, 1028–1044.

33 Hendrik Boer in Ettre and Zlatkis, *75 Years of Chromatography*, p. 14. In his essay, Boer states his visit took place in the autumn of 1955, but this is not compatible with Lovelock's use of argon in late 1956 (which appears to be correct).

34 R. P. W. Scott in Ettre and Zlatkis, *75 Years of Chromatography*, pp. 399–400.

35 See the essay by Lipsky in Ettre and Zlatkis, *75 Years of Chromatography*, pp. 265–276.

36 James E. Lovelock, 'Electron Capture Detector', in Robert Bud and Deborah Jean Warner (eds.), *Instruments of Science: An Historical Encyclopedia*, Garland, New York and London, 1998, pp. 213–214.

37 Lovelock, *Homage to Gaia*, p. 186.

38 For the history of DDT, see Thomas R. Dunlap, *DDT: Scientists, Citizens and Public Policy*, Princeton University Press, Princeton, 1981; Kenneth Mellanby, *The DDT Story*, British Crop Protection Council, Farnham, 1992; Christian Simon, DDT; *Kulturgeschichte einer chemischen Verbindung*, Christoph Merian Verlag, Basel, 1999; and Edmund P. Russell III, 'The Strange Career of DDT: Experts, Federal Capacity and Environmentalism in World War II', *Technology and Culture*, 1999, **40**, 770–796. For broader surveys, see John Sheail, *Pesticides and Nature Conservation: The British Experience, 1950–1975*, Oxford University Press, Oxford, 1985; and Christopher J. Bosso, *Pesticides and Politics: The Life Cycle of a Public Issue*, University of Pittsburgh Press, Pittsburgh, 1987. Adam Curtis produced a brilliant documentary history of DDT, 'Goodbye Mrs Ant' as part of his television series 'Pandora's Box'. This programme was transmitted by the BBC on 2 July 1992. I wish to thank Michael Beasley of the National Museum of Photography, Film and Television, Bradford, for his help with this citation.

39 Gustave K. Kohn, 'Agriculture, Pesticides and the American Chemical Industry' in Gino J. Marco, Robert M. Hollingsworth and William Durham (eds.), *Silent Spring Revisited*, American Chemical Society, Washington DC, 1987, p. 163.

40 Lloyd Henderson, 'Comparison of Laboratory Techniques for the Determination of Pesticides Residues in Milk', *Journal of the AOAC*, 1963, **46**, 209–215; Joseph D. Rosen and Fred M. Gretch, 'Analysis of Pesticides: Evolution and Impact' in Marco, Hollin-

gsworth and Durham (eds.), *Silent Sprint Revisited*, p. 129. For a thorough survey of pesticide analysis, see Francis A. Gunther and Roger C. Blinn, *Analysis of Insecticides and Acaricides: A Treatise on Sampling, Isolation and Determination Including Residue Methods*, Volume VI of *Chemical Analysis*, Interscience, New York, 1955; and J. H. Ruzicka, 'Methods and Problems in Analysing for Pesticides Residues in the Environment' in C. A. Edwards, *Environmental Pollution by Pesticides*, Plenum Press, London and New York, 1973, pp. 11–56. For summaries of the appropriate techniques for each pesticide, also see Hubert Martin (ed.), *Pesticide Manual*; and William Horwitz (ed.), *Official Methods of Analysis of the Association of Official Agricultural Chemists*, 10th edition, Association of Official Agricultural Chemists, Washington DC, 1965 and earlier editions.

41 Quoted in Rita Gray Beatty, *The DDT Myth: Triumph of the Amateurs*, John Day, New York, 1973, p. 26. Francis Gunther was the leading figure in the field of pesticide residue analysis for over twenty years and he would have been an important oral source for this paper, but sadly he died of lung cancer in the late 1980s.

42 The long-established Gutzeit method was very sensitive for arsenic. In food, the detection limit laid down by the AOAC in 1925 was 0.001 mg, and as the initial sample (for vegetables) was 25 g, this equates to 1 ppm (as the original solution of the digested sample would be divided into four aliquots). By 1950, the AOAC laid down that the concentration should be recorded to not more than three decimal places as grains per pound. This implies a detection limit of around 0.2 ppm. See R. E. Doolittle (chairman), *Official and Tentative Methods of Analysis of the Association of Official Agricultural Chemists*, 2nd edition, Association of Official Agricultural Chemists, Washington DC, 1925, pp. 171–173; and Henry A. Lepper (chairman), *Official Methods of Analysis of the Association of Official Agricultural Chemists*, 7th edition, Association of Official Agricultural Chemists, Washington DC, 1950, pp. 369–373.

43 The history of organic toxicology is sadly underdeveloped, but see Katherine D. Watson, Philip Wexler and Janet M. Everitt, 'History' in Philip Wexler, Pertti J. Hakkinen, Gerald Kennedy and Frederick W. Stoss (eds.), *Information Resources in Toxicology*, 3rd edition, Academic Press, San Diego and London, 1999; and P. J. T. Morris and W. A. Campbell, 'Laboratory Practice and Chemical Analysis in the Nineteenth Century' in David Knight (ed.), *History of Nineteenth Century Chemistry*, a volume of the *Encyclopedia Italiana*, forthcoming.

44 Gunther and Blinn, *Analysis of Insecticides and Acaricides*, p. 12.

45 Jules Duboscq introduced a comparative colorimeter in 1854; a later variant was the Klett colorimeter. J. W. Lovibond introduced his 'Tintometer' in 1886 and it is still used today in a practically unaltered form.

46 Paul A. Mills, 'Detection and Semiquantitative Estimation of Chlorinated Pesticides Residues in Foods by Paper Chromatography', *Journal of the AOAC*, 1959, **42**, 734–740; also see Lloyd Henderson, 'Comparison of Laboratory Techniques'.

47 Rachel Carson, *Silent Spring*, Hamish Hamilton, London, 1963, p. 124, citing a private communication from John C. Pearson. Also see L. D. Newson, 'Some Ecological Implications of Two Decades of Use of Synthetic Organic Insecticides for Control of Agricultural Pests in Louisiana' in M. Taghi Farvar and John P. Milton (eds.), *The Careless Technology*, p. 451 and Table 24.7. I am indebted to George Twigg for the latter reference.

48 Assembled from Hubert Martin (ed.), *Pesticide Manual: Basic Information on the Chemicals Used as Active Components of Pesticides*, first edition, British Crop Protection Council, Worcester, 1968, supplemented by information from Erik Verg, Gottfried Plumpe and Heinz Schultheis, *Milestones*, Bayer AG, Leverkusen, 1988. Lin-

dane was actually developed in 1942, but kept secret until the end of the war, see W. J. Reader, *Imperial Chemical Industries, A History*, Volume Two, *The First Quarter-Century, 1926–1952*, Oxford University Press, London, 1975, pp. 456–457.

49 Francis A. Gunther, 'Advances in Analytical Detection of Pesticides', in *Scientific Aspects of Pest Control*, National Academy of Sciences/National Research Council, Washington DC, Publication no. 1402, 1966, pp. 276–302, on p. 293.

50 This claim first appeared in Lovelock, 'Midwife to the Greens', p. 17 and was repeated in Lovelock, 'Travels with an Electron Capture Detector', p. 9 and recently in Lovelock, *Homage to Gaia*, p. 189. *Silent Spring* was published in the United States by Houghton Mifflin in 1962 and in Britain by Hamish Hamilton a year later.

51 Linda Lear, *Rachel Carson: Witness for Nature*, Allen Lane, London, 1998.

52 William F. Durham, *et al.*, 'Insecticide Content of Diet and Body Fat of Alaskan Natives' *Science*, 1961, **134**, 1880–1881.

53 Dunlap, *DDT*, pp. 107–8; Carson, *Silent Spring*, pp. 183–4; Beatty, *The DDT Myth*, pp. 153–155.

54 Henderson, p. 210.

55 P. H. Needham, 'An Investigation into the Use of Bioassay for Pesticide Residues in Foodstuffs', *The Analyst*, 1960, **85**, 792–809.

56 E. S. Goodwin, R. Goulden, A. Richardson and J. G. Reynolds, 'The Analysis of Crop Extracts for Traces of Chlorinated Pesticides by Gas-Liquid Partition Chromatography', *Chemistry and Industry*, 24 September 1960, 1220–1. Also see E. S. Goodwin, R. Goulden and J. G. Reynolds, 'Rapid Identification and Determination of Residues of Chlorinated Pesticides in Crops by Gas-Liquid Chromatography', *The Analyst*, 1961, **86**, 697–709.

57 J. O. Watts and A. K. Klein, 'Determination of Chlorinated Pesticide Residues by Electron-Capture Gas Chromatography', *Journal of the AOAC*, 1962, **45**, 102–108; J. E. Lovelock and S. R. Lipsky, 'Electron Affinity Spectroscopy – A New Method for the Identification of Functional Groups in Chemical Compounds Separated by Gas Chromatography', *Journal of the American Chemical Society*, 1960, **82**, 431–433.

58 A. D. Moore, 'Electron Capture with an Argon Ionization Detector in Gas Chromatographic Analysis of Insecticides', *Journal of Economic Entomology*, 1962, **55**, 271–2. In a footnote, Moore remarks 'After this manuscript was submitted it was brought to the authors [*sic*] attention that E. S. Goodwin, R. Goulden, A. Richardson and J. G. Reynolds have also used the Shandon argon detector as an electron capture detector for studies of insecticide residues'. One wonders if Goulden was one of the referees.

59 A. K. Klein, J. O. Watts and J. A. Damico, 'Electron Capture Gas Chromatography for the Determination of DDT in Butter and Some Vegetable Oils', *Journal of the AOAC*, 1963, **46**, 165–171.

60 Lloyd Henderson, 'Comparison of Laboratory Techniques', quote on p. 215.

61 The other two methods were the Mills technique, already mentioned, and the Coulson method (see below).

62 Advisory Committee on Poisonous Substances Used in Agriculture and Food Storage, *Review of the Persistent Organochlorine Pesticides*, HMSO, London, 1964, p. 13 and Appendix F.

63 D. C. Abbott and J. Thomson, 'Analytical Aspects of Water Pollution by Some Domestic and Agricultural Chemicals', *Proceedings of the Society of Water Treatment and Examination*, 1965, **14**, 70–80, on p. 74.

64 Frank Graham, Jr., *Since Silent Spring*, Hamish Hamilton, 1970, pp. 96–108; A. S. Travis, 'Contaminated Earth and Water: A Legacy of the Synthetic Dye Industry', *Ambix*, forthcoming.

65 Dale M. Coulson, Leonard A. Cavanagh and Janet Stuart, 'Gas Chromatography of Pesticides', *Agricultural and Food Chemistry*, 1959, **7**, 250–1; and Dale M. Coulson, Leonard A. Cavanagh, John E. de Vries and Barbara Walther, 'Microcoulometric Gas Chromatography of Pesticides', *Agricultural and Food Chemistry*, 1960, **8**, 399–402.

66 E. D. Peters, G. C. Rounds and E. J. Agazzi, 'Determination of Sulfur and Halogens: Improved Quartz Tube Combustion Analysis', *Analytical Chemistry*, 1952, **24**, 710–714 and E. J. Agazzi, E. D. Peters and F. R. Brooks, 'Combustion Techniques for the Determination of Residues of Highly Chlorinated Pesticides', *Analytical Chemistry*, 1953, **25**, 237–240.

67 Dale M. Coulson and Leonard A. Cavanagh, 'Automatic Chloride Analyzer', *Analytical Chemistry*, 1960, **32**, 1245–1247.

68 Dale M. Coulson 'Electrolytic Conductivity Detector for Gas Chromatography', *Journal of Gas Chromatography*, April 1965, **3**, 134–137. According to Coulson's entry in *American Men of Science, The Physical & Biological Sciences*, eleventh edition, R. R. Bowker, New York, 1965, the Coulson Instrument Company was founded in 1964. I have not located any instruments made by this company and cannot find any trade literature published by it. Curiously, Coulson makes no reference to his company or even his chromatograph in his Harvey Wiley Award speech in 1974, reproduced in *Journal of the AOAC*, 1975, **58**, 174–183. It appears that his company may have been absorbed into SRI Co, the commercial wing of the Stanford Research Institute.

69 Randall C. Hall, 'A Highly Sensitive and Selective Microelectrolytic Conductivity Detector for Gas Chromatography', *Journal of Chromatographic Science*, March 1974, **12**, 152–160.

70 For instance, EPA Method 8081, for which see http://www.speclab.com/compound/m8081.htm

71 http://tmqaustin.com/Detectors/t2mdetspecs.htm

72 Lovelock, 'Midwife to the Greens', p. 18. The proper quotation (which is rarely seen) is 'Alle ding sind gift und nichts ohn gift; alein die dosis macht das ein ding kein gift ist'. (Everything is poisonous and nothing is not poisonous; it is merely the dose that makes something non-poisonous.) Quoted in Walter Pagel, *Paracelsus: An Introduction to Philosophical Medicine in the Era of the Renaissance*, 2nd edition, Karger, Basel, 1982, p. 363, citing Paracelsus, *Defensiones III*, Sudhoff Volume XI, p. 138. The objection is sometimes made that it is a misuse of Paracelsus's comment to apply it outside its intended context of the medicinal use of poisons. It is unlikely that Paracelsus intended his remark to be circumscribed in this way and indeed, contemporary commentators objected to his rejection of the concept that poisons were intrinsically harmful, which seems close to the viewpoint of the more extreme present-day environmentalists, see Henry M. Pachter, *Paracelsus, Magic into Science*, Henry Schuman, New York, 1951, p. 311.

73 Lovelock, *Homage to Gaia*, p. 189.

74 Robert N. Proctor, *Cancer Wars: How Politics Shapes What We Know and Don't Know About Cancer*, Basic Books, New York, 1995, p. 155.

75 T. F. West and G. A. Campbell, *DDT: The Synthetic Insecticide*, Chapman and Hall, London, 1946, p. 56.

76 'Toxic Manifestations', Chapter Four of West and Campbell, *DDT*. Also see 'FDA Hearings Offer Dual Opportunity', editorial by Robert L. Taylor in *Chemical Industries*, November 1949, 703. For later studies on the mammalian safety of DDT see Beatty, *DDT Myth*, pp. 24–44. The ability to eat dessert-spoonfuls of DDT and remain (apparently) unharmed was strikingly demonstrated by an elderly propagandist for DDT on 'Goodbye Mrs Ant' (see note 38).

77 Gunther and Blinn, *Analysis of Insecticides and Acaricides*, p. 501 and Martin, *Pesticide Manual*, p. 321 (see note 40). Significantly, even in 1950 there was no provision for the testing of nicotine residues in vegetables in the AOAC's *Methods of Analysis* (see note 42) although the colometric method had been published in 1939 in the AOAC's own journal (J. N. Markwood, 'The Photometric Determination of Nicotine on Apples without Distillation', *Journal of the AOAC*, **22**, 1939, 427–436). For the history of the pre-DDT era, see James C. Whorton, *Before Silent Spring: Pesticides and Public Health in pre-DDT America*, Princeton University Press, Princeton, 1975 and Paul W. Riegert, *From Arsenic to DDT: A History of Entomology in Western Canada*, University of Toronto Press, Toronto and Buffalo, 1980.

78 This topic has been analysed brilliantly by Robert Proctor in his books, *Cancer Wars* and its sequel, *The Nazi War on Cancer*, Princeton University Press, Princeton, New Jersey, 1999.

79 Carson, *Silent Spring*, p. 151.

80 The issue of threshold limits has hitherto received scant attention from historians, with the partial exception of asbestos. For tantalising hints of such an analysis see 'The Political Morphology of Dose–Response Curves', Chapter Seven of Proctor, *Cancer Wars*, pp. 153–173; Barry I. Castleman, *Asbestos: Medical and Legal Aspects*, 2nd edition, Law and Business, Clifton, NJ, 1986, Chapter 4, 'Thresholds and Standards', pp. 213–296; and Geoffrey Tweedale, *Magic Mineral to Killer Dust: Turner & Newall and the Asbestos Hazard*, Oxford University Press, Oxford, 2000, *passim*.

81 The only study on the application of the Delaney Clause to pesticides is: Board on Agriculture. Committee on Scientific and Regulatory Issues Underlying Pesticide Use, Patterns and Agricultural Innovation, *Regulating Pesticides in Food: The Delaney Paradox*, National Academy Press, Washington DC, 1987.

82 Milton S. Schechter, '1962 Wiley Award Address: Comments on the Pesticide Residue Situation', 1963, **46**, 1063–1069. This point is also made (with respect to blood alcohol testing) by Hugh D. Crone, *Chemicals and Society: A Guide to the New Chemical Age*, Cambridge University Press, Cambridge, 1986, p. 55.

83 For discussions of the impact of the Delaney clause on pesticide residues, see http://www.ewg.org/pub/home/reports/newpestlaw/NewPestLaw.html and http://cnie.org/nle/pest-1.html

84 For the banning of DDT at state and national levels in the United States, see Dunlap, *DDT*, pp. 129–245; Beatty, *DDT Myth*, pp. 135–185; and Bosso, *Pesticides and Politics: The Life Cycle of a Public Issue*, University of Pittsburgh Press, Pittsburgh, 1987.

85 See Rachel's Environment and Health Weekly #46 at http://www.monitor.net/rachel/r46.html or *via* http://www.rachel.org/home_eng.htm. The International Agency for Research on Cancer (IARC) currently classifies DDT, lindane, chlordane and heptachlor as possibly carcinogenic in man, but strong evidence for such a link is wanting. For IARC rulings on pesticides, visit the massive IARC website: http://www.iarc.fr/. For DDT see http://193.51.164.11/htdocs/Monographs/Vol53/04-DDT.HTM and for other agents (*e.g.* dieldrin) use the powerful search engine: http://193.51.164.11/cgi/iHound/Chem/iH_Chem_Frames.html. For a general survey of the carcinogenic hazards of chemicals and the rôle of the IARC, see Corbett McDonald and Pier Bertazzi, Chapter 2, 'Occupational Cancer: Metals and Chemicals', in Corbett McDonald (ed.), *Epidemiology of Work Related Diseases*, 2nd edition, BMJ Books, London, 2000, pp. 7–39.

86 Personal communication from David Carter, LGC (Teddington).

87 Quoted in Beatty, *DDT Myth*, p. 26.

88 See Barry Commoner, *Science and Survival*, Victor Gollancz, London, 1966 (but first

published in 1963) p. 27 and pp. 88–90. Also see Barry Commoner, 'Fallout and Water Pollution – Parallel Cases', *Scientist and Citizen*, December 1964, **7**, 2–7. For the development of anti-chemical hysteria in the 1960s and 1970s, see Edith Efron, *The Apocalyptics: How Environmental Politics Controls What We Know About Cancer*, Simon and Schuster, New York, 1984, for Commoner see pp. 35–37. For a different approach to the development of environmentalism, see Donald Worster, *Nature's Economy: A History of Ecological Ideas*, 2nd edition, Cambridge University Press, New York, 1994; see p. 354 for Commoner.

89 The World-wide Web is a fertile breeding ground for these fears (for better or worse), see for example: General: http://www.parentstrackingthenet.com/pesticid.htm and http://www.nrdc.org/health/kids/ocar/chap5.asp; Breast milk: http://www.eap.mcgill.ca/MagRack/JPR/JPR_07.htm; Baby food: http://www.ewg.org/pub/home/reports/Baby_food/baby_home.html

Instrumentation in Environmental Analysis, 1935–1975

ANTHONY S. TRAVIS

Introduction

The large-scale manufacture of aromatic organic chemicals based on coal-tar products that began during the second half of the nineteenth century contributed immensely to improvements in material life. The products included dyestuffs ('aniline' or coal-tar dyes), pharmaceuticals, explosives, and, in the early 1900s, the phenol-derived plastic Bakelite. However, the waste and other releases from their manufacture proved detrimental to the environment, particularly the soil, and surface and ground water. Concern had been aroused in Britain, France, Switzerland, and Germany during the 1860s when arsenic-containing wastes from the manufacture of dyestuffs that had permeated soil and entered well water were found to be responsible for severe health problems, and even deaths. Similar situations arose when the waste from coal gas works infiltrated ground water. By the 1890s, it was clear that when nature was overwhelmed with industrial waste then even deep subsurface-waters were contaminated, and also that certain of the novel organic compounds did not easily degrade. Early in the twentieth century, a new class of poorly or non-biodegradable industrial compounds appeared, the chlorinated hydrocarbons, whose members increasingly served as solvents in manufacturing and cleaning processes, particularly from the 1930s. Their various uses stemmed, in part, from their great stability, but this also meant that they persisted when released into nature. The rapid growth of the United States chemical industry during and after World War I ensured that waste disposal problems were common to both sides of the Atlantic. From 1930, and 1941, respectively, the aromatic chlorinated hydrocarbons PCBs (PolyChlorinated Biphenyls), and DDT (DichloroDiphenylTrichloroethane) joined the large group of inert substances.[1]

Though tests were introduced for small amounts of phenolic compounds in waters, measurements of organic matter were generally indirect and non-specific,

Table 1   *Desirable detection limits for organic chemicals, 1930–1965*[2]

Year	Sensitivity (minimum detectability required)
1930–40	10 ppm, 1:100 000
1945	1.0 ppm, 1:1 000 000
1950	0.1 ppm, 1:10 000 000
1955	0.02 ppm, 2:100 000 000
1960	1.0 ppb, 1:1 000 000 000
1965	0.1 ppb, 1:10 000 000 000

particularly biological oxygen demand (BOD), which measured only the carbon compounds that underwent aerobic decomposition over a fixed period (generally five days), and chemical oxygen demand (COD), which, by oxidation with permanganate (later dichromate), measured the total amount of organic matter. With increasing health concerns, fish kills, and foaming of river and tap water, there was, by the 1950s, a considerable demand for rapid, routine and reliable methods of analyses for trace amounts of individual substances (see Table 1), even though chemists could not achieve these levels at that time. This encouraged new applications of instruments and sensitive detectors, and the amalgamation of instrumental and classical methods. Instruments enabled not only the hitherto impossible species identification in complex mixtures but also measurement at the low parts per million (ppm) level, and in the 1960s at the low parts per (American) billion (ppb) level.

The earliest applications of analytical instruments in general industrial problem solving took place in the United States, where, as Yakov Rabkin has demonstrated, physicists and chemists at American Cyanamid played major rôles in the promotion and widespread adoption of instruments, particularly after World War II.[3] It was in the 1930s, however, that American Cyanamid medical officers, drawing upon the wide expertise available in the firm's laboratories, first demonstrated the power of instrumental analysis in undertaking studies of industrial hygiene and environmental significance. Accordingly, the first sections of this paper highlight the rôle of American Cyanamid scientists in both developing and extending the use of analytical instruments, especially during the 1940s. Federal laboratories, particularly those in Cincinnati, Ohio, were at the forefront of the application and promotion of instrumental methods, starting with infrared spectroscopy, to water pollution studies from around 1950. Their target audience was made up of analysts, sanitary engineers, toxicologists, and wastewater experts.

American Cyanamid and Infrared Spectrophotometry

The insalubrious and noxious properties of the aromatic compounds nitrobenzene and aniline have been known since the 1860s, when they and their congeners became the principal intermediates employed in the coal-tar dye industry. In the twentieth century, nitrobenzene was an important solvent used, for example, in the manufacture of anthraquinonoid vat dyes, while other aromatic nitro compounds became the basis of the modern explosives industry. Well water contamination, caused by migration of chemicals through soil from factories, and high concentrations of aromatics in workplace atmospheres stimulated the development of tests for their detection and estimation. In particular, the prevalence of cyanosis, or 'blue lip' (methemoglobinema), among workers handling nitrobenzene and aromatic amines in dye and intermediate factories was a source of considerable concern. The presence of these compounds could often be established by that highly sensitive instrument, the nose. Nitrobenzene for example has a characteristic shoe polish odour.

By the 1930s, factory inspectors, especially in Britain, were engaged in the development of robust and reliable devices suitable for quantitative determinations of specific contaminants, including aniline, in workplace atmospheres. These relied upon traditional 'wet-and-dry' chemical analysis, and were suited to measurements in low parts per hundred thousand. From 1937, official methods and specifications were based on work done at the Laboratory of the Government Chemist in London. Ethel Browning of HM Factory Inspectorate gathered together data on toxic chemicals found in industry (much of it gleaned from inspectors' reports), which became the basis of publications that on both sides of the Atlantic were considered to be authoritative.[4]

With the advance during the same decade of novel optical systems suited to physical instrumentation, synthetic dyestuffs and dye intermediates (coloured and other aromatic molecules that interact with radiation) were obvious choices for studies involving colorimetry and, later, spectrophotometry. The industrial leader in the development and application of instrumental methods for the study of dyes and related compounds was American Cyanamid. This company is a good starting point here, because it was the first to successfully employ newly available instrumental methods in the understanding of industrial hygiene and environmental problems. Thus at the Bound Brook, New Jersey, dye factory of American Cyanamid the medical officers Donald Hamblin and Arthur Mangelsdorff used the costly Hardy recording spectrophotometer to undertake what was to become the definitive study of how nitrobenzene and aniline caused methemoglobinemia.[5] Their results were published in 1938, one year after the corporation's new research laboratories were opened at Stamford, Connnecticut.

For almost three decades, American Cyanamid would remain at the forefront of spectrophotometry as applied to industrial aromatic, as well as other, chemicals, including applications to environmental problems. Publications in *The Review of Scientific Instruments, Journal of Applied Physics* and *Analytical Chemistry*, as well as numerous editorials, review articles and chapters in books on the newer instrumental techniques of analytical chemistry, attested to the cutting-edge studies carried out at American Cyanamid.[6] Significantly, the first factory of Perkin–Elmer, founded by Richard S. Perkin and Charles W. Elmer in 1938 to manufacture advanced optical systems, was almost adjacent to the Stamford laboratories of American Cyanamid, and publications by American Cyanamid scientists acknowledged contributions from Perkin–Elmer.

During the 1940s, the practical application of spectrophotometry was advanced more than in any academic laboratory by the distinguished American

Calco factory at Bound Brook, New Jersey, 1940. Reproduced from [Williams Haynes], 'Dyes Made in America, 1915–1940: Calco Chemical Division, American Cyanamid Company' (Bound Brook, American Cyanamid, 1940)

Cyanamid scientists Edwin I. Stearns and R. Bowling Barnes, at the Bound Brook and Stamford laboratories, respectively.[7] Both were present at the October 1943 meeting of the Optical Society of America in which the rôle of the National Bureau of Standards in arriving at consensus was emphasised, and subsequently were among the principal participants in the development and use of spectrophotometers and colorimeters. In 1945, Barnes and colleagues, jointly with Richard F. Kinnaird of Perkin–Elmer, for the first time described the latter firm's model 12 infrared spectrophotometer,[8] while during 1945–46, 'Barnes and his co-workers published two papers in America which were to have a profound effect on the design of flame photometers'.[9]

Barnes was co-author of *Infrared Spectrosopy: Industrial Applications, and Bibliography* (1944), which was adopted by 1950 and later recommended by Francis M. Middleton, Chief, Chemistry and Physics Section, Water Supply and Research Branch, US Public Health Service, Robert A. Taft Sanitary Engineering Center, Cincinnati, Ohio.[10] This clearly demonstrates that instrumentation was soon appreciated by both chemical and environmental communities. At a time when water analysis still meant measurement of BOD, COD, suspended solids, pH, turbidity, and colour, Middleton's group employed functional group and fingerprint analysis based on infrared to successfully detect organic industrial waste in water supplies.[11] In 1952, Middleton was joined by Aaron A. Rosen, who had received his PhD in organic chemistry at Cincinnati in 1938, and served for five years with the US Chemical Warfare Service. In 1954, Middleton and Rosen described their instrument-aided research based on infrared before the division of water, sewage and sanitation chemistry at the annual meeting of the American Chemical Society, held in Kansas City.[12]

In August 1960, The Taft Sanitary Engineering Center hosted the first major

*Perkin-Elmer12C infrared spectrometer, presented to Harold Thompson by Richard S.
Perkin of Perkin-Elmer in 1945. Science Museum collections, inventory number 1980–993.
Science Museum/Science and Society Picture Library*

gathering of federal, academic and industrial scientists, as well as instrument
manufacturers, totalling 42 participants, that emphasised the rôle of analytical
instruments, as well as instruments for data transmission, in the field of water
quality measurement. Middleton's paper drew attention to the use of gas
chromatography and infrared spectroscopy, and in the discussion session named
DDT, aldrin, and derivatives of benzene among the contaminants found and
identified in waste waters.[13] The conference was timely, for in the following year
the US Water Pollution Control Act was introduced (this followed an earlier act
of 1948, amended in 1956), requiring the Secretary of Health, Education, and
Welfare to undertake research into improved methods for identifying and
measuring pollution. Also in 1961, the Public Health Service organised a major
symposium dealing with ground water contamination.[14]

These initiatives had an immediate influence on activities at many academic
and industrial laboratories and afforded Rosen and his colleague M. B. Ettinger
much-enlarged audiences at gatherings of specialist groups. The Taft centre
analysts emphasised the value of infrared in wastewater analysis when addressing the 1962 annual meeting of the Water Pollution Control Federation (held in
Toronto) on studies of contaminants in the Kanawha River, at Nitro, West
Virginia, which, like northern New Jersey, was a major chemical producing
area.[15] The federation, which published a monthly journal, consisted of fifty
technical societies, whose members included chemists and engineers engaged in
examination and treatment of wastewater. It was also co-publisher of *Standard
Methods for the Examination of Water and Wastewater Including Bottom Sediments and Sludges* ('Standard Methods'), with the American Public Health
Association and the American Water Works Association, which first appeared in
1905.

Beckman DU quartz (prism) ultraviolet–visible spectrometer, 1940s. Science Museum collections, inventory number 1979-238. When it was launched in 1941, the DU revolutionised the use of ultraviolet spectroscopy by chemists and biochemists. Science Museum/Science and Society Picture Library

American Cyanamid and Ultraviolet Spectrophotometry

Robert C. Hirt was another notable American Cyanamid contributor at Stamford, particularly to ultraviolet spectrophotometry, which offered more accuracy than IR for quantitative analysis. Ultraviolet analysis had been adopted from 1951 at Du Pont's Chambers Works, another important US dye-manufacturing site, for industrial hygiene studies involving detection of aromatic nitro compounds in the 5–50 ppm range. In this case, the nitro compound was reduced to the corresponding amino derivative, which was then diazotized to afford a coloured substance that gave a characteristic ultraviolet spectral curve.[16] Hirt and his colleagues modified ultraviolet spectrophotometers, namely the Beckman DU and a Cary machine, for more direct use in identifying individual substances in the ultraviolet region.[17] Their publications included instrumental analysis of chemicals in the industrial environment. Thus they analysed two-component mixtures that included nitrobenzene and aniline, 'a combination which had been of interest in an industrial hygiene investigation', and that was characteristic of dye manufacture, which in the case of American Cyanamid meant Bound Brook. Notably, '[s]peed and ease in setting up an analysis are obtained, as knowledge of only three absorptivities is needed for a two-component analysis if an isoabsorptive point is used; these values may be readily obtained from the user's files, the literature, or another laboratory'. Details of this absorbance-ratio method were presented at the 4th Pittsburgh Conference on Analytical Chemistry and Applied Spectroscopy held in March 1953.[18] The benefit of novel instrumental applications at, and data banks held within, a central research laboratory, such as American Cyanamid Stamford, was that important and accurate information could be rapidly transmitted to other company-owned sites:

> This allows a laboratory with a large library of spectra to transmit a minimum of information – by telephone, for example – to a smaller laboratory or plant control unit to enable it to set up and use a plot of absorbance ratio *vs.* relative concentration.[19]

By 1957, moderately priced ultraviolet spectrophotometers were available from Beckman, Cary, and Perkin–Elmer, and improvements in sensitivity and detectability enabled rapid trace analyses, including in industrial hygiene and wastewater studies, where it was desirable to monitor compounds specific to a manufacturing site or process. Absolute concentrations were read off from a calibration graph in which, for a given compound, absorbance (on a scale of 0 to 100%) was plotted against concentration. Thus in 1959 at American Cyanamid's Bound Brook plant, ultraviolet spectrophotometric quantitative analysis demonstrated that nitrobenzene was present in a newly driven water well at dilutions of less than one part per million.[20]

Instrumental Analysis of Halocarbons

The determination of organic halogen by instrumental techniques was also developed in industrial laboratories where manufacturing, product control, and environmental considerations applied. This is also a good example of the adaptation of classical methods of analysis, such as the Fujiwara pyridine–sodium hydroxide colour reaction (1914) that was used to detect minute, or trace, amounts of trichloroethylene (TCE) and other chlorinated organics. This included in soils, using the Lovibond Tintometer to provide quantitative information.[21] If TCE was the only, or principal, solvent in contaminated well water then its measurement in low ppm was possible in 1949, and 'it should be possible to estimate 5 parts per million with ease'.[22] Du Pont, following its investigation of worker exposure to aromatic nitro compounds, funded a 1950s study of TCE and other chlorinated compounds in urine that enabled identification and quantitative determination of specific compounds through the Fujiwara test, followed by measurement of absorption with a Beckman DU spectrophotometer.[23]

A second way of detecting organic halogen was by conversion into halide ion, through combustion and sodium biphenyl reduction, which was widely applied in the 1950s and 60s. It was also adapted to instrumental measurement, particularly in the petrochemical industry, and chloride concentration could be determined potentiometrically down to 10 ppm.[24] To avoid interference from atmospheric chloride in salt, it was recommended that the apparatus be placed in an air-conditioned room. This is how Atlantic Refining Co. of Philadelphia detected trace halides in petroleum naphthas and catalysts.[25] Atlantic, incidentally, had demonstrated a long record of environmental concern, much of it associated with its waste disposal expert Wilson B. Hart, who published a series of papers on pollution prevention in *Petroleum Processing* in the 1940s. The Atlantic

method for organic halogen was adapted for environmental analysis in agriculture by Roger C. Blinn (who later joined American Cyanamid) at the Department of Entomology, University of California, Riverside.[26]

The use of mass spectra to determine volatile 'industrial solvents and cleaning
agents' in water (in addition to hydrocarbons) was suggested in 1953 by Frank
Melpolder and colleagues, also at Atlantic Refining Co.[27] Three years later, at
the 7th Pittsburgh Conference on Analytical Chemistry and Applied Spectroscopy, chemists from the research laboratories of Eastman Kodak Co, Rochester,
NY, described mass spectrometric determination of trace amounts of volatile
organic solvents in industrial waste water. The volatiles included chlorinated
hydrocarbons, whose presence was detected in water below 1 ppm, even in
mixtures containing five test solvents. To achieve this, the vapour above the
liquid in an air-free system, which prevented interference from dissolved or
suspended solids, was analysed (the solvent vapours were concentrated by
freezing in liquid nitrogen). To check the results, experiments were undertaken
with 100 ppm of chlorinated hydrocarbon added to a plant's industrial sewer half
a mile from the sampling point. Chlorinated hydrocarbon was detected, peaking
to 95 ppm, before dropping off. This method offered rapid identification with
negligible interference from a wide variety of other organics.[28] Notably, the
spectrometer, a 60-degree sector-type instrument, was made in the Eastman
Kodak laboratories. The in-house design and construction of instruments at
industrial laboratories continued to be widely practised, and from the mid-1950s
included gas chromatographs (sometimes referred to as chromatograms). It was
gas chromatography that became the most important technique available from
the late 1950s for qualitative and quantitative determination of chlorinated
hydrocarbons and many aromatics.

Gas Chromatography

Though use of gas chromatography for detecting hydrocarbon vapours in gas
streams was first carried out during the early 1940s, the greatest progress was
made after Archer Martin and Anthony James in 1952 published the practical
details of what was soon to be known as gas-liquid chromatography, or gas
chromatography (GC). Their early experiments at the National Institute for
Medical Research, Mill Hill, near London, included a distinctly environment-
related analysis, the measurement of the constituents in motor car exhausts. In
1956, James published a comprehensive paper on the 'separation and microestimation of volatile aromatic amines', and discussed the effects of substituents in
aniline and its congeners on retention volumes (retention times). James observed
that:

> [t]he sensitivity of the titration technique is such that as little as 0.1 mg of
> aniline is detectable (corresponding to a step height of 1 mm). The later
> development of the gas-density meter by Martin has allowed the detection
> limit to be lowered even further. This device, unlike the automatic titrator,
> records the instantaneous concentration of the material emerging from the

Archer Martin, 1950s. Photograph courtesy of NIMR

chromatogram.... Each step denotes a separate substance and the horizontal line denotes a period in which no titratable material is emerging from the columns.[29]

By this time the first commercial gas chromatographs had appeared, notably the Perkin–Elmer Vapor Fractometer 154, introduced in 1955, that soon proved its value in trace analysis and identification of halocarbons. In 1958, chemists at the Ethyl Corp, Baton Rouge, Louisiana, reported that in trace analysis '[i]t has now become common practice to collect selected fractions for further study by infrared, mass spectrometric, or repeated gas chromatographic techniques'.[30] They used the Model 154 to achieve measurements down to 1 ppm. One year later, the development department of Union Carbide Chemicals Co, Division of Union Carbide Corp, South Charleston, West Virginia, showed that it was possible to differentiate between TCE, PCE (perchloroethylene), and other chlorinated aliphatic hydrocarbons, again with a Model 154.[31]

In 1957, Roland S. Gohlke, at the Dow, Midland, Spectroscopy Laboratory, was one of the first to hyphenate gas chromatography with mass spectrometry (GC–MS). At the 17th Pittsburgh Conference in 1966, Ragnar Ryhage of the Karolinska Institute, Stockholm, described the hyphenation of a gas chromatograph with a rapid-scanning mass spectrometer and observed that the technique held out new potential in the development of air and water pollution research.[32] In the following year, Varian introduced an advanced mass spectrometer, the Model CH 5, suitable for coupling with a gas chromatograph. The importance of this innovation was that though GC brought about separation, in common with, but superior to, other chromatographic methods, it did not (unlike the mass spectrometer) provide unambiguous identification of components in mixtures.

An Hewlett-Packard 5973 mass spectrometer, hyphenated with an Hewlett-Packard 6890 gas chromatograph and an Hewlett-Packard 7683 injected system, used for drug analysis at the Drug Control Centre, King's College, London, 2000. Nowadays, GC–MS is used for a wide range of analytical applications. Science Museum/Science and Society Picture Library

The widespread combination of gas chromatography (for separation) with other instrumental methods (for identification) soon led to trace analysis studies at levels and separations never previously achievable with many organic substances. This, along with rapidity, was the major breakthrough in environmental analysis.

Detection Limits and Detectors

From the 1940s, and particularly after the introduction of gas chromatography in 1955, instrumental methods greatly enhanced the capabilities of quantitative chemical analysis. During the 1950s, trace analyses for residues, using precision colorimeters, infrared and ultraviolet spectrophotometers, and gas chromatographs, was possible down to 0.1 ppm.[33] This had significant implications for environmental work. However, since waste mixtures were invariably inhomogenous, techniques that showed sensitivity to particular types of compounds were required. The performance of early GC thermal conductivity instruments proved to be no match for the flame ionisation detector (FID). From 1960 sensitivity to specific compounds was improved greatly, including in environmental work, with the electron capture detector (ECD) that was well suited to detection of nitrobenzene and halogenated hydrocarbons. American Cyanamid's Stamford laboratories manufactured FID detectors for in-house use, and through a strong emphasis on electrochemistry the company contributed towards detectors suited to chlorinated hydrocarbons, including DDT.[34]

By 1965, there was a sense of urgency among analysts to achieve even better results, created by newly perceived needs in measurement of water quality, particularly identification and trace analysis of contaminants. National and international organisations interested in trace analysis, particularly of refractory pesticide residues in soil and water, demanded minimum sensitivity levels of fractions of a part per billion (see Table 1), which in turn drove the development of GC detector technology.[35] The setting up of the annual Joint Food and Agriculture Organisation (FAO)/World Health Organisation (WHO) Meeting on Pesticide Residues (JMPR) in 1963 and the establishment, two years later, of the US Pesticide Residues Committee of the National Academy of Sciences and the National Research Council gave instrumental environmental analysis a major boost, as did publication of the 12th edition of 'Standard Methods', which included 'for the first time ultraviolet spectrophotometric, spectrographic, and gas chromatographic methods'.[36] Middleton and Rosen were members of the American Water Works Association's committee that jointly prepared this publication.

Applications of GC–MS to atmospheric pollutants extended work reported in 1973 and 1974 on 1,1,1-trichloroethane (Tri) and freons (fluorochloromethanes) to other chlorinated chlorocarbons, including methyl chloride, and related volatile organic compounds (VOCs). Parts per trillion measurements were also reported.[37] By this time GC–MS was computerised, and rapid examination of trace amounts of contaminants in waste waters had become routine, particularly since a wide range of data had been published. Trace analysis of waste water by liquid chromatography was also available.[38] The methods of sample preparation and concentration, however, relied on techniques developed during the 1950s and 1960s.

Polluted Waters

While water pollution, particularly from non-biodegradable industrial products, including chlorinated hydrocarbons, grew during the 1940s, so too did instrumental methods for its analysis. In a 1951 review article, S. Kenneth Love of the US Geological Survey, Washington, DC, observed:

> Increasing use of instruments in the analysis of water is apparent. Photometers and spectrophotometers of various kinds have largely displaced visual observation in colorimetric analysis. Better instruments, making possible greater precision, are being placed on the market. Flame photometry is finding wider application, and, under carefully controlled conditions, good results are being reported.... Several of the analytical procedures ... relate to the analysis of water-borne industrial wastes. With the increasing emphasis on the control and abatement of pollution of natural waters, there is a growing need for methods of analysis for determining constituents frequently present in waste but not ordinarily found in unpolluted waters.[39]

New constituents in surface and ground waters around 1950 included the refractory alkylbenzenesulfonate detergents. Infrared spectroscopy was used in their analysis at low ppm levels, such as in the investigation of biochemical breakdown in sewage, part of a 1958 initiative by the British Standing Technical Committee on Synthetic Detergents.[40]

The collection of contaminants in water on activated carbon, prior to undertaking instrumental analysis, was widely investigated. This originated in the US, where in 1951:

> Braus isolated 133 μg /l^{-1} of organics from Cincinnati tap water by passing 8000 l through a 500 g carbon filter, and then eluting the organics from the carbon by Soxhlet extraction with diethyl ether. This extract was separated into groups of acidic, basic and neutral compounds.

Harry Braus was a colleague of Middleton at the Taft Center, who applied the same method in 1956 to investigate organics in the Kanawha River:

> This water was particularly heavily polluted with industrial waste. The extract was first separated into neutral, weak acid, strong acid, amine and ether insoluble fractions which composed 56, 1.5, 2.5, 3.0 and 12% of the extract, respectively. The neutrals were then further separated using an alumina column into aliphatic, aromatic and oxygenated materials. The aromatic and oxygenated fractions were the most odorous.

Around 1960, a number of conference papers demonstrated a high level of awareness of ground water pollution problems that existed and required attention, while others identified the range of chemical and instrumental methods for analysis of contaminated and waste waters.[41] Chemists at the Taft Center and elsewhere drew attention to the potential of gas chromatography in water analysis.[42] This was a time when routine analysis was becoming available at moderate cost through the availability of commercial instruments and novel detectors. It was possible to rapidly isolate and identify individual components of complex mixtures by GC–MS and/or GC–infrared. Instrumental colorimetry had displaced or supplemented earlier non-instrumental colorimetric methods. In connection with analysis of water, carbon chloroform extracts (that is, organic components extracted into chloroform) were, after reduction in volume, subjected to instrumental analysis. Middleton noted that a:

> recent and powerful technique involves steam distillation of the chloroform extract, followed by a group separation, then further separation by gas chromatography with collection of samples from the outlet. Infrared examination of these fractions may identify the [contaminating] materials.[43]

In 1963, Rosen reported his analyses of Kanawha River water at Nitro also using carbon chloroform extracts, followed by gas chromatography, sample collection and infrared identification. He identified several products of local industry, including naphthalene, styrene, acetophenone and 2-methyl-5-ethylpyridine:

> Over 1 million litres of water were passed through large filters yielding 1600 g of organics on elution with chloroform. The steam-volatile material from this extract was separated into acidic, basic and neutral groups and

then further separated by gas chromatography. Infrared spectroscopy was used to identify the separated materials.[44]

Extraction of organics was also undertaken with diethyl ether:

> The solvent [was] reduced by vacuum evaporation . . . portions are injected into a gas chromatograph. The idea is not original; the general scheme has reportedly been used by the US Public Health Service in their studies.[45]

These and similar methods were to receive increasing prominence at major conferences of water experts and analytical chemists. Thus in 1964, L. G. Cochran and F. D. Bess of Union Carbide gave an account of analysis of undiluted waste at the annual meeting of the Water Pollution Control Federation. The results were published in 1966 in the *Journal of the Water Pollution Control Federation*. In the same year, R. A. Baker's analyses of phenols in water by direct aqueous injection gas chromatography with a flame ionisation detector system appeared in the *Journal of the American Water Works Association*.[46]

When the Federal Water Pollution Control Administration was established in 1965, Rosen was appointed acting chief of Waste Identification and Analysis Activities at its Cincinnati Water Research Laboratory. (The administration was transferred to the Department of the Interior in 1966.) Rosen pointed out that '[g]as chromatography is the most powerful separation technique available to the water pollution chemist'. And though he still preferred to identify separated components by IR, he noted that:

> the development of specific detector systems such as electron capture, microcoulometry, and thermionic emmision, adds to the identification capabilities of gas chromatography.[47]

In 1970, the Federal Water Pollution Control Administration was renamed the Federal Water Quality Administration, which in 1971 became part of the newly founded Environmental Protection Agency (EPA). The EPA retained its Analytical Quality Control Laboratory in Cincinnati, in addition to nine water analysis laboratories in other states. The Cincinnati laboratory, part of the National Environmental Research Center, maintained a computer-accessed literature file on analytical methods for determining water quality.

By 1970, gas chromatography and other instrumental methods were widely applied to trace determination of many contaminants in water, following extraction by methods such as activated carbon/charcoal adsorption, as well as liquid–liquid (chloroform–water, and diethyl ether–water). These were included in Brian Croll's review, published in 1972, that covered both domestic and industrial organic pollutants:[48]

> At the present time these methods [analysis of organic compounds in water] are being developed and attempts are being made to separate and identify all the organic compounds present in water . . . concentration and separation techniques for the analysis of organic and inorganic compounds in water covers precipitation techniques, liquid–liquid extraction, freeze

concentration, adsorption bubble separators, chromatography, ion exchange, membrane techniques, carbon adsorption, distillation, evaporation and sublimation. Two techniques not covered are the use of ion-exchange resins for the concentration of humic substances and the use of direct injection of water and head-space gas chromatographic analysis.[49]

Instrumental methods were now standard for separation and identification of contaminants in water:

> Caruso and Koslow have advocated the use of solvent extraction using diethyl ether followed by gas liquid chromatography to trace sources of river pollution. They also advocated the use of infrared and ultraviolet spectroscopy and mass spectrometry to identify the components separated by gas chromatography. Using these combination techniques they were able to identify phenol and naphthalene in water contaminated by an industrial spillage, quinaldine in an industrial effluent, and hydrocarbons in water from Lake Michigan.[50]

Fred K. Kawahara, also at Cincinnati, employed gas chromatography in analysis of chlorinated hydrocarbons in waste water during the 1960s, and in 1970 published a useful modification suited to phenolic and other acidic organic compounds in carbon chloroform extracts obtained from the Ohio River. The acids were converted to halogenated derivatives with bromo-2,3,4,5,6-pentafluorotoluene and analysed by electron-capture gas chromatography:

> The selectivity of the detector to halogenated compounds enables the acids to be identified in the presence of much larger quantities of non-electron capturing materials, and at lower levels than when using flame-ionisation detectors.

Similar studies to those carried out in the US were performed in Germany and Holland on the River Rhine. In 1970, 'Kölle, Koppe and Sontheimer desorbed the organics from carbon filters on the Rhine using carbon tetrachloride and benzene… Using thin layer chromatography and gas chromatography they were able to isolate [ten-carbon] chlorinated hydrocarbons'. Adriaan Meijers, at Delft Polytechnic:

> working on the Dutch Rhine, investigated recovery of organics using activated carbon filters, batch adsorption onto activated carbon, a strongly basic anion-exchange resin and freeze concentration followed by chloroform extraction. Some of the extracts were subjected to group separations, and then further separation by column and gas-liquid chromatography. One carbon chloroform extract was separated into groups and many compounds in each group were identified using a mass spectrometer linked to a gas chromatograph. [Identified organics included tetrachloroaliphatics].

By 1972, John McGuire, Chief, Mass Spectrometry and Gas Chromatography Section, at the EPA's Southeast Environmental Research Laboratory, had adopted computerised matching of the mass spectra obtained from GC separation, in preference to using the electron capture detector alone (due to its restricted selectivity),

> [a] system [that] has been developed for the rapid identification of volatile organic water pollutants.... Application of this system to the analysis of industrial waste effluents revealed a significant number of pollutants.[51]

Volatile organic compounds in water were separated by purge-and-trap methods, in which concentration took place on an adsorbent. This greatly aided ppb analysis by GC–MS. In this manner:

> evidence has accumulated for the presence of a number of simple aliphatic chlorohydrocarbons in the environment. Among those commonly found are chloroform, carbon tetrachloride, tri- and perchloroethylene and 1,1,1-trichloroethane which are widely distributed in the atmosphere, fresh and marine waters, foodstuffs and animal (including human) tissues.[52]

Pesticides

Pesticide analysis by GC, particularly detection and measurement of chlorinated aromatic compounds such as DDT, at first proved difficult due to the fact that these compounds decomposed in the spectrometers, and there was little or no standardisation of conditions. Also, the earliest detectors available from the mid-1950s were unsuited to this work. Initially, Rosen and Middleton preferred IR analysis, which in 1958 they described to the American Chemical Society's division of water, sewage and sanitation chemistry at the 133rd annual meeting, held in San Francisco. Working with eight chlorinated insecticides, they detected amounts of less than 10 ppb in surface waters. Though conservative in their claims, they did emphasise that the results gave a satisfactory measure of the magnitude of trace amounts.[53]

The analysis of pesticides changed in the same year, 1958, when the Dohrmann (Coulson method) microcoulometric gas chromatograph, specific for carbon–halogen and carbon–sulfur bonds, was installed at the Stanford Research Institute. Two years later:

> [t]he introduction of electron capture detection for the gas-liquid chromatography of organochlorine pesticide residues was widely welcomed and this sensitive technique is now the method of choice.... Electron capture detection gives great sensitivity but only limited selectivity.[54]

In 1960, the British Pesticides Residues Foodstuffs Sub-Committee, set up by the Analytical Methods Committee of the Society of Analytical Chemistry, commissioned an investigation by Needhams into bioassay methods for determination of pesticide residues in foodstuffs (from this initiative the Laboratory of the Government Chemist established the Pesticide Residue Analysis Service in 1964). Techniques based on GC were by then acknowledged to show considerable promise, and there was much interest in the development of a rapid 'sorting test' for identifying traces of chlorinated pesticides in crops by means of GC with electron capture detectors.

Application to residue analysis on the microgram scale with more sensitive detection, for example, argon ionisation, was made difficult by chromatographic interference resulting from material co-extracted from the crop, unless a 'clean-up' stage was included. Coulson, however, overcame the clean-up by employing a GC–combustion–coulometric titration procedure, applicable to both chlorinated and thiophosphate residues, as did Zweig, Archer, and Rubenstein, who used GC–infrared.[55] In 1961, Goodwin, Goulden, and Reynolds, observed that:

> [t]o avoid the necessity for these somewhat complex combination methods, a simpler means of detection is required that possesses not only great sensitivity but also a high degree of selectivity towards the pesticides to be identified and determined. These requirements are met to a considerable extent by the electron capture detector of Lovelock and Lipsky, which can be made to exhibit exceptional response to halogenated compounds. This selective response, which is such that nanogram (10^{-9} g) amounts of chlorinated compounds can readily be determined, permits the identification and determination of traces of chlorinated pesticides in crop extracts without the need either for prior 'clean up' or for the preliminary concentration of the extract solution.[56]

One year earlier they had provided evidence that 0.035 ppm of aldrin and 0.07 ppm of dieldrin could be detected without prior clean up.[57] Through this work there became available:

> a simple rapid 'sorting test' for identifying and determining residues of chlorinated pesticides in crops. At the same time, the wide scope of the electron capture gas-liquid chromatographic technique in the analysis of agricultural, atmospheric and industrial samples for trace halogenated pesticides is indicated.[58]

In 1965, D. C. Abbott and J. Thomson of the Laboratory of the Government Chemist in London, reviewed the range of available analytical methods for domestic and agricultural chemicals, again drawing attention to the value of GC, which had lowered the detection limit for pesticides and their metabolic products to 0.0001 ppm:

> As analytical techniques have progressed, so the older colorimetric methods have been largely replaced by chromatographic systems of greater selectivity and sensitivity. Paper and thin-layer chromatographic techniques are applicable to many pesticides at the 0.1 μg level, and below this in favourable cases, for both qualitative and semi-quantitative purposes, gas-liquid chromatography offers still greater sensitivity.... The use of a sample in the range from 200 ml to 2 litres allows the detection and determination of pesticides down to about 0.0001 ppm by chromatographic methods.... The use of a liquid–liquid extraction apparatus has

been recommended for organochlorine pesticides.[59]

They also discussed trace contamination of water, drawing attention to earlier problematic releases of refractories, in this case at Montebello, California, in 1945:

> The determination of traces of herbicidal compounds in water has attracted considerable attention. This is partly because they may be applied directly to waters and also because of taste problems which could be caused by the presence of phenolic bodies which are associated with chlorophenoxy acids in technical products. Swenson has described the effects of an industrial accident involving 2,4-D at Montebello, California in 1945; the tastes and odours in water persisting for four to five years. A colorimetric method using 4-aminoantipyrene has been suggested [1962] for the determination of 2,4-D phenol: on a one litre water sample a sensitivity of 0.007 ppm was obtained. Halogenated nitrophenols have also been determined colorimetrically [1960].[60]

By the mid-1960s, the Velsicol Chemical Company's pesticide-manufacturing plant at Memphis, on the Mississippi River, was analysing pesticide residues in spent cooling waters and process wastes using an electron capture gas chromatograph. As reported in 1966, the company had earlier rejected results of analyses using microcoulometric gas chromatography carried out by the Public Health Service following a massive fish-kill in the river. Subsequently, the PHS, drawing on the work of Middleton and Rosen, developed a new test based on IR spectrometry.[61] Also in 1966, Francis A. Gunther, professor of entomology at the University of California, Riverside, and the leading expert on pesticide analysis (including by bioassay methods), reviewed progress in instrumental trace analysis at a 'Symposium on Scientific Aspects of Pest Control', held under the auspices of the National Academy of Sciences, National Research Council.[62]

Though the annual Pittsburgh Conference on Analytical Chemistry and Applied Spectroscopy was the main venue for dissemination of knowledge about novel applications of instrumental methods, including water pollution studies, this topic also received close attention at the annual Purdue Industrial Waste Conference. Thus in 1968, H. A. Clarke's paper on 'Characterization of industrial wastes by instrumental analysis', included discussion of GC (for 'water samples....., particularly for chlorinated hydrocarbons and especially for DDT and its metabolites') and the use of carbon adsorption.[63] Croll, in his 1972 review paper, discussed chlorinated organic compounds, including polychlorinated biphenyls (PCBs), organomercury compounds and polyaromatic [polynuclear] hydrocarbons (PAHs):

> The specific known organic toxins which have been detected in water cover a wide range of classes of compounds, but it is convenient to discuss them all together. The polychlorinated biphenyl compounds, organomercury compounds, carcinogenic polynuclear hydrocarbons, mineral oils and detergents have all been the subject of extensive reviews.... [Polychlorinated

biphenyl compounds] [b]ecause of their very low solubility in water ... are normally detected in bottom muds and sludges and not in water itself.... These compounds have similar toxicities and toxic properties to the more familiar organochlorine insecticides and are analysed by electron capture gas chromatography in a similar manner to the insecticides.[64]

By the 1970s, the analytical sensitivity and selectivity of GC was such that 'it permits detection of traces of halogenated pesticides in the presence of hydrocarbon background materials, to which it is relatively insensitive'.[65]

PCBs

Gas-liquid chromatography was at first used to identify trace amounts of PCBs by finger-print technique, using reference samples. Partly this was due to the fact that problems were encountered in distinguishing PCBs from other chlorinated compounds, such as pesticide residues. Also, though they were measured in the atmosphere, and in marine species, they were not detected in aqueous environments. This anomaly was resolved in 1970 by Alan V. Holden at the Freshwater Fisheries Laboratory, Pitlochry, Scotland, who discovered PCBs in sludges and river muds, by extractions using a mixture of *n*-hexane and *iso*propanol, and GC analysis. However, though the media holding the PCBs was now identified, the origins, in this case somewhere from Glasgow and district, were not: 'It has proved impossible to locate the origins of the PCB discharges to the sewage plants concerned, because several hundred factories are linked to the sewers'.[66]

Consensus and Analytical Protocol

In the development and use of instrumental methods, industry, as American Cyanamid demonstrated, was the leader, setting standards based on its long-standing capabilities and knowledge of its products, backed up with in-house refinements in instrumentation. This was soon applied to hygiene and environmental problems in US industrial and federal laboratories. Typical were investigations of aromatic compounds such as nitrobenzene and aliphatic chlorinated hydrocarbons and, later, aromatic chlorinated hydrocarbons, such as DDT and PCBs.

During the early 1950s, sensitivity for organic substances was 0.1 ppm, and anything less became 'none detectable'. The demand for lowering the limits of detection (instrument sensitivity) in environmental monitoring brought about the need to investigate standardisation of experimental conditions, particularly in the analysis of water and agricultural products. Among the earliest official bodies to investigate standardisation and reproducibility was the 1965 US Pesticide Residues Committee. At that time, trace analysis of pesticides was required at 0.1 ppb, which was possible since GC had lowered the detection limit to this level. Initiatives by WHO and FAO and various other national and international bodies encouraged even greater sensitivity.

Subsequently, and after taking over from the Association of Official Analytical

Chemists, the US Environmental Protection Agency has set the international standards for water analysis, as well as of contaminants in soil, by instrumental methods through 'EPA Methods', the abbreviated title for *Methods for Chemical Analysis of Water and Wastes*. The EPA's method of choice in 1973 for organic chlorinated compounds and pesticides was GC, based on interim procedures available from the EPA's Methods Development and Quality Assurance Laboratory in Cincinnati.[67] Consensus regarding precision and accuracy followed the review of results from assigned analysts and laboratories.

Changes in legislation, such as the US Federal Water Pollution Control Acts, and the need to comply with regulations concerning soil, water and air contamination for so-called priority pollutants, brought about consistency in methodology, and reduced analytical uncertainties. Analytical protocols are based on agreement over the available techniques, required sensitivity, accuracy, reliability, precision, interferences, matrix effects, *etc.* These all contribute towards quality assurance. GC–MS still remains the technique of choice in environmental analysis, though high pressure liquid chromatography (HPLC) and other instrumental methods are also widely used. Except for greater sensitivity and sophistication, there is little difference with the principles behind the instrumental procedures employed in industrial hygiene and environmental investigations carried out since the 1950s.

The rapid advance and expansion in chemical technology drove the implementation of precision analytical instruments. These in turn were applied to the understanding of problems created by the impact of that technology, both where it happened and was applied, and made available evidence about the environmental contaminants and their identities. This is surely one of the landmarks of modern chemistry, as well as being the basis of legislative pressures, regulatory control, modelling and clean-up programmes, which in turn drive much of the present-day development in instrumentation. Along with other areas of instrumental analysis, monitoring the environment, especially from the 1950s, transformed the practice of analytical chemistry.

Notes and References

1 For DDT see Christian Simon, *DDT: Kulturgeschichte einer chemischen Verbindung*, Christian Merian Verlag, Basel, 1999 and references therein.
2 Francis A. Gunther, 'Advances in Analytical Detection of Pesticides', in *Scientific Aspects of Pest Control*, National Academy of Sciences/National Research Council, Washington DC, Publication no. 1402, 1966, pp. 276–302, on p. 293.
3 Yakov M. Rabkin, 'Technological Innovation in Science: The Adoption of Infrared Spectroscopy by Chemists', *Isis*, 1987, **78**, 31–54.
4 Ethel Browning, *The Toxicity of Industrial Organic Solvents*, Industrial Health Research Board Reports, no. 80, Medical Research Council, London, 1937 and Ethel Browning, *Toxicity and Metabolism of Industrial Solvents*, Elsevier, Amsterdam, 2nd ed., 1965 (first edition, 1953).
5 D. O. Hamblin and A. F. Mangelsdorff, 'Methemoglobinemia and its Measurement', *Journal of Industrial Hygiene and Toxicology*, 1938, **20**, 523–530.
6 Typical review articles by American Cyanamid Stamford chemists include Robert C.

Gore, 'Infrared Spectroscopy', *Analytical Chemistry*, 1958, **30**, 570–579 and Robert C. Hirt, 'Ultraviolet Spectrophotometry', *Analytical Chemistry*, 1958, **30**, 589–593.

7 See for example, E. I. Stearns, 'Applications of Ultraviolet and Visible Spectrophotometric Data', in M. G. Mellon (ed.), *Analytical Absorption Spectroscopy: Absorptimetry and Colorimetry*, Wiley, New York, 1950, pp. 306–438. For Barnes and the American Cyanamid–Perkin–Elmer connection see Rabkin, 'Technological Innovation in Science: The Adoption of Infrared Spectroscopy by Chemists', pp. 38 and 44.

8 R. Bowling Barnes, Robert S. McDonald, Van Zandt Williams and Richard F. Kinnaird, 'Small Prism Infrared Spectrometry', *Journal of Applied Physics*, 1945, **16**, 77–86.

9 A. G. Jones, *Analytical Chemistry: Some New Techniques*, Butterworths, London, 1959, p. 1. On the leading rôle played by the chemical and allied industries in the development of instrumental analysis see R. Clark Chirnside, 'Analytical Chemistry in Great Britain', *Analytical Chemistry*, 1961, **33**(12), 25A–33A.

10 R. B. Barnes, R. C. Gore, V. Liddel and V. Z. Williams, *Infrared Spectroscopy: Industrial Applications and Bibliography*, Reinhold, New York, 1944, and R. B. Barnes, R. C. Gore, R. W. Stafford and V. Z. Williams, 'Qualitative Organic Analysis and Infrared Spectometry', *Analytical Chemistry*, 1948, **20**, 402–410, were both listed by Francis M. Middleton, A. A. Rosen and R. M. Burttachell, in their *Manual for Recovery and Identification of Organic Chemicals in Water*, National Technical Information Service, US Department of Health, Education and Welfare/Public Health Service/Robert A. Taft Sanitary Engineering Center, Cincinnati, 1957, p. VII–5. In 1948, Barnes and Van Zandt Williams moved, respectively, from American Cyanamid to American Optical Co, Southbridge, Massachusetts and Perkin–Elmer. In 1964, Middleton was described as Chief, Advanced Waste Treatment Research Program, Basic and Applied Sciences Branch, Division of Water Supply and Pollution Control.

11 F. M. Middleton, Harry Braus and C. C. Ruchhoft, 'Fundamental Studies of Taste and Odor in Water Supplies', *Journal of the American Water Works Association*, 1952, **43**, 538–546, on pp. 543–545.

12 A. A. Rosen and F. M. Middleton, 'Identification of Petroleum Refinery Waste', *Analytical Chemistry*, 1955, **27**, 790–794.

13 F. M. Middleton, 'Detection and Measurement of Organic Chemicals in Water and Waste', in *Water Quality Measurement and Instrumentation. Proceedings of a Symposium held at Cincinnati, Ohio, August 29–31, 1960*, US Department of Health, Education and Welfare/Public Health Service/Robert A. Taft Sanitary Engineering Center, 1961, SEC TR W61-2, pp. 50–54, 69.

14 *Ground Water Contamination: Proceedings of the 1961 Symposium*, Public Health Service, US Department of Health, Education and Welfare, Technical Report. W61-5. 1961.

15 A. A. Rosen, R. T. Skeel and M. B. Ettinger, 'Relationship of River Water Odor to Specific Organic Contaminants', *Journal of the Water Pollution Control Federation*, 1963, **35**, 777–782. In 1964, Rosen was in charge of the Organic Contaminants Unit, Chemistry and Physics Section, Basic and Applied Sciences Branch, Division of Water Supply and Pollution Control. Ettinger was head of the Chemistry and Physics Section. See *Training Course Manual: Water Quality Management*, US Department of Health, Education and Welfare/Public Health Service/Robert A. Taft Sanitary Engineering Center, March 1964.

16 W. B. Konieki and A. L. Linch, 'Determination of Aromatic Nitro Compounds', *Analytical Chemistry*, 1958, **30**, 1134–1137.

17 Robert C. Hirt and Frank T. King, 'Use of Micrometer Baly Cells with Beckman and

Cary Ultraviolet Spectrophotometers', *Analytical Chemistry*, 1952, **24**, 1545–1554. For American Cyanamid Stamford see R. P. Chapman, 'Organisation and Functions of an Analytical and Testing Group', *Chemical Industries*, 1949, **65**, 718–721.

18 Robert C. Hirt, Frank T. King and R. G. Schmitt, 'Graphical Absorbance-ratio Method for Rapid Two-component Spectrophotometric Analysis', *Analytical Chemistry*, 1954, **26**, 1270–1273. Data were provided for nitrophenols, nitrobenzene, aniline, and mixtures of melamine, ammeline, and trimethylolmelamine (the latter three compounds were relevant to production of amino resins).

19 *Ibid.*, p. 1272.

20 I thank former American Cyanamid Bound Brook chemist Dr Erwin Klingsberg for discussions on analytical capabilities of the facility.

21 R. P. Daroga and A. G. Pollard, 'Colorimetric Method for the Determination of Minute Quantities of Carbon Tetrachloride and Chloroform in Air and in Soil', *Chemistry and Industry*, 1941, **60**, 218–222.

22 F. A. Lyne and T. McLachlan, 'Contamination of Water by Trichloroethylene', *The Analyst*, 1949, **74**, 513. In 1936, Barrett used the Fujiwara colour reaction for determining TCE in air, and detection to 20 ppm was claimed. H. M. Barrett, *Journal of Industrial Hygiene and Toxicology*, 1936, **18**, 341. See also F. H. Brain, 'The Estimation of Trichloroethylene in Air', *The Analyst*, 1949, **74**, 555–559.

23 Y. A. Seto and M. O. Schultze, 'Determination of Trichloroethylene, Trichloroacetic Acid and Trichloroethanol in Urine', *Analytical Chemistry*, 1956, **28**, 1625–1628.

24 J. G. Bergmann and John Sanik, Jr. (Research Department, Standard Oil Co (Indiana), Whiting, Ind.), 'Determination of Trace Amounts of Chlorine in Naphtha', *Analytical Chemistry*, 1957, **29**, 241–243.

25 F. W. Chapman, Jr. and R. M. Sherwood (Atlantic Refining Co, Phil.), 'Spectrophotometric Determination of Chloride, Bromide and Iodide', *Analytical Chemistry*, 1957, **29**, 172–176.

26 R. C. Blinn 'Techniques Useful for Sodium Biphenyl Determination of Micro Quantities of Organic Chlorine', *Analytical Chemistry*, 1960, **32**, 292–293.

27 F. W. Melpolder, C. W. Warfield and C. E. Headington, 'Mass Spectrometer Determination of Volatile Contaminants in Water', *Analytical Chemistry*, 1953, **25**, 1453–1456.

28 G. P. Happ, D. W. Stewart and H. C. Cooper, 'Mass Spectrometric Determination of Volatile Solvents in Industrial Waste Water', *Analytical Chemistry*, 1957, **29**, 68–71.

29 A. T. James, 'Gas–liquid Chromatography: Separation and Microestimation of Volatile Aromatic Amines', *Analytical Chemistry*, 1956, **28**, 1564–1567.

30 James D. Boggus and N. G. Adams 'Gas Chromatography for Trace Analysis', *Analytical Chemistry*, 1958, **30**, 1471–1473, on p. 1471.

31 G. W. Warren, L. Priestley, Jr., J. F. Haskin and V. A. Yarborough, 'Gas Chromatographic Analysis of Various Mixtures of Compounds Containing Chlorine', *Analytical Chemistry*, 1959, **31**, 1013–1016.

32 Judith Wright, *Vision, Venture and Volunteers: 50 Years of History of the Pittsburgh Conference on Analytical Chemistry and Applied Spectroscopy*, The Pittsburgh Conference/Chemical Heritage Foundation, Pittsburgh, Philadelphia, 1999, pp. 36–37.

33 'Pesticide Analyses Demand Sensitive Methods', *Chemical & Engineering News*, 14 February 1966, pp. 40–44.

34 I thank Dr. Paul Stonehart, of Stonehart Associates, Inc, for sharing with me his experiences of American Cyanamid Stamford in the 1960s.

35 Francis A. Gunther, 'Advances in Analytical Detection of Pesticides', 1966, p. 293.

36 *Standard Methods for the Examination of Water and Wastewater Including Bottom Sediments and Sludges*, American Public Health Association, New York, 12th ed., p. X.

37 Bennett J. Tyson, 'Chlorinated Hydrocarbons in the Atmosphere: Analysis at the Parts-per-trillion Level by GC–MS', *Analytical Letters*, 1975, **8**, 807–813 and Dagmar Rais Cronn and David E. Harsch, 'Rapid Determination of Methyl Chloride in Ambient Air Samples by GC–MS', *Analytical Letters*, 1976, **9**, 1015–1023.

38 For state of the art in the 1970s, see, for example, Harry S. Hertz and Stephen N. Chesler, *Trace Organic Analysis: A New Frontier in Analytical Chemistry*, National Bureau of Standards Special Publication, no. 519, US Department of Commerce, Washington, 1979.

39 S. K. Love, 'Water Analysis', *Analytical Chemistry*, 1951, **23**, 253–257.

40 H. L. Bolton, H. L. Webster and J. Hilton, 'Collaborative Work on the Determination of Alkylbenzenesulfonates in Sewage, Sewage Effluent, River Waters and Surface Waters', *The Analyst*, 1961, **86**, 719–723.

41 Sources on ground water pollution include: *Ground Water Contamination: Proceedings of the 1961 Symposium*, Public Health Service, US Department of Health, Education and Welfare, Technical Report, W61-5. 1961. 'This is the earliest comprehensive analysis of the many facets of ground water pollution'. (p. 9); W. K. Summers and Zane Spiegel, *Ground Water Pollution: A Bibliography*, Ann Arbor Science Publishers, Ann Arbor, 1974; *Polluted Ground Water: A Review of the Significant Literature*, General Electric Company, prepared for the EPA, Office of Water Resources Research, March 1974, PB-235 556. See also James F. Pankow, Stan Feenstra, John A. Cherry and Cathryn Ryan, 'Dense Chlorinated Solvents in Groundwater: Background and History of the Problem' in *Dense Chlorinated Solvents and Other DNAPLs in Groundwater: History, Behaviour, and Remediation*, James F. Pankow and John A. Cherry (eds.), Waterloo Press, Ontario, 1996, pp. 1–52.

42 Middleton, Rosen and Burttachell, 'Manual for Recovery and Identification of Organic Chemicals in Water'; and F. M. Middleton, 'Detection and Measurement of Organic Chemicals in Water and Waste', in *Water Quality Measurement and Instrumentation. Proceedings of a Symposium held at Cincinnati, Ohio, August 29–31, 1960*, US Department of Health, Education and Welfare/Public Health Service/Robert A. Taft Sanitary Engineering Center, 1961, SEC TR W61-2, pp. 50–54.

43 Middleton, 1961, p. 52.

44 B. T. Croll (Water Research Association), 'Organic Pollutants in Water', *Water Treatment and Examination*, 1972, **21**, 213–238, on p. 226, referring to A. A. Rosen, *et al.*, 'Relationship of River Water Odor to Specific Organic Contaminants', *Journal of the Water Pollution Control Federation*, 1963, **35**, 268–274.

45 S. C. Caruso, H. C. Bramer and R. P. Hoak (Mellon Institute, Pittsburgh), 'Tracing Organic Compounds in Surface Streams', *Air and Water Pollution*, 1966, **10**, 41–48.

46 L. G. Cochran and F. D. Bess, 'Waste Monitoring by Gas Chromatography', *Journal of the Water Pollution Control Federation*, 1966, **38**, 2002–2008; R. A. Baker, 'Phenolic Analysis by Direct Aqueous Injection Gas Chromatography', *Journal of the American Water Works Association*, 1966, **58**, 751–760.

47 A. A. Rosen, 'Chemical Analysis: Weapon Against Water Pollution', *Analytical Chemistry*, 1967, **39**(12), 26A–33A, on 30A.

48 B. T. Croll (Water Research Association), 'Organic Pollutants in Water', *Water Treatment and Examination*, 1972, **21**, 213–238.

49 *Ibid.*, pp. 213–214.

50 *Ibid.*, p. 225.

51 John M. McGuire, Ann Alford and Mike Carter (Environmental Protection Agency, Southeast Environmental Research Laboratory, Athens, Georgia), 'Organic Pollutant Identification Utilizing Mass Spectrometry', abstract in 'Report on the Third Annual

[CIBA] Symposium on Recent Advances in the Analytical Chemistry of Pollutants', May 14–16, 1973, Holiday Inn Convention Center, Athens, Georgia. See also the comprehensive survey *Instrumentation for Environmental Monitoring: Water*, University of California, Berkeley, 1973, H2O-PES Methods, p. 30. This was one of five volumes, but the only one dealing with water, published during 1972–74. Particularly valuable are tables of instrument costs and sensitivities, and instrument specifications.

52 George McConnell, 'Halo Organics in Water Supplies', *Journal of the Institution of Water Engineers and Scientists*, 1976, **30**, 431–445, on p. 433. See also C. R. Pearson, and G. McConnell, 'Chlorinated C$_1$ and C$_2$ Hydrocarbons in the Marine Environment', *Proceedings of the Royal Society B*, 1975, **189**, 305–332.

53 A. A. Rosen and F. A. Middleton, 'Chlorinated Insecticides in Surface Water', *Analytical Chemistry*, 1959, **31**, 1729–1731.

54 D. C. Abbott and J. Thomson, 'Analytical Aspects of Water Pollution by Some Domestic and Agricultural Chemicals', *Proceedings of the Society of Water Treatment and Examination*, 1965, **14**, 70–80, on p. 74.

55 G. Zweig, T. E. Archer and D. Rubinstein, 'Residue Analysis of Endosulfan by Combination of Gas Chromatography and Infrared Spectrophotometry', *Journal of Agricultural and Food Chemistry*, 1960, **8**, 403–405.

56 E. S. Goodwin, R. Goulden and J. G. Reynolds, 'Rapid Identification and Determination of Residues of Chlorinated Pesticides in Crops by Gas-Liquid Chromatography', *The Analyst*, 1961, **86**, 697–709.

57 E. S. Goodwin, R. Goulden, A. Richardson and J. G. Reynolds, 'The Analysis of Crop Extracts for Traces of Chlorinated Pesticides by Gas–Liquid Partition Chromatography', *Chemistry and Industry*, September 1960, **24**, pp. 1220–1221.

58 Goodwin, Goulden and Reynolds, 'Rapid Identification', pp. 697–698. These authors used an unmodified Shandon universal gas chromatograph fitted with an electron capture detector, since it was typical of detectors readily available commercially. Insecticides were detected down to 0.5 ppm. Also described was use of the argon-ionisation detector, which when used as an electron capture ionisation detector was similar to Lovelock's small-diode argon-ionisation detector, and which had a sensitivity about 10 times that of the standard Shandon model.

59 D. C. Abbott and J. Thomson, 'Analytical Aspects of Water Pollution by Some Domestic and Agricultural Chemicals', on p. 74.

60 *Ibid.*, p. 75.

61 'New Moves in the Wake of the Pesticides–Pollution Squabble', *Chemical Engineering*, 3 January 1966, pp. 32–33. The Public Health Service had earlier considered that Velsicol's endrin was responsible for the death of 5.2 million fish in the lower Mississippi River during the autumn and winter of 1963.

62 'Pesticide Analyses Demand Sensitive Methods', *Chemical & Engineering News*, 14 February 1966, pp. 40–44; and Francis A. Gunther, 'Advances in Analytical Detection of Pesticides', 1966, pp. 276–302.

63 H. A. Clarke, 'Characterization of Industrial Wastes by Instrumental Analysis', *Purdue, Proceedings of the 23rd Industrial Waste Conference*, 1968, Part 1, pp. 31–34.

64 Croll, 'Organic Pollutants in Water', p. 222.

65 F. W. Karasek, 'Detection Limits in Instrumental Analyses', *Research/Development*, 1975, **24**, 20–24.

66 A. V. Holden, 'Source of Polychlorinated Biphenyl Contamination in the Marine Environment', *Nature*, 1970, **228**, 1220–1221.

67 'Title 40, Protection of Environment, Chapter 1, Environmental Protection Agency, Subchapter D, Water Programs, Part 136, Guidelines Establishing Procedures for the

Analysis of Pollutants', *Federal Register*, **38**, no. 199 (October 16, 1973), 28758–28760. For analysis of water and waste water by GC–MS, see Dennis Schuetzle (ed.), *Monitoring Toxic Substances*, American Chemical Society, Washington DC, 1979, especially Thomas A. Bellar, William L. Buddle and James W. Eichelberger, 'The Identification and Measurement of Volatile Organic Compounds in Aqueous Environmental Samples', pp. 49–62; and Ronald A. Hites, G. A. Jungclaus and V. Lopez-Avila, 'Potentially Toxic Organic Compounds in Industrial Wastewater: Two Case Studies', pp. 63–90.

The Impact of Chromatographic and Electrophoretic Techniques on Biochemistry and Life Sciences

LUIGI CERRUTI

Dipartimento di Chimica Generale, C.so M.D'Azeglio 48, 10125 Turin, Italy, Email: 1cerruti@ch.unito.it

Introduction

Chromatography and electrophoresis are two of the most important separation techniques in biochemistry. They are therefore ideal for demonstrating the fundamental problem of separation techniques, namely the search for 'the best *system* from the point of view of resolution', to quote Michael Lederer.[1] The key word is *system*, and we will see how complex the apparently simple chromatographic and electrophoretic experimental devices actually are. The exceptional flexibility of these methods is a consequence of their particular (chemical) complexity. Most of the technical innovations were introduced in the course of research on proteins, or on their ubiquitous presence in the living world, thus it was an obvious choice to take research into the primary structure of proteins as the main theme of this paper. The influence of this new kind of knowledge will be discussed in the fields of human genetics and population genetics.

The period covered has a precise *terminus a quo* in the fundamental article of 1944 by Raphael Consden Hugh Gordon and Archer Martin on paper chromatography.[2] The *terminus ad quem* is much broader, depending on the particular topic. As sources, I have used almost exclusively the scientific and technical literature, in which the researcher's knowledge procedure is offered openly and critically to the scientific audience. The pertinent historiographic material is rather abundant, albeit heterogeneous and often scattered in technical publications. However, a few contributions are of outstanding value as historical essays or as primary autobiographical sources.[3–6]

Paper and Thin-Layer Chromatography

Archer Martin once remarked that he had been fascinated since 1933 by 'the relationship between [...] chromatogram and distillation columns', so he constructed several machines to be used for solving separation problems. He assembled one of the most notorious of these machines with Richard Synge at the laboratory of the Wool Industries Research Association in Leeds. It was a 'fiendish piece of apparatus', and a separation took a week.[7] Little wonder that the two chemists looked for a alternative, which they found in the 'invention' of partition chromatography. The paper was sent to the *Biochemical Journal* in November 1941, and was quickly printed in the same volume that had hosted the article on the separation machine. Their article is justly famous, and often quoted not only for the important part on the theory of chromatography, but also for its *en passant* mention of the possibility of gas chromatography.[8] Martin continued to strive towards a further simplification of the procedure, and thus 'invented' paper chromatography.

Martin's group had to be resourceful in the face of the inevitable wartime shortages,[8a] and at the beginning of the experimental part of their well-known paper they described the chromatographic apparatus in these terms:

> The essentials of the apparatus consist of a filter-paper strip, the upper end of which is immersed in a trough containing the water-saturated solvent. The strip hangs in an airtight chamber in which is maintained an atmosphere saturated with water and solvent.

This general description was followed by the description of the chamber:

> *Chamber.* For one-dimensional experiments, stoneware drain-pipes ... have been found convenient. The pipe stands in a close-fitting tray hammered from a sheet of lead.[9]

The use of drain-pipes for one-dimensional chromatograms was confirmed by detailed figures,[10] but the new procedure produced the most amazing results in the two-dimensional version, which used two different solvents. Consden, Gordon and Martin demonstrated that a theoretical estimate of the positions of the amino acid on a two-dimensional chromatogram was possible, by using particular diagrams based on an important new kind of experimental data. They proposed a connection with Martin's preceding theoretical elaboration on partition chromatography, by relating the partition coefficient to the rate of movement of the bands. This second quantity was defined as the movement of the band divided by the movement of the advancing front of the solvent and given the symbol R_F.

The important practical significance of R_F was clearly shown in the diagrams. Actually, they developed the discussion in terms of R_F,[11] and stressed that:

> the reproducibility of R_F values depends on the constancy of the following *six* factors: paper, temperature, quantity of amino acids, extraneous sub-

Paper chromatography. Courtesy of Charles D. Winters/Science Photo Library

stances, degree of saturation with water, supply of solvent and distance between starting-point and source of solvent.[12]

This list of experimental variables which can influence R_F demonstrates that even an apparently simple chromatographic device is a complex system.[13] Obviously a crucial point of the technique was the type of paper used in the experiments; for the first time in its long history, filter paper became the object of research in order to determine its properties *as an instrument*. In spite of the experimental research on many types of filter paper no generalisation was possible, and still in 1955 the experts' advice was plain: 'it is necessary [...] to try several grades of paper in preliminary experiments before undertaking a long-range chromatographic study'.[14]

Several different factors assisted the success of paper chromatography, amongst which at least three stand out: the affordability of the technique; the style of professional communication, which was extremely circumstantial regarding experimental details; and the flexibility of the method, which permitted numerous variations. The development and diffusion of paper chromatography also occurred in a period of shortages, particularly in Europe. However, the most important reason for the success of paper chromatography (obviously!) its great value as a separation technique. In 1988 Frederick Sanger wrote:

With the insulin chain we were again lucky in that the method of paper chromatography had just been developed by Martin and his colleagues...
The fractionation of small peptides was superior to anything that had been achieved previously, and it seemed something of a miracle at the time to see

these hitherto intractable products separated from one another on a simple sheet of filter paper.[15]

An important component of Sanger's surprise was being able *to see* the separated substances; the theme of the dramatic effect of the *immediate visibility* of the separated components will come up again in this paper.

Simple as it might seem, paper chromatography evolved in several different directions. For purposes of this paper, we can stop in the early 1960s, by which time paper chromatography was fully developed. In June 1961, a symposium was held at Liblice Castle, in central Bohemia, on the 'relation between paper chromatographic behaviour and chemical structure'.[16] Consden gave the introductory lecture.[17] He paid little attention to the different spatial arrangements, because he preferred to try a classification of the 'methods and varieties' under three principal variables: paper treatment, stationary and mobile phases. Consden discussed 13 different 'forms of paper chromatogram'.[18] In particular, he stressed 'the wide variety of paper chromatographic methods'; 'these in conjunction with a large number of sensitive physical and chemical location methods' enable paper chromatography 'to be a powerful tool in many investigations'. He gave 'a brief selection of advances which have been made possible by paper chromatography', and this list of six research fields allows us to estimate the general impact of paper chromatography on biochemistry.

Consden's choice of the fields of biochemical application of paper chromatography is as follows. 'Structural study of large molecules': paper chromatography is one of the essential fractionation procedures for the analysis of the breakdown products of proteins, oligopeptides, polysaccharides and nucleic acids. 'Discovery of unusual or uncommon constituents in mixtures': in this field Consden's inventory is particularly varied; 'Various clinical applications'; 'Studies involving trace labelled compounds'; and 'Origin of life and alterations in macromolecules'. On the final theme, Consden referred to Vernon Ingram's research on 'the mixture of peptides obtained by enzymolysis of haemoglobin, by a two-dimensional separation (electrophoresis/chromatography)'.[19]

At the Liblice symposium several problems remained unresolved. These problems not only concerned applications, as is usual in scientific meetings, but also on the more unusual aspect of the technique itself, because the issue of standardisation was still unsettled. In the preface to the proceedings published in 1962, the editors wrote: 'Other obvious tasks for the future concern *standardisation* which always brings about restriction of variety by proper selection: limitation of experimental conditions, of the number of systems, reference substances, nomenclatures, symbols, *etc*'. The standardisation was an 'obvious task' and it was also a necessary one because paper chromatography had found a strong rival in thin-layer chromatography. An interesting process from the point of view of the complementary function of techniques had now begun. This process was described by Kurt Randerrath in 1963:

> When the first glass plates began to appear in our biochemistry department about two and half years ago [January 1961] and some of the paper chromatography tanks began to disappear, there were few adherents of the

new method and many who were sceptical. Soon, however, the obvious success of thin-layer chromatography attracted more and more disciples. Meanwhile, a state of peaceful coexistence has been established between paper chromatography and thin-layer chromatography. We know the possibilities of both methods and use each to the best advantage.[20]

Thus, at the beginning of the 1960s, some fifteen years after its introduction, paper chromatography began to be replaced by thin-layer chromatography. While the origin of modern paper chromatography can be traced back only to the group of researchers led by Martin, the success of thin-layer chromatography can be linked to at least two different sources, whose contributions were rather different, namely the research of Justus Kirchner and Egon Stahl.

'It occurred to the writer [Kirchner] that by suitable modifications a technique could be evolved which would combine the advantages of column and paper chromatography'.[21] In his understandable (but pointless) claim of priority Kirchner refers to almost forty different laboratories, in which thin-layer chromatography was in use before the crucial year of 1958, when Stahl presented his *Grundausrüstung* (basic equipment) for thin-layer chromatography. During his research at the US Department of Agriculture, Kirchner also described a home-made coating device,[22] and other authors also proposed different devices, but only the commercial types of apparatus came into general use.[23] This critical instrumental aspect of the experimental practice brings us to Stahl's contribution.

In 1958, the Ausstellung chemischer Apparate (ACHEMA) ran from 31 May until 8 June, in Frankfurt am Main, and was the occasion for several meetings of European and German scientific societies, including the Gesellschaft deutscher Chemiker. The feverish atmosphere of the West German *Wirtschaftswunder* was an appropriate background for Stahl's presentation of his industrialised version of thin-layer chromatography; moreover the show was performed again a year later in Basle, at the ILMAC.[24] To be sure, the contribution of Egon Stahl to thin-layer chromatography went much more beyond the launching of experimental equipment, marketed by the Desaga firm of Heidelberg, and of the Silica gel G, 'specially prepared for this purpose'[25] by Merck of Darmstadt. His greatest contributions to the field were

(a) the standardisation of materials, procedures, nomenclature; and
(b) the description of selective solvent systems for the resolution of many important classes of organic compounds.[26]

It is precisely by classes of compounds that he organised his famous *Handbuch* published in 1962.[27] Stahl gave an interesting analysis of the current situation when he attended a symposium held in Rome in 1963. He showed the rapid diffusion of thin-layer chromatography by tabulating the number of publications: about 20 papers in 1958 and 1959, 60 in 1960, 150 in 1961, and over 300 in 1962. He also gave a quantitative classification of the fields of application, on which he commented from an 'intrinsic' point of view: 'From the graphic representation of the area of application […] we see that the greatest utilisation lies in the sphere of lipophilic materials: a similar analysis for paper chromatography would probably show the same for hydrophilic materials'.[28] This is the principal reason why the thin-layer technique has not been important in the particular field of biochemistry covered in this paper. I have discussed it here in order to stress two points, the competition between techniques and the different degrees of connection with the instruments industry, weak for paper chromatography, and strong for thin-layer chromatography.

Paper, Agar, Starch and PAGE Electrophoresis

Hugh McDonald, professor of biochemistry at the Stritch School of Medicine of Loyola University in Chicago carried out a very thorough study in 1954 of the 'origins' of the numerous electrophoretic techniques.[29] His work is particularly interesting because he collected about 800 references on electrophoresis in stabilised media, a valuable resource for any historian interested in the development of these techniques. Furthermore, his personal epistemological and disciplinary position is intriguing, as he defended the autonomy of electrophoresis in stabilised media from the established fields of chromatography and traditional electrophoresis *à la* Tiselius. The alleged independence of electrophoresis in stabilised media of chromatographic techniques was questioned by one of the founders of paper electrophoresis, Emmett Durrum. In his introduction to a report on paper electrophoresis, he wrote: 'it seems certain that the subsequent "rediscovery" of the method was influenced to a large extent by experience with paper chromatography, as well as with the silica gel ionophoresis of Consden, Gordon and Martin'.[30] The use of paper as the solid medium for electrophoresis had been pioneered before the Second World War, but it only entered in the field of separation techniques in the period 1948–1950, when no less than ten different laboratories published papers on the subject.[31]

The clinical applications of paper electrophoresis aroused strong interest among the medical community. The CIBA Foundation organised a symposium in London at the end of July 1955, 'on an even more informal basis than usual'.[32] Martin took the Chair, and opened the symposium with a self-ironic, 'historical' remark:

> I gather that I have been chosen as Chairman principally because I have published nothing on the subject and may, therefore, be held less likely to have a bias than perhaps others. I should like however [to say] that I started playing with this method in 1942, but I got rather busy on other subjects and never developed it. I can see now just how much I missed by dropping things too soon.[33]

From the rich proceedings of the symposium, it is possible to draw two assessments on the value of the technique. The first is a quantitative estimation of the success of paper electrophoresis. One of the 'inventors' of the technique, Wolfgang Grassmann, stated that in five years about 4000 articles had been published on 'the methods of paper electrophoresis and its manifold applications'.[34] Secondly, the entire contents of the symposium demonstrate the social making of scientific knowledge. For example, in a contribution on the analysis of human haemoglobin by paper electrophoresis, John White, Gilbert Beaven and Margaret Ellis described in minute detail, with the aid of photographs, the two sets of apparatus used in their analyses, and referred to four different groups of researchers whose published experiences served them as guidelines. Their contribution is also interesting because they explicitly connect the qualitative and quantitative aspects of biochemical and biomedical research:

> Zonal electrophoresis on filter paper strips has proved invaluable in studying the human haemoglobin variants. The relative simplicity of the method, and the possibility of examining large numbers of samples from various racial and family groups, has resulted in a rapid increase in understanding of the hitherto obscure group of anaemias, comprising sickle-cell anaemia, thalassaemia and inter-related hereditary haemolitic disorders.[35]

In addition to paper, many other solid media were tried for zone electrophoresis. In 1949 Gordon, Borivoj Keil and Karel Sebesta replaced the silica gel medium used in 1946 with agar gel; Gordon's group was thereby able to achieve zonal separation of proteins.[36] However, the first real success with agar electrophoresis was obtained at the Service de Chimie Microbienne of the Institut Pasteur, Paris. In 1953, Pierre Grabar and Curtis A. Williams Jr. published a brief but important note in *Biochimica et Biophysica Acta*. On one hand it proposed an electrophoretic procedure, which quickly became standard in protein research; on the other it introduced a very important variation which marked the birth of immunoelectrophoresis. The authors were able to conjoin the advantages of the two methods:

> We have attempted both (1) to confirm the individuality of the electrophoretic constituents *via* immunochemical methods; (2) conversely, to classify those constituents detectable by immunochemical methods among the electrophoretic fractions, as defined by their migration rates.[37]

Another versatile method was proposed in 1955 by Oliver Smithies, who described a technique of zone electrophoresis, which used a starch gel as supporting medium.[38] Smithies stressed that the high resolving power of the method probably depended on the pore size of the medium. This technique employed very simple apparatus. Because of the sieving effect of the starch gel, molecules of equal charge but different molecular size could now be separated, so the method gave a dramatic improvement in resolution compared with paper electrophoresis. In many cases, the resolving power of the technique was found to be equal, if not superior, to that of the classical Tiselius method.[39] It is not surprising that the method was soon widely adopted. Almost immediately, its use led to the discovery of genetic variation in blood serum proteins.[40]

'Unfortunately being a natural product the composition of starch can vary, and can affect its gelling ability and resolution. The optimum starch concentration therefore has to be determined in preliminary experiments for each batch of

Polyacrylamide gel electrophoresis. Courtesy of Digital Imagery © copyright 2001 PhotoDisc, Inc

material'.[41] This negative appraisal appeared in 1981, by which time starch gel electrophoresis had been superseded by techniques using polyacrylamide gels (PAGE). Larry Sherman and James Goodrich have outlined the history of this electrophoretic technique and the following observations are drawn from their article. After the introduction of starch gels in 1955, experiments with other forms of gels were carried out to find a more reproducible medium. In 1959, Samuel Raymond and Lewis Weintraub proposed a polyacrylamide gel as a useful gel for zone electrophoresis, and a particular gelling agent called 'Cyanogum' was used. According to Sherman and Goodrich, the resulting gel was 'optically clear and colourless, flexible and elastic, stable, and completely insoluble in water'.[42] These six adjectives describe optical, mechanical and chemical properties. They refer to a fundamental aspect of laboratory research, the availability of special materials for specific purposes. Apropos of this aspect, it has to be stressed that Raymond's material was homemade (and not acquired from a technologically sophisticated firm), so it pertains strictly to the chemical tradition.[43] It is also to be added that the list of the 'miraculous' properties of polyacrylamide gels required some time to be completed. For example, only in 1966 was it realised that it was possible to estimate the molecular sizes of polypeptides from the relative mobilities of their complexes with sodium dodecyl sulfate (SDS) on polyacrylamide gels.[44]

Electrophoresis theory long remained backward compared with the great extension of experimental practice. As late as 1967, complaints were being made concerning the 'lack of experimental data that can be used to test theoretical predictions', with only four investigations that had been carried out systematically with a view to checking theoretical results for the most important model (rigid spherical particles!).[45] In a sense, the empirical attitude that went hand in hand with the development and diffusion of chromatography was simply being maintained. In 1963, Kurt Randerath stressed that the theoretical principles of chromatography had not been worked out in all details, and that most analysts were unacquainted with the great number of factors involved. In this critical vein, Randerath quoted a somewhat disconsolate statement, picked from an important textbook by Harold Cassidy, *Fundamentals of Chromatography*. At the end of the chapter on 'General Theory', Cassidy concluded: 'However, when new separation problem are faced, many of the above parameters remain unknown until after chromatography has been carried out, when usually any interest in them is likely to be small'.[46]

Before leaving the long timeline of electrophoretic techniques, I have to mention two 'singular points' on this line: a crossing point and a terminal one. As we will see later, in 1956 Vernon Ingram was able to identify the molecular 'defect' of haemoglobin responsible for sickle-cell anaemia. He obtained this important result with a particular technique, which 'blended' zone electrophoresis and chromatography, which he called 'fingerprinting'. He gave a detailed description of this method in *Biochimica et Biophysica Acta* in 1958. The section which dealt with the 'fingerprinting' procedure began with a simple statement, which depicts a key idea in the history of separation techniques: 'The separation of the peptides obtained by tryptic digestion was achieved by a two-dimensional combination of paper electrophoresis and chromatography'. The description of the result is more emotional:

> The most striking feature of the fingerprints of haemoglobin A and S[47]
> […] is the faithful reproduction of the intricate and highly characteristic
> pattern of peptide spots in both proteins. All *except one*; this is the No. 4
> peptide which appears amongst the uncharged peptides of a haemoglobin
> A fingerprint, but it is positively charged in haemoglobin S and appears in
> a new position nearer to the cathode.[48]

I have emphasised the phrase *except one* because it delimits the border between 'normality' and 'abnormality', in this case between life and death.

With their innumerable variations, electrophoretic techniques invaded biochemical laboratories in the 1960s and 1970s. Anthony Andrews could write in 1981 that 'it is probably no exaggeration to say that well over half and perhaps as many as three-quarters of all research papers in the whole field of biochemistry make some use of electrophoresis'.[49]

An 'Automatic Recording Apparatus' and Other Machines

The progress in the analytical perfection of separation techniques was not limited to the relatively simple 'flat-bed' techniques I have described in the preceding two sections, and the successful attack on protein sequences followed several different paths.

Stanford Moore and W. H. Stein with the bench model of the amino acid analyser, Rockefeller University, 1965. Photograph by Sam Vandivirt, courtesy of The Rockefeller University Archives

The path that Stanford Moore and William Stein followed was that of the exact determination of the amino acid composition of proteins. In some recollections published in 1979, they referred to the contributions to partition chromatography and paper chromatography made by Martin's group.[50] These contributions stimulated Moore and Stein to explore the best conditions for a quantitative separation and recovery of any single amino acid found in a protein hydrolysate. Their first relevant result was obtained in 1948, when they presented what appeared to be a complete system for the determination of protein composition to a symposium of the New York Academy of Science.[51] It is possible to speak of a system because Moore and Stein not only proposed the use of partition chromatography on starch column, but also the use of an automatic fraction collector and the quantitative colorimetric estimation of the content of each fraction with an improved ninhydrin reagent. Emil Smith has written that 'The impact of this paper and the subsequent detailed publication of these methods produced a small shock wave'. He adds an important annotation: 'Now [in 1979] it is difficult to say which had the greater impact, the analytical method or the fraction collector'.[52] Anyway, the method was short-lived, because the 'starch columns worked, but they were not rapid. It took two weeks to complete a full analysis'.[53] Just after the New York symposium, Moore and Stein's attention was attracted by published experiments on the chromatographic use of synthetic organic ion exchange resins. Following this lead, they 'finally arrived to a perfect solution by using a sulfonated polystyrene resin neutralised by Na ions'.[54] This 'perfect solution' was essentially concerned with the new columns, however, '[a]s experience showed the utility of the approach, it seemed worthwhile to invest the effort necessary to make the analysis more rapid and more automatic'. Therefore, with the help of Darrel Spackman, they began to design and produce a completely automated procedure.

The result of the joint efforts of Spackman, Stein and Moore was a very useful machine, and a key paper, published in *Analytical Chemistry* in July 1958. The Rockefeller Institute group described their apparatus in minute detail, with assistance of graceful and naturalistic drawings. The absorbance curves obtained for almost all the individual amino acids were perfect, and the few difficult cases solved in an appropriate way.[55] The worn pages of the issue shelved in the Library of Turin University's Istituto Chimico show that the paper has been read many, many times. With the 'automatic recording apparatus' of Spackman, Stein and Moore, a hydrolysate of a protein could be analysed in less than 24 hours; the most complex mixtures characteristic of blood plasma, urine and mammalian tissue could be analysed in two days.

The scientific contributions of Moore and Stein were diverse. Since their introduction of ion exchange resins and citrate buffers for amino acid analysis, virtually all column chromatographic techniques have used these materials for the separation of mixtures of amino acids; their system became the standard for comparison of all other amino acid analysis methods.[56] Lastly, the introduction of the automatic apparatus was followed by the marketing of commercial instruments. As early as 1963, Christian Anfinsen could write: 'Techniques for the determination of amino acid sequences will become increasingly automatic over the coming years, and such efforts may eventually become the province of well-trained machine operators'.[57]

With the story of Stein and Moore, I have concluded my brief account of the development of chromatographic and electrophoretic techniques. However, I also wish to mention the contribution from Pehr Edman of Lund University, because his work demonstrates the possibility of obtaining information at the molecular level with different chemical methods. In 1950, Edman described a method for the successive chemical removal of individual amino acids from the amino-terminus of a peptide chain.[58] Fruton has stressed that the chemical reaction on which Edman's degradation was based had been known since 1926. For the rest of his life the Swedish researcher was dedicated to the development of a fully automatic machine for the sequential degradation of proteins.[59] The machine, called the 'sequenator' was ready just in time for a description of the machine to be published in the first issue of the *European Journal of Biochemistry*, in 1967.[60] The process had been applied to the whole molecule of apomyoglobin from the humpback whale, and it had been possible to establish the sequence of the first 60 amino acids from the nitrogen-terminal end. The machine is intrinsically unfriendly, because it uses very dangerous and very expensive reagents, and it is constructed with special and equally expensive materials. In any event, Edman's machine was developed by Beckman Instruments, who marketed it under the name 'Sequencer', in 1969. In 1972 a group led by Gerhard Braunitzer published the first sequence of a protein (162 residues) obtained by a completely automated procedure.[61]

From several points of view, the different techniques described in this section and in the preceding two may be perceived as complementary. Hence, in many laboratories several techniques were in use in order to tackle the diverse faces or phases of the same problem. A case in point, at the beginning of the 1960s, was Ingram's laboratory at MIT. In research published in February 1961, Ingram and Corrado Baglioni used five different techniques;[62] the following paper[63] also described the use of five techniques, but with two differences from the preceding list. The complexity of the knowledge procedure is evident. Any actual chromatographic or electrophoretic application involves only apparently simple sets of apparatus, which, together with solvents, buffers and reagents, constitute complex systems. The systems of instruments and procedures applied in leading laboratories are thus 'systems of systems'. In the next two sections, I examine the impact of chromatography and electrophoresis on the life sciences, specifically their influence on the development of human genetics and population genetics.

Haemoglobins, or Human Genetics and the Competition Between Separation Techniques

The discovery of differences between the haemoglobins of different animal species may be traced back to E. von Körber's inaugural dissertation at Dorpat in 1866, but a reliable method of differentiation was only discovered in 1925, when Friedrich von Krüger proposed the criterion of a 'decomposition time' of haemoglobin solutions treated with 0.05 M NaOH. Shortly afterwards, in 1929, Felix Haurowitz, in Prague, investigated the rate of denaturation spectrophotometrically, and suggested the presence of a homogeneous haemoglobin in six different species. A year later, Haurowitz made an important discovery, when he used denaturation kinetics to show the presence of two different haemoglobins in the umbilical blood of new-born children. In 1935, he succeeded in separating the two haemoglobins and obtaining crystals of the neonate haemoglobin. In a historical paper published in 1979, he stressed that 'The discovery of the foetal Hb was the first definite proof for the presence of more than one Hb in an individual'.[64]

The next advance was the well-known discovery of the sickle-cell haemoglobin by Linus Pauling, Harvey Itano, John Singer and Ibert Wells.[65] As the history of this important stride has already been written from different vantage points,[66,67] I can confine my remarks to the original paper published in *Science* in November 1949. Pauling's group used a modified Tiselius electrophoresis apparatus, which enabled it to distinguish between the haemoglobin obtained from normal individuals and that from patients suffering from sickle-cell anaemia. The group also demonstrated that individuals with sicklemia[68] had a mixture of the two haemoglobins in their blood. On the molecular level, Pauling's group knew that the two haemoglobins had the same sedimentation and diffusion constants, so the difference in electrophoretic behaviour was put down to 'a difference in the number or kind of ionisable groups'. From the differences in isoelectric points they deduced that 'sickle-cell anaemia haemoglobin [had] 2–4 more net charges per molecule than normal haemoglobin'.[69] Sickle-cell anaemia had been recognised as 'a clinical entity' in 1917, and in 1923 it was observed that the ability of the red cells to sickle had a genetic basis. A few months before Pauling's paper, James Neel of the Heredity Clinic, University of Michigan, had established that the gene responsible for the sickling characteristic was in heterozygous condition in individuals with sicklemia, and homozygous in those with sickle-cell anaemia.[70] Pauling's group fully validated Neel's results. In its paper on the 'molecular disease' the group stated: 'The results of our investigation are compatible with a direct quantitative effect of ⌊the⌋ gene pair [...]. This investigation reveals [...] a clear change produced in a protein molecule by an allelic change in a single gene involved in synthesis'.[71] At the end of the paper, the Caltech-based researchers defined their future interest in the field: 'The results obtained in the present study suggest that the erythrocytes of other hereditary haemolitic anaemias be examined for the presence of abnormal haemoglobins. This we propose to do'.[72]

To be precise, the research was continued by Itano and Neel himself. They followed the thread of apparent exceptions to the general pattern of inheritance of the sickle-cell trait and sickle-cell anaemia in certain families. Thus, in 1950, they discovered that a new haemoglobin, termed haemoglobin C, was responsible for a distinct haematological syndrome, in general less severe than 'classical' sickle-cell anaemia. Following the same pattern of haematological research along genetic lines, in 1951 Itano discovered a new 'molecular species of haemoglobin', which was called haemoglobin D. During the 1950s, several other members were added to the human haemoglobin 'molecular family', which at the beginning of 1958 had 12 different members.[73] By this time a new turning point had been reached, thanks to Ingram's use of his own 'fingerprint' technique described above. In 1956, Ingram had discovered 'a specific chemical difference between the globins of normal and sickle-cell anaemia haemoglobin'. Using a two-dimensional combination of paper electrophoresis and chromatography he had examined the peptides obtained by hydrolysis of the proteins with trypsin. From the amino acid composition, the number of peptides was expected to be about thirty, with an average chain length of ten amino acids. The experimental data confirmed this prediction: both haemoglobins yielded around thirty peptides whose analytical behaviour was very similar with one exception. The 'anomalous' peptide in the sickle-cell haemoglobin was rather more positively charged at pH 6.5 than the 'anomalous' peptide in the normal haemoglobin. Ingram concluded that sickle-cell anaemia was due to 'a difference in the amino acids sequence in one small part of one of the polypeptide chains'.[74] Subsequently, Ingram was able to isolate the two anomalous peptides, and determine their amino acid sequences. He discovered that the only difference between them was the exchange of a glutamic acid residue in the normal adult haemoglobin peptide by a valine residue in the same position in the sickle-cell peptide.[75]

In 1958, when Harry Harris published his book on 'Human Biochemical Genetics', he made several important statements, which foreshadowed a new research program:

'It is clear that a whole series of different abnormal genes exists in human populations which may in one way or another influence the character of haemoglobin synthesis'.

Linus Pauling. Courtesy of the Science Photo Library

'One of the most striking things about the genes which determine the different types of haemoglobin is their remarkable variation in incidence from population to population'.

'A single gene difference such as that between the sickle-cell gene and its normal allele is presumably the result of a single mutation step. This is the smallest unit of inherited variation'.

'[S]uch work as that of Ingram on the structural difference between haemoglobins A, S and C ... has opened an entirely new chapter in human genetics'.[76]

I will trace the development of this research program in the next section; but here we should note that the discovery of the *structural* difference between the haemoglobins was permitted only by the shift from the classical Tiselius moving boundary electrophoresis to zone electrophoresis, appropriately associated with enzymatic digestion and paper chromatography.

The diffusion of the new technique alarmed experts in moving boundary electrophoresis such as Harvey Itano, who, since 1956, had drawn attention to the fact that analyses on paper in acidic buffers were 'handicapped by increased denaturation and trailing'. In an extensive review of 1957 on properties and genetic control of human haemoglobins, Itano stressed that '[i]n general the method [of zone electrophoresis] is not as effective as moving boundary electrophoresis for separating components of closely similar charge', and conceded, with an obvious understatement, only that 'its simplicity lends itself to screening and diagnostic procedures'.[77] In actual fact, zone electrophoresis in its variants was extensively used for screening purposes. The development and systematisation of screening for inborn errors of metabolism is reported in several chapters of the four editions (between 1958 and 1976) of the volumes edited by Ivor Smith and James Seakins on *Chromatographic and Electrophoretic Techniques*, with a particular emphasis on urine chromatography.[78] However, as we will see shortly, the knowledge horizon opened by zone electrophoresis coupled to paper chromatography was much wider. While it was evident that 'Competent techniques of moving boundary electrophoresis require[d] too much and expensive equipment for regular laboratory use',[79] the experts had no doubts about the general usefulness of the technique:

> Unlike classical electrophoresis, so far practically restricted to substances with high molecular weight, paper electrophoresis is employed in the entire fields of inorganic and organic substances with both high and low molecular weights, and there is hardly any branch of chemistry in which this method cannot be, or has not been, applied successfully.[80]

I will now turn to the development of Harry Harris's research programme, and to the consequences of the discovery of 'molecular polymorphism' for population genetics.

The 'Romantic Period' in the Study of Genetic Variations

Cambridge University Press reprinted Harris's book on human biochemical genetics in 1962. By this time, he had experienced a radical change in his attitude towards the study of proteins in the context of human genetics; indeed, he had shifted the focus of his research from the analysis of the 'inborn errors of metabolism', with their associated clinical or metabolic disturbances, to screening for possible enzyme variations in normal populations. Harris explained his motivation in the opening section of his seminal paper of 1966.[81] According to Harris, the enzyme deficiencies associated with inborn errors 'represent a highly selected group of mutants and cannot be expected *per se* to provide any clear picture of how extensively genetically determined enzyme variation which does not result in overt pathological manifestations may occur in the general population'. 'In attempting to tackle this rather general problem', Harris added, 'we have adopted a quite empirical and perhaps somewhat simple-minded approach':

> 'Our idea was to see whether, if we examined a series of arbitrarily chosen enzymes in normal individuals in *sufficient detail*, we would find genetically determined differences, and if so whether such differences were common or rare, and whether they were peculiar to one class of enzyme rather than another'.

The staggering scope of this research program is apparent. The constraints of the experiments he envisaged were manifold, including the availability of sufficient quantities of enzymes in an acceptable state of purity, and the feasibility of 'family studies on any enzyme differences that turned up'. An essential issue was the degree of molecular knowledge required to obtain the desired *sufficient detail* about enzymes. Harris's answer was as follows:

> We had, of course, also to make some decision about the kind of techniques
> we would utilise in looking for such differences. A wide variety of methods
> suitable for examining the many different properties of enzyme proteins are
> available... In practice we have mainly relied in the first instance on the
> technique of starch gel electrophoresis... Despite [these] limitations, we
> have found, during the course of examining in varying degrees of detail
> some ten arbitrarily chosen enzymes, three quite striking examples of
> genetically-determined polymorphism.[82]

Starch gel electrophoresis yielded data which was highly appropriate for Harris's
experimental research. The differences in electrophoretic mobilities were sufficient to identify the molecular variants, and simple differences were enough to
start the family study.

The first important results were published in 1963 and 1964, in two papers on
red cell acid phosphatase. As we have seen, Harris used starch gel electrophoresis, coupled with suitable chemical treatment of the surface to obtain coloured
compounds. In the case of red cell acid phosphatase, he found a set of five
isoenzymes. When haemolysates from normal individuals were examined, it was
found that every sample showed the presence of more than one enzyme: 'Furthermore, there were clear-cut person-to-person differences in the number, the
mobilities, and the relative activities of these isoenzyme components'. In a note
published in *Nature*, Harris announced the discovery of 'a new human polymorphism'. Five distinct phenotypes were soon identified, and a sixth one was
predicted. Harris – and the new research perspective – was soon vindicated by
the observation of a small number of cases of the expected phenotype in 1964, by
an independent research group.[83]

Harris's research on enzyme polymorphism in the British population was
mirrored on the other side of Atlantic by Richard Lewontin's research on enzyme
polymorphism in *Drosophila pseudoobscura*. Lewontin started his investigation
after Harris had published his first important results. Lewontin and John Hubby
used gel electrophoresis to characterise genetic variations in five natural populations of *Drosophila pseudoobscura*, a classical choice in experimental genetics.
They studied 18 different loci, and arrived at a surprising conclusion: about 30%
of the structural gene loci within any population were segregating for elec-
trophoretically detectable alternative alleles (allozymes). The two researchers
estimated that in natural populations of *D. pseudoobscura* 12% of loci in each
individual were heterozygous. They demonstrated that – even after leaving to
one side the fact that not all amino acid substitutions may cause detectable
changes in electrophoretic mobility of the enzyme – the amount of genetic
diversity in a typical population of *D. pseudoobscura* was immense.[84]

The appearance in 1966 of Harris's and Lewontin's papers led immediately to
similar research on a wide variety of other organisms and to improved estimates
of genetic variations in *D. pseudoobscura* and man. As Lewontin stated in 1973:

From a broad and diverse science, attempting to establish a theory and phenomenology of all aspects of evolutionary change, population genetics has, in the last few years, turned into one possessed by a single problem, the problem of protein polymorphism and protein evolution.[85]

Lewontin, as a professional population geneticist, united into *a single problem* the experimental aspect of quantitatively determining and qualitatively specifying protein polymorphism on the one hand, and the theoretical side of describing protein evolution on the other. From the very outset, it was evident that the electrophoretic techniques could not detect all genetic variations in a population. The actual proportion detected was unknown, but in the middle of the 1970s, the estimate was about one third. To render the situation less certain, a new kind of heterogeneity *within* isoenzymes was detected by electrophoresis. In an exemplary piece of research, Hubby, Shelly Bernstein and Lynn Throckmorton reexamined the amount of genetic variation for an enzyme of the *D. virilis* species using electrophoresis and the obsolescent technique of heat denaturation. Electrophoresis revealed 11 'electromorphs', but heat denaturation increased the number to 32 different 'thermoelectromorphs'.[86] The very title of their paper illustrates their scientific discomfort: 'Still more genetic variability in natural populations'. In 1977, in a review on *Drosophila* systematics and biochemical evolution, one of the preceding authors (Throckmorton) headed a section dealing with these discoveries with a blunt question: 'Twelve years of work down the drain?' He collated the enormous amount of research carried out during these twelve years under the title: 'The bandwagon era'.[87] A sense of discomfort is also evident in a paper by Francisco J. Ayala, published in 1982:

> Because of the pervasive enthusiasm originally generated by electrophoretic studies, the decade from the mid-1960s to the mid-1970s may be called the 'romantic period' in the study of genetic variation. One might also call it the 'delusive period', because the anticipated expectations remain largely unfulfilled'.[88]

However, in the course of the same essay, Ayala, one of most renowned biologist of modern times, lauds electrophoresis:

> 'Electrophoresis makes it possible to study gene loci independently of whether they are variable or not'.
> 'The application of electrophoretic techniques to the study of genetic variation generated enormous enthusiasm among evolutionists for one additional reason: it provides a method for organisms not suitable for breeding experiments'.
> 'Before the electrophoretic revolution, genetic data existed for only a few dozen multicellular organisms. Now, hundreds of different species have been studied by electrophoresis. The number of loci sampled in many species is sufficiently large, 15 or more, so that average estimates of genetic variation can be advanced with some degree of confidence'.[89]

The last statement is in contrast with a preceding one in the same article, where the author wonders 'whether formulae can be found (or even whether they are worth finding) to transform electrophoretic measures into "true" estimates of genetic variation'.[90] The two sections in which these two contrasting statements are found are entitled 'From the hazy to the delusive', and 'The electrophoresis revolution'. Clearly, Ayala was not at all convinced that the revolution had ended.

Conclusions

The main thrust of my paper was encapsulated by Emil L. Smith more than twenty years ago, in a very interesting contribution to the New York Academy discussion on the origin of modern biochemistry. Addressing his audience directly, Smith wrote:

> Perhaps some of you may feel that I am overemphasising the tools of the trade rather than the intellectual developments of protein chemistry. Be that as it may, this is the reality of the period ... most of the conceptual notions regarding the importance of proteins and the general view of their covalent structure were already part of our intellectual heritage. What was needed were the methods and the tools to tackle the structure of proteins in order to begin the task of trying to understand how the proteins accomplished their functions.[91]

I have included many instrumental details in my account, tedious lists of experimental variables, and a lengthy lineage of electrophoretic techniques. I have adopted a 'rhetoric of quantity', in order to emphasise that science is a social construct. However, I developed the first part of my paper with a view to stressing the relationship between the formidable technical development and the practical epistemology of chemists. Any chemist adopts a sensible epistemology, attentive to the minute variation of experimental conditions. This attitude is found in an extreme form in any biochemist. Thus, the chemical complexity of chromatography and zone electrophoresis appealed to analytical chemists and biochemists alike. It is fair to point out that many technical breakthroughs originated in biochemical laboratories. In several points in my paper, I have spoken of experimental 'systems'. My use of this term is similar to Robert Kohler's 'systems of production of knowledge'. In his essay, Kohler noted that the 'language of scientific experiment and "practice" [had] become fashionable' for sociologists and historians.[92] However, I would argue that this 'fashion' has not yet influenced the 'spontaneous' philosophy of scientists, even less the 'spontaneous' pyramidal hierarchy between theoretical and experimental disciplines.

The multitude of techniques supports not only a 'rhetoric of quantity', but also a 'rhetoric of quality', in a twofold sense. The first refers to the molecular individuality of many substances of biological interest. Proteins excel in this sense, so the search for new 'adaptive' techniques was in many cases a necessity. The second qualitative sense depends on the quantitative one. The molecular individuality of proteins for many years created a 'personal relationship' between the biochemist and the substance he was studying. I will cite only two authoritative appraisals. In 1960, J. Ieuan Harris and Vernon Ingram wrote: 'Each protein or peptide presents a different problem which is treated individually'.[93] Fifteen years later, an analogous attitude may be found in Saul Needleman's overview: 'one is encouraged to design instrumental approaches tailored to one's own particular requirements'.[94] From this point of view, Edman's machine wipes out the individuality of proteins.

In the second part of my paper, I tried to demonstrate the effect of the new biochemical knowledge on life science. The experimental knowledge of proteins' individuality changed human genetics and population genetics into really different disciplines, but the overall result went beyond pure academic progress in scientific knowledge. In fact, the very concept of 'normality' became questionable because data on the variations in human populations made it impossible to define any kind of 'enzymatic norm'. Every human being has her or his own norm, at least at the molecular level.

Notes and References

1 M. Lederer, *Chromatography for Inorganic Chemistry*, Wiley, New York, 1994, p. 27.
2 R. Consden, A. H. Gordon and A. J. P. Martin, 'Qualitative Analysis of Proteins: A Partition Chromatographic Method Using Paper', *Biochem. J.*, 1944, **38**, 224–232.
3 J. S. Fruton, *Molecules and Life. Historical Essays on the Interplay of Chemistry and Biology*, Wiley, New York, 1972; J. S. Fruton, *A Skeptical Biochemist*, Harvard University Press, Cambridge, MA, 1992; J. S. Fruton, *Proteins, Enzymes, Genes. The Interplay of Chemistry and Biology*, Yale University Press, New Haven, 1999.
4 P. Laszlo, *A History of Biochemistry. Molecular Correlates of Biological Concepts*, Amsterdam, Elsevier, 1986; it is Volume 34A of A. Neuberger and L. L. M. van Deenen, *Comprehensive Biochemistry*.
5 P. R. Srinivasan, J. S. Fruton and J. T. Edsall (eds.), *The Origins of Modern Biochemistry: A Retrospect on Proteins*, New York Academy of Science, New York, 1979.
6 L. S. Ettre and A. Zlatkis, *75 Years of Chromatography – A Historical Dialogue*, Elsevier, Amsterdam, 1979.
7 A. J. P. Martin, personal reminiscences, in Ettre and Zlatkis, *75 Years of Chromatography*, pp. 286–296, on p. 288. In the published paper the two authors soberly state that 'the procedure is still in a crude state'; A. J. P. Martin and R. L. M. Synge, 'Separation of the Higher Monoamino acids by Counter-current Liquid–Liquid extraction: The Amino Acid Composition of Wool', *Biochem. J.*, 1941, **35**, 91–121, on p. 119.
8 A. J. P. Martin and R. L. M. Synge, 'A New Form of Chromatogram Employing Two Liquid Phases. (1) A Theory of Chromatography. (2) Application to the Micro-determination of the Higher Monoamino Acids in Proteins', *Biochem. J.*, 1941, **35**, 1358–1368. The portent of gas chromatography is in these words: 'very refined separations of volatile substances should therefore be possible in a column in which permanent gas is made to flow over gel with a non-volatile solvent in which the substances to be separated approximately obey Raoult's law', on p. 1359.
8a Note added by the editor: It should be noted, however, that Lovelock displayed a similar frugality and ingenuity long after the war was over. There was a long-standing (British?) tradition of making-do in chemistry laboratories. See the papers by Morris and Knight in this volume.
9 Consden, Gordon and Martin, 'Qualitative Analysis of Proteins', p. 226.

10	Consden, Gordon and Martin, 'Qualitative Analysis of Proteins', Figures 2 and 3, p. 227.

11	The authors reported the R_F values for the 22 amino acids and 17 solvents; Consden, Gordon and Martin, 'Qualitative Analysis of Proteins', Table 2, p. 228.

12	Consden, Gordon and Martin, 'Qualitative Analysis of Proteins', p. 230; my italics.

13	It began a research on the experimental variables which can influence R_F. A later list by P. W. Zimmermann, 1953, quotes *nine* different variables; see R. J. Block, E. L. Durrum and G. Zweig, *A Manual of Paper Chromatography and Paper Electrophoresis*, Academic Press, New York, 1955, p. 11.

14	Block, Durrum and Zweig, *A Manual of Paper Chromatography and Paper Electrophoresis*, p. 34.

15	F. Sanger, 'Sequences, Sequences, Sequences', *Ann. Rev. Biochem.*, 1988, **57**, 1–28, on p. 9.

16	I. M. Hais and K. Macek, *Some General Problems of Paper Chromatography. Relation Between Paper Chromatographic Behaviour and Chemical Structure. Attempts at Systematic Analysis*, Czechoslovak Academy of Science, Prague, 1962.

17	R. Consden, 'Origin, Development and Future of Paper Chromatography', in Hais and Macek, *Some General Problems of Paper Chromatography*, pp. 13–21.

18	Consden, 'Origin, Development and Future of Paper Chromatography', Table 1 and pp. 15–17.

19	Consden, 'Origin, Development and Future of Paper Chromatography', p. 19; V. M. Ingram, 'Abnormal Human Haemoglobins, I. The Comparison of Normal Human and Sickle-cell Hemoglobins by "Fingerprinting"', *Bioch. Biophys. Acta*, 1958, **28**, 539–543.

20	K. Randerath, *Thin-Layer Chromatography*, Verlag Chemie, Weinheim, 1963, on p. V.

21	J. G. Kirchner, *Thin-Layer Chromatography*, Interscience, New York, 1967; when the book was published, the author was working at the Coca Cola Company.

22	J. M. Miller and J. G. Kirchner, 'Apparatus for the Preparation of Chromatostrips', *Anal. Chem.*, 1954, **26**, 2002; by then the authors were working at the Fruit and Vegetable Chemistry Laboratory, Pasadena.

23	Randerath, *Thin-Layer Chromatography*, p. 27.

24	E. Stahl, 'Development and Application of Thin-Layer Chromatography', in G. B. Marini-Bettòlo (ed.), *Thin-Layer Chromatography*, Elsevier, Amsterdam, 1964, pp. 3–12, on p. 3. The ILMAC was an international exhibition on chemical and analytical technology; the Swiss version of ACHEMA.

25	Stahl, 'Development and Application of Thin-Layer Chromatography', p. 9.

26	B. Fried and J. Shema, *Thin-Layer Chromatography. Techniques and Applications*, Dekker, New York, 1982, pp. 1–6.

27	E. Stahl, *Dunnschicht-Chromatographie, eine Labotatoriumshandbuch*, Springer, Berlin, 1962.

28	Stahl, 'Development and Application of Thin-Layer Chromatography', p. 12.

29	H. J. McDonald, *Ionography. Electrophoresis in Stabilized Media*, Year Books Publishers, Chicago, 1955; the preface is dated November, 1954.

30	Block, Durrum and Zweig, *A Manual of Paper Chromatography and Paper Electrophoresis*, p. 333. Here Durrum makes reference to an important article published in 1946: R. Consden, A. H. Gordon and A. J. P. Martin, 'Ionophoresis in Silica Jelly: A Method for the Separation of Amino Acids and Peptides', *Biochem. J.*, 1946, **40**, 33–41.

31	J. Grégoire, and J. Reynaud, *Électrophorèse sur papier. Méthodes et Résultat*, Paris, Vigot, 1956, p. 11.

32	Preface to G. E. W. Wolstenholme and E. C. P. Millar (eds.), *Ciba Foundation*

Symposium on Paper Electrophoresis, Churchill, London, 1956; p. V.

33 Wolstenholme and Millar, *Ciba Foundation Symposium on Paper Electrophoresis*, p. 1.

34 W. Grassmann, 'General Methods of Paper Electrophoresis with Examples of its use in Medical and Biochemical Problems', in Wolstenholme and Millar, *Ciba Foundation Symposium on Paper Electrophoresis*, pp. 2–21, on p. 3.

35 J. C. White, G. II. Beaven and M. Ellis, 'Analysis of Human Haemoglobins by Paper Electrophoresis', in Wolstenholme and Millar, *Ciba Foundation Symposium on Paper Electrophoresis*, 43–57; quoted on p. 45.

36 A. H. Gordon, B. Keil and K. Sebesta, 'Electrophoresis of Proteins in Agar Jelly', *Nature*, 1949, **164**, pp. 498–499.

37 P. Grabar and C. A. Williams, 'Méthode Permettant L'étude Conjuguée des Propriétés Électrophorétiques et Immunochimiques d'un Mélange de Protéines. Application au Sérum Sanguin', *Biochim. Biophys. Acta*, 1953, **10**, 193–194. Even McDonald did not find any predecessor of Grabar and Williams; *cf.* McDonald, *Ionography. Electrophoresis in Stabilized Media*, p. 129. Translated from the French by Pierre Laszlo.

38 O. Smithies, 'Zone Electrophoresis in Starch Gels: Group Variations in the Serum Proteins of Normal Human Adults', *Biochem. J.*, 1955, **61**, 629–641; the author was working at the University of Toronto.

39 L. R. Sherman and J. A. Goodrich, 'The Historical Development of Sodium Dodecyl Sulfate–Polyacrylamide Gel Electrophoresis', *Chem. Soc. Rev.*, 1985, **14**, 225–236, on p. 226.

40 A. T. Andrews, *Electrophoresis: Theory, Techniques and Biochemical and Clinical Applications*, Oxford, Clarendon, 1981, p. 230.

41 Andrews, *Electrophoresis: Theory, Techniques and Biochemical and Clinical Applications*, p. 230.

42 L. R. Sherman and J. A. Goodrich, 'The Historical Development of Sodium Dodecyl Sulfate', p. 226.

43 I am thinking, for example, of the search for the most appropriate solvent for isolating a substance, or of efforts to obtain the most active and stable catalyst for a certain reaction. On these themes I refer the reader to: L. Cerruti, 'Chemicals as Instruments. A Language Game', *Hyle*, 1998, **4**, 39–61; L. Cerruti, 'Historical and Philosophical Remarks on Ziegler–Natta Catalysts. A Discourse on Industrial Catalysis', *Hyle*, 1999, **5**, 3–41. Nevertheless, a chemical process may need expensive and exotic material; *vide infra*, on Edman's machine.

44 L. R. Sherman and J. A. Goodrich, 'The Historical Development of Sodium Dodecyl Sulfate', p. 229.

45 J. Th. G. Overbeek and P. H. Wiersema, 'The Interpretation of Electrophoretic Mobilities', in M. Bier, *Electrophoresis. Theory, Methods and Applications*, Vol. II, Academic Press, New York, 1967, pp. 1–52, on p. 20.

46 H. G. Cassidy, *Fundamentals of Chromatography. Technique of Organic Chemistry*, Vol. 10, Interscience, New York, 1957, quoted in Randerath, *Thin-Layer Chromatography*, p. 7.

47 A and S refer to normal and sickle-cell haemoglobin, respectively.

48 V. M. Ingram, 'Abnormal Human Haemoglobins, I. The Comparison of Normal Human and Sickle-cell Haemoglobins by "Fingerprinting"', *Bioch. Biophys. Acta*, 1958, **28**, 539–543.

49 Andrews, *Electrophoresis: Theory, Techniques and Biochemical and Clinical Applications*, p. I.

50 S. Moore and W. H. Stein, personal recollections in Ettre and Zlatkis, *75 Years of*

Chromatography, pp. 299–308, on p. 299.

51 S. Moore and W. H. Stein, 'Partition Chromatography of Amino Acids on Starch', *Ann. N. Y. Acad. Sci.*, 1948, **49**, 265–278.

52 E. L. Smith, 'Amino Acids Sequences of Proteins – The Beginnings', in Srinivasan, Fruton, and Edsall (eds.), *The Origins of Modern Biochemistry*, pp. 107–118, on p. 111.

53 S. Moore and W. H. Stein, personal recollections in Ettre and Zlatkis, *75 Years of Chromatography*, on p. 302.

54 T. Wieland and M. Bodanszky, *The World of Peptides. A Brief History of Peptide Chemistry*, Springer, Berlin, 1991, on p. 50.

55 D. H. Spackman, W. H. Stein and S. Moore, 'Automatic Recording Apparatus for Use in the Chromatography of Amino Acids', *Anal. Chem.*, 1958, **30**, 1190–1206.

56 P. E. Hare, 'Amino Acid Composition by Column Chromatography', in S. B. Needleman (ed.), *Protein Sequence Determination. A Sourcebook of Methods and Techniques*, Springer, Berlin, 1975, pp. 204–231, on p. 205.

57 R. E. Canfield and C. B. Anfinsen, 'Concepts and Experimental Approaches in the Determination of the Primary Structure of Proteins', in H. Neurath (ed.), *The Proteins. Composition, Structure and Function*, Vol. I, Academic Press, New York, 1963, pp. 311–378, on p. 312.

58 P. Edman, 'A Method for Determination of the Amino Acid Sequence in Peptides', *Acta Chem. Scand.*, 1950, **4**, 283–293. Fruton has stressed that the chemical reaction on which Edman's degradation was based had been known since 1926; Fruton, *Molecules and Life. Historical Essays on the Interplay of Chemistry and Biology*, p. 173.

59 L. R. Croft, *Handbook of Protein Sequence Analysis. A Compilation of Amino Acid Sequences of Proteins with an Introduction to the Methodology*, Wiley, New York, 1980, p. 47.

60 P. Edman and G. Begg, 'A Protein Sequenator', *Europ. J. Biochem.*, 1967, **1**, 80–91.

61 S. Neufeldt, *Chronologie Chemie, 1800–1980*, VCH, Weinheim, 1987, p. 280.

62 C. Baglioni and V. M. Ingram, 'Four Adult Haemoglobin Types in One Person', *Nature*, 1961, **189**, 465–467. The techniques were starch-gel electrophoresis; electrophoresis on starch block; an *ad hoc* modified fingerprinting technique; 'Spinco Model Moore and Stein Automatic Analyser'; stepwise degradation with the Edman procedure.

63 C. Baglioni, V. M. Ingram and E. Sullivan, 'Genetic Control of Foetal and Adult Human Haemoglobin', *Nature*, 1961, **189**, 467–469. The new inventory is: starch-gel electrophoresis; electrophoresis on starch block; column ion-exchange chromatography, according to T. H. J. Huisman; fingerprinting, as usual and 'by an improved method'.

64 F. Haurowitz, 'Protein Heterogeneity: Its History, Its Bases and Its Limits', in Srinivasan, Fruton and Edsall (eds.), *The Origins of Modern Biochemistry*, pp. 37–47; quotation on p. 40.

65 L. Pauling, H. A. Itano, S. J. Singer and I. C. Wells, 'Sickle Cell Anemia, a Molecular Disease', *Science*, 1949, **110**, 543–548.

66 Laszlo, *A History of Biochemistry*, pp. 392–397.

67 P. Heller, 'Historic Reflections on the Clinical Roots of Molecular Biology', in D. A. Chambers (ed.), *DNA: The Double Helix Perspective and Prospective at Forty Years*, New York Academy of Science, New York, 1995, pp. 83–93. Heller speaks of a 'legendary story of the discovery of haemoglobin S', and gives (on p. 89) a version of the origins of Pauling's interest in sickle-cell anaemia somewhat different from that told by Pauling himself.

68 People with sicklemia have cells capable of sickling, but exhibit no pathological

consequences although they have the 'sickle-cell trait'. People with sickle-cell anaemia have cells which are capable of sickling and also suffer from severe chronic anaemia.

69 Pauling, Itano, Singer and Wells, 'Sickle Cell Anemia, a Molecular Disease', p. 546.

70 J. V. Neel, 'The Inheritance of Sickle Cell Anemia', *Science*, 1949, **110**, 64–66.

71 Pauling, Itano, Singer and Wells, 'Sickle Cell Anemia, a Molecular Disease', p. 547.

72 Pauling, Itano, Singer and Wells, 'Sickle Cell Anemia, a Molecular Disease', p. 548.

73 H. Harris, *Human Biochemical Genetics*, Cambridge University Press, Cambridge, 1959, p. 146–151. Since 1953 a group of researchers in the field convened by the US National Institutes of Health had recommended a nomenclature by which normal haemoglobin was named haemoglobin A, the foetal one F, the sickle-cell one S, and the others by letters following the order of discovery and the English alphabet.

74 V. M. Ingram, 'A Specific Chemical Difference between the Globins of Normal and Sickle-cell Anaemia Hemoglobin', *Nature*, 1956, **178**, 792–794, on p. 794.

75 Harris, *Human Biochemical Genetics*, p. 154.

76 Harris, *Human Biochemical Genetics*, pp. 157, 161, 290 and 298.

77 H. A. Itano, 'The Human Haemoglobins: Their Properties and Genetic Control', *Adv. Prot. Chem.*, 1957, **12**, 215–268, on p. 228.

78 J. W. T. Seakins and R. S. Erseser, 'Chromatography and Screening for Inborn Errors of Metabolism', in I. Smith, J. W. T. Seakins (eds.), *Chromatographic and Electrophoretic Techniques*, Vol. I, *Paper and Thin Layer Chromatography*, Heinemann, London, 1976, pp. 11–17.

79 Sherman and Goodrich, 'The Historical Development of Sodium Dodecyl Sulfate', p. 225.

80 W. Grassmann, 'General Methods of Paper Electrophoresis with Examples of its Use in Medical and Biochemical Problems', in Wolstenholme and Millar, *Ciba Foundation Symposium on Paper Electrophoresis*, pp. 2–21, on p. 13.

81 H. Harris, 'Enzyme Polymorphism in Man', *Proc. R. Soc. London (B)*, 1966, **164**, pp. 298–310.

82 H. Harris, 'Enzyme Polymorphism in Man', on p. 299; my italics.

83 H. Harris, 'Enzyme Polymorphism in Man', on pp. 299–300.

84 J. L. Hubby and R. C. Lewontin, 'A Molecular Approach to the Study of Genetic Heterozygosity in Natural Populations, I. The Number of Alleles at Different Loci in *Drosophila pseudoobscura*', *Genetics*, 1966, **54**, 577–594; R. C. Lewontin and J. L. Hubby, 'A Molecular Approach to the Study of Genetic Heterozygosity in Natural Populations, II. Amount of Variation and Degree of Heterozygosity in Natural Populations of *Drosophila pseudoobscura*', *Genetics*, 1966, **54**, 595–609.

85 R. C. Lewontin, 'Population Genetics', *Ann. Rev. Genetics*, 1973, **7**, 1–17.

86 S. C. Bernstein, L. H. Throckmorton, J. L. Hubby 'Still More Genetic Variability in Natural Populations', *Proc. Nat. Acad. Sci.*, 1973, **70**, 3928–3931.

87 D. J. Merrell, *Ecological Genetics*, University of Minnesota Press, Minneapolis, 1981, p. 347.

88 F. J. Ayala, 'The Genetic Structure of Species', in R. Milkman (ed.), *Perspectives on Evolution*, Sinauer Associates, Sunderland, 1982, pp. 60–82, on p. 63.

89 Ayala, 'The Genetic Structure of Species', pp. 66–67.

90 Ayala, 'The Genetic Structure of Species', p. 63.

91 Smith, 'Amino Acids Sequences of Proteins – The Beginnings', in Srinivasan, Fruton and Edsall (eds.), *The Origins of Modern Biochemistry*, p. 112.

92 R. E. Kohler, 'Systems of Production: Drosophila, Neurospora, and Biochemical Genetics', *Hist. Stud. Phys. Sci.*, 1991, **22**, 88–91.

93 J. Ieuan Harris and V. M. Ingram, 'Methods of Sequence Analysis in Proteins' in P.

Alexander and R. J. Block (eds.), *A Laboratory Manual of Analytical Methods of Protein Chemistry*, Pergamon, Oxford, 1960, pp. 424–499, on p. 424.

94 S. B. Needleman, 'General Considerations', in S. B. Needleman (ed.), *Protein Sequence Determination. A Sourcebook of Methods and Techniques*, pp. 1–4, on p. 3.

Index

Bold page numbers indicate illustrations.